Modeling Forest Response to Climate Change

Modeling Forest Response to Climate Change

Editors

Daniela Dalmonech
Gina Marano
Alessio Collalti

MDPI

Basel • Beijing • Wuhan • Barcelona • Belgrade • Novi Sad • Cluj • Manchester

Editors

Daniela Dalmonech
Forest Modelling Laboratory, Institute for Agriculture and Forestry Systems in the Mediterranean
National Research Council of Italy (CNR-ISAFOM)
Perugia
Italy

Gina Marano
Environmental Systems Science
ETH Zürich
Zürich
Switzerland

Alessio Collalti
Forest Modelling Laboratory, Institute for Agriculture and Forestry Systems in the Mediterranean
National Research Council of Italy (CNR-ISAFOM)
Perugia
Italy

Editorial Office
MDPI AG
Grosspeteranlage 5
4052 Basel, Switzerland

This is a reprint of articles from the Special Issue published online in the open access journal *Forests* (ISSN 1999-4907) (available at: https://www.mdpi.com/journal/forests/special_issues/6F1AD5HPH3).

For citation purposes, cite each article independently as indicated on the article page online and as indicated below:

Lastname, A.A.; Lastname, B.B. Article Title. *Journal Name* **Year**, *Volume Number*, Page Range.

ISBN 978-3-7258-1757-3 (Hbk)
ISBN 978-3-7258-1758-0 (PDF)
doi.org/10.3390/books978-3-7258-1758-0

Cover image courtesy of Gina Marano

Contents

About the Editors

Daniela Dalmonech

Daniela Dalmonech is a researcher at the Institute for Agricultural and Forestry Systems in the Mediterranean (ISAFOM-CNR) at the National Research Council of Italy. She holds a Master's in environmental engineering from the University of Trento with a focus on numerical modelling applied to the environment; she holds a PhD in Forest Ecology attained at the Tuscia University and a Marie Skłodowska-Curie postdoctoral fellowship at the Max Planck Institute for Biogeochemistry in Jena (Germany). She is interested in the carbon and nitrogen cycle dynamics at different spatial scales and under anthropogenic and natural stress through numerical modelling and statistical analyses. In 2019, she started collaborating with the Forest Modeling Laboratory, where she is responsible for the 3D-CMCC-FEM (Three Dimensional—Coupled Model Carbon Cycle—Forest Ecosystem Module) code development and output data analyses.

Gina Marano

Gina Marano is a PhD student in Forest Ecology in the Stand and Landscape Dynamics group at ETH Zurich and a collaborator of the Forest Modeling Laboratory. Her research focuses on developing and testing formulations of drought stress to better capture the effects of drought on tree growth and mortality in stand-scale dynamic forest models.

Alessio Collalti

Alessio Collalti is a senior researcher at the Institute for Agricultural and Forestry Systems in the Mediterranean (ISAFOM-CNR) and the Forest Modelling Laboratory Head at the National Research Council of Italy. He has a Master's in Natural Sciences and a Ph.D. in Forest Ecology. His background concerns Forest Ecology, Carbon and Nitrogen Cycle, and Forest and Vegetation Modelling, particularly concerning vegetation numerical modelling and response under natural and anthropogenic stress, including climate change impacts and forest management scenarios. He has been working at the University of Tuscia, at the Foundation euro-Mediterranean Centre on Climate Change, and at the National Research Council of Italy, where he developed the 3D-CMCC-FEM (Three Dimensional-Coupled Model Carbon Cycle-Forest Ecosystem Module).

forests

Editorial

Modeling Forest Response to Climate Change

Gina Marano [1,2,*,†], Daniela Dalmonech [1,†] and Alessio Collalti [1,†]

1 Forest Modelling Laboratory, Institute for Agriculture and Forestry Systems in the Mediterranean, National Research Council of Italy (CNR–ISAFOM), Via Madonna Alta 128, 06128 Perugia, Italy; daniela.dalmonech@cnr.it (D.D.); alessio.collalti@cnr.it (A.C.)

2 Forest Ecology, Institute of Terrestrial Ecosystems, Department Environmental Systems Science, ETH Zurich, 8092 Zurich, Switzerland

* Correspondence: gina.marano@usys.ethz.ch

† These authors contributed equally to this work.

Citation: Marano, G.; Dalmonech, D.; Collalti, A. Modeling Forest Response to Climate Change. *Forests* **2024**, *15*, 1194. https://doi.org/10.3390/f15071194

Received: 2 July 2024
Accepted: 5 July 2024
Published: 10 July 2024

In an era marked by unprecedented climate shifts, understanding the intricate responses of forest ecosystems to these changes is of paramount importance. The research presented in this Special Issue delves deeply into various dimensions of forest dynamics under the influence of climate change, offering critical insights that can guide effective conservation and management strategies.

Vegetation seasonality, a crucial component of ecological systems, is under significant stress due to global warming. Nooni et al.'s study [1] highlights how Normalized Difference Vegetation Index (NDVI) trends in Equatorial Africa (EQA) have been influenced by changes in precipitation and temperature over the past four decades. The research reveals that while forest and cropland areas have experienced declining NDVI trends, shrubland and grassland areas have tended to increase, suggesting that there is a complex interplay between climate factors and vegetation types. This nuanced understanding is essential for ecological conservation and resource management in the face of ongoing climate change.

Similarly, the capacity of forests to act as carbon sinks is under threat. In their study, Morichetti et al. [2] examine carbon fluxes within forest ecosystems using the 3D-CMCC-FEM model. Their analysis of five contrasting European forest sites under current and future climate scenarios demonstrates the model's robust ability to estimate net ecosystem exchange (NEE). The study predicts a consistent reduction in the carbon sink capabilities of forests due to climate change and forest aging. Despite an increase in the number of days that evergreen forests act as carbon sinks, their overall annual capacity is projected to decrease. Similarly, deciduous forests maintain stable carbon sink days but also show a reduction in their annual capacity. This highlights the need for the implantation of adaptive forest management practices that mitigate the anticipated decline in carbon sequestration.

The same model was employed by Vangi et al. [3] by simulating carbon stocks and wood production across different forest ages and climate scenarios. Their findings indicate a pronounced decline in biomass for older coniferous stands, such as spruce, under warming conditions; meanwhile, beech forests may sustain or even enhance their carbon storage capacity. Scots pine forests display intermediate behavior, with a stable stock capacity but decreasing annual increment. These insights highlight the variable resilience of different forest types to climate change, necessitating tailored management approaches and, most importantly, underscoring the differential impacts of climate change on coniferous and broadleaf forests; in addition, they highlight the necessity of species-specific management practices.

An important component of the carbon cycle and its dynamics is soil respiration; therefore, its influence on the carbon cycle was explored by Kivalov et al. [4]. The authors developed empirical models to better understand soil respiration in different forest ecosystems. Their research highlights the importance of soil's organic carbon and water-holding capacity in predicting soil respiration, providing a foundation for the enhanced modeling of the carbon cycle in terrestrial ecosystems. Moving to a management-oriented perspective, several studies investigated how management practices, namely restoration, plantation and thinning techniques, can alleviate forest ecosystems from the future pressure of environmental stressors.

The restoration and conservation of native forests also emerge as critical themes in Yong et al.'s study [5]. The authors employ a joint species distribution model to analyze the distribution of tree species in China's Jilin Province. The study identifies climate, site, and soil as the key environmental factors influencing tree species niches, with the model demonstrating strong explanatory power. Their work emphasizes the importance of environmental factors—climate, site, and soil—in shaping tree species niches, thus providing a robust framework for forest restoration and proactive forest management.

The impact of climate change on economically significant timber trees is a crucial aspect of timber-based bioeconomies. In their work, Feng et al. [6] focus on *Cunninghamia lanceolata* by using the MaxEnt model to project its distribution under future climate scenarios. Their research identifies the key environmental variables affecting its growth and suggests that suitable habitats will shift to higher latitudes as the climate warms. This predictive modeling is crucial for the planning of future planting strategies and conservation efforts to ensure the survival of this valuable species.

Innovative methodologies also play a pivotal role in forest management. In their study, Liu et al. [7] integrate remote sensing, deep learning, and statistical modeling to monitor forest changes and carbon storage dynamics in China. Their approach demonstrates high accuracy in mapping forest types and quantifying carbon storage, offering a valuable tool in local forest management and the achievement of carbon neutrality.

On the same level, predictive models of species distribution under various climate scenarios offer critical insights into conservation planning. For instance, rare and endangered species such as *Magnolia wufengensis* 'Jiaolian' are projected to experience significant habitat shifts due to climate change, as reported by Shi et al. [8]. According to their study, the suitable habitats for such species will move to higher elevations and latitudes, highlighting the need for dynamic conservation strategies that can adapt to these changes. Understanding these shifts is crucial for the protection and sustainable management of biodiversity.

Thinning practices, which are an essential technique in sylviculture and the optimization of its management, were examined by Qin et al. [9] through a hybrid modeling approach; this combined the 3-PG process model and a long short-term memory neural network. Their study offers practical guidelines for thinning practices that enhance forest growth and carbon sequestration, demonstrating the significance of adaptive management in response to climate and anthropogenic pressures.

The conservation of endemic ornamental species was explored by Shi et al. [10] who reported that, under more severe scenarios of climate change, the populations of *Helleborus tibetanus* Franchet, are at high risk of destruction. These insights are critical for the conservation and sustainable utilization of this species in China.

Similarly, Korznikov et al. [11] employed Random Forest models to explore changes in the distribution of Jezo spruce (*Picea jezoensis* (Siebold and Zucc.) Carrière) in Northeast Asia under climate change scenarios. For this species, however, the key refugia are predicted to remain suitable; hence, the establishment of artificial stands in these future climate-acceptable regions may be vital for preserving genetic diversity.

The potential ability of forest plantations to mitigate climate change was also explored by Altamirano-Fernandez et al. [12], who developed a mathematical model to optimize carbon capture in forest plantations. Their work underscores the importance of strategic planning in reforestation, thinning, and fire prevention to maximize carbon sequestration and combat global warming.

Climate change impacts the productivity of sites differently across tree species and regions. For example, in Ontario, Canada, the effects of climate on site productivity vary among jack pine, black spruce, red pine, and white spruce plantations [13]. Sharma reports that while jack pine shows positive climate effects in western Ontario, black spruce, red pine, and white spruce exhibit negative impacts, especially under high-emission scenarios. These findings highlight the need for localized management strategies that account for species-specific and regional climate responses in order to sustain forest productivity.

In conclusion, the collective research presented in this Special Issue underscores the multifaceted responses of forest ecosystems to climate change by means of both statistical and process-based models. Through modeling techniques and comprehensive analyses, these studies provide critical insights and practical solutions regarding the management and conservation of forests in a warming world. The knowledge gained from these investigations is vital for informing policy and guiding actions that will help sustain forest ecosystems and their invaluable services for future generations.

Conflicts of Interest: The authors declare no conflicts of interest.

References

1. Nooni, I.K.; Ogou, F.K.; Prempeh, N.A.; Saidou Chaibou, A.A.; Hagan, D.F.T.; Jin, Z.; Lu, J. Analysis of Long-Term Vegetation Trends and Their Climatic Driving Factors in Equatorial Africa. *Forests* **2024**, *15*, 1129. [CrossRef]
2. Morichetti, M.; Vangi, E.; Collalti, A. Predicted Future Changes in the Mean Seasonal Carbon Cycle Due to Climate Change. *Forests* **2024**, *15*, 1124. [CrossRef]
3. Vangi, E.; Dalmonech, D.; Morichetti, M.; Grieco, E.; Giannetti, F.; D'Amico, G.; Nakhavali, M.; Chirici, G.; Collalti, A. Stand Age and Climate Change Effects on Carbon Increments and Stock Dynamics. *Forests* **2024**, *15*, 1120. [CrossRef]
4. Kivalov, S.; Lopes de Gerenyu, V.; Khoroshaev, D.; Myakshina, T.; Sapronov, D.; Ivashchenko, K.; Kurganova, I. Soil Temperature, Organic-Carbon Storage, and Water-Holding Ability Should Be Accounted for the Empirical Soil Respiration Model Selection in Two Forest Ecosystems. *Forests* **2023**, *14*, 1568. [CrossRef]
5. Yong, J.; Duan, G.; Chen, S.; Lei, X. Environmental Response of Tree Species Distribution in Northeast China with the Joint Species Distribution Model. *Forests* **2024**, *15*, 1026. [CrossRef]
6. Feng, J.; Cao, Y.; Manda, T.; Hwarari, D.; Chen, J.; Yang, L. Effects of Environment Change Scenarios on the Potential Geographical Distribution of *Cunninghamia lanceolata* (Lamb.) Hook. in China. *Forests* **2024**, *15*, 830. [CrossRef]
7. Liu, J.; Yang, B.; Li, M.; Xu, D. Assessing Forest-Change-Induced Carbon Storage Dynamics by Integrating GF-1 Image and Localized Allometric Growth Equations in Jiangning District, Nanjing, Eastern China (2017–2020). *Forests* **2024**, *15*, 506. [CrossRef]
8. Shi, X.; Yin, Q.; Sang, Z.; Zhu, Z.; Jia, Z.; Ma, L. Habitat Distribution Pattern of Rare and Endangered Plant *Magnolia wufengensis* in China under Climate Change. *Forests* **2023**, *14*, 1767. [CrossRef]
9. Qin, J.; Ma, M.; Zhu, Y.; Wu, B.; Su, X. 3PG-MT-LSTM: A Hybrid Model under Biomass Compatibility Constraints for the Prediction of Long-Term Forest Growth to Support Sustainable Management. *Forests* **2023**, *14*, 1482. [CrossRef]
10. Shi, X.; Mao, L.; Sun, M.; Ma, G.; Zhu, K. Paleo Distribution and Habitat Risks under Climate Change of *Helleborus thibetanus*. *Forests* **2023**, *14*, 630. [CrossRef]
11. Korznikov, K.; Petrenko, T.; Kislov, D.; Krestov, P.; Doležal, J. Predicting Spruce Taiga Distribution in Northeast Asia Using Species Distribution Models: Glacial Refugia, Mid-Holocene Expansion and Future Predictions for Global Warming. *Forests* **2023**, *14*, 219. [CrossRef]
12. Altamirano-Fernández, A.; Rojas-Palma, A.; Espinoza-Meza, S. Optimal Management Strategies to Maximize Carbon Capture in Forest Plantations: A Case Study with *Pinus radiata* D. Don. *Forests* **2023**, *14*, 82. [CrossRef]
13. Sharma, M. Modeling Climate Effects on Site Productivity of Plantation Grown Jack Pine, Black Spruce, Red Pine, and White Spruce Using Annual/Seasonal Climate Values. *Forests* **2022**, *13*, 1600. [CrossRef]

forests MDPI

Article

Analysis of Long-Term Vegetation Trends and Their Climatic Driving Factors in Equatorial Africa

Isaac Kwesi Nooni [1], Faustin Katchele Ogou [2], Nana Agyemang Prempeh [3], Abdoul Aziz Saidou Chaibou [4], Daniel Fiifi Tawiah Hagan [5], Zhongfang Jin [6] and Jiao Lu [1,*]

1 School of Atmospheric Science and Remote Sensing, Wuxi University, Wuxi 214105, China; nooni25593@alumni.itc.nl
2 Laboratory of Atmospheric Physics, Department of Physics, University of Abomey-Calavi, Cotonou 01 BP 526, Benin; ogofaustin@gmail.com
3 School of Geosciences, Department of Atmospheric and Climate Science, University of Energy and Natural Resources, Sunyani P.O. Box 214, Ghana; agyemang.prempeh@uenr.edu.gh
4 Département de Physique, Faculté des Sciences et Techniques, Université Abdou Moumouni, Niamey BP 10662, Niger; abdoulaziz.saidou@uam.edu.ne
5 Hydro-Climate Extremes Lab, Ghent University, 9000 Ghent, Belgium
6 School of Electronic and Information Engineering, Wuxi University, Wuxi 214105, China; jinzhongfang@cwxu.edu.cn
* Correspondence: jiao_lu@cwxu.edu.cn

Abstract: Understanding vegetation seasonality and its driving mechanisms improves decision-making in the management of ecological systems in a warming global climate. Using multiple statistical methods (i.e., trend analysis, abrupt changes, and partial correlation analysis), this study analyzed the spatiotemporal variations in the Normalized Difference Vegetation Index (NDVI) in the Equatorial Africa (EQA) region and their responses to climate factors from 1982 to 2021. The NDVI values declined at a rate of 0.00023 year^{-1}, while the precipitation (P) and mean temperature (TMEAN) values increased at rates of 0.22 mm year^{-1} and 0.22 °C year^{-1}, respectively. The mean minimum temperature (TMIN) had a higher rate of 0.2 °C year^{-1} than the mean maximum temperature (TMAX) at 0.02 °C year^{-1}. An abrupt change analysis showed that the TMAX, P, and NDVI breakpoints occurred in 2000, 2002, and 2009, respectively; TMEAN and TMIN breakpoints occurred in 2001. The NDVI trends declined in forest and cropland areas but increased in shrubland and grassland areas. The summer NDVI trends declined for all vegetation types and were reversed in the winter season. The NDVI positively correlated with the P (r = 0.50) and TMEAN (r = 0.60). All seasonal analyses varied across four seasons. A temporal analysis was conducted using partial correlation analysis (PCR), and the results revealed that TMIN had a greater impact on the NDVI (PCR = −0.45), followed by the TMAX (PCR = 0.31) and then the P (PCR = −0.19). The annual trend showed that areas with significant greening were consistent with stronger wetter and weaker warming trends. Both precipitation and temperature showed a positive relationship with vegetation in semi-arid and arid regions but a negative relationship with humid regions. Our findings improve our insight into scientific knowledge on ecological conservation.

Keywords: NDVI; vegetation dynamics; climate change; precipitation; temperature; Equatorial Africa

Citation: Nooni, I.K.; Ogou, F.K.; Prempeh, N.A.; Saidou Chaibou, A.A.; Hagan, D.F.T.; Jin, Z.; Lu, J. Analysis of Long-Term Vegetation Trends and Their Climatic Driving Factors in Equatorial Africa. *Forests* **2024**, *15*, 1129. https://doi.org/10.3390/f15071129

Academic Editors: Daniela Dalmonech, Alessio Collalti and Gina Marano

Received: 9 May 2024
Revised: 23 June 2024
Accepted: 27 June 2024
Published: 28 June 2024

1. Introduction

Terrestrial vegetation is a dominant component of terrestrial ecosystems on Earth. Vegetation is an intermediary in the biosphere that influences energy–water–carbon cycles [1,2]. Therefore, monitoring and tracking vegetation dynamics are essential practices in the management of multiple sectors of ecological systems in a warming global climate [3,4]. Recent observation and climate modeling studies have indicated that global warming significantly influences climate patterns and vegetation dynamics [5,6]. For example, changes in the hydrological cycle affect soil moisture and vegetation growth. Vegetation, in turn, impacts the climate by altering energy and biogeochemical cycles [7,8]. According to the sixth assessment report of the Intergovernmental Panel on Climate Change (IPCC AR6), global warming is impacting our ecosystems, and these changes are mainly due to natural and anthropogenic causes [7]. The impacts on ecosystems in a given region affect communities whose livelihoods are directly tied to local ecosystems. However, the factors that explain global ecosystem changes vary across space and time [9]. For example, it is well known that intense social pressures on certain global land areas can suppress or promote major ecosystem richness [4]. Typical empirical evidence includes land degradation in the Amazon and Congo rainforests [10,11] and restoration programs (e.g., in China and India [12,13] and greening in the Sahel region [14,15]). Thus, the study of vegetation dynamics (i.e., patterns, seasonality, and relationships) has gained substantial attention in climate change studies [3,4].

With advances in observational remote sensing, it has become more convenient to disentangle the issue of climate impacts on greening. Remote sensing (RS) data offer scalable multi-temporal and multi-spatial analysis solutions. Many RS products are being developed to measure vegetation health, plant phenology, productivity, etc. [16–18]. The NDVI (Normalized Difference Vegetation Index) is a commonly used index based on the ratio of red (R) to near-infrared (NIR) reflectance (i.e., NDVI = (NIR − R)/(NIR + R)) [19]. The NDVI is primarily used to determine the health and density of vegetation. This indicator measures vegetated and non-vegetated terrain within the range of +1 and −1, where high NDVI values (near +1) indicate dense green vegetation and low values indicate moisture-stressed vegetation [20]. Multiple NDVI datasets are available for global and regional studies [17]. The Global Inventory Modeling and Mapping Studies of the National Oceanic and Atmospheric Administration/Advanced Very High-Resolution Radiometer Normalized Difference Vegetation Index third generation (GIMMS NOAA/AVHRR NDVI3g) dataset has been widely used to study vegetation greening since the 1980s. Over the past three decades, the GIMMS NOAA/AVHRR NDVI3g dataset has significantly improved our understanding of intra- and inter-annual variations in vegetation activity from regional to global scales [21,22]. Most of these studies have reported large patterns and trends in the magnitude and timing of vegetation activity in the Northern Hemisphere (NH) [23–25]. Additionally, increasing ("greening") or decreasing ("browning") vegetation trends have been documented over multiple timescales [26]. The trajectories of these vegetation trends have been reported to be gradual or abrupt for global and regional studies [27]. Vegetation greenness in relation to trends in climate has been investigated at different spatial (i.e., global, regional, and watershed) and temporal scales [28].

Climate drivers, such as precipitation and temperature, generally influence vegetation growth. Precipitation and temperature are the two most widely used climate variables in different regions [29]. Despite these numerous studies, the relationship between precipitation or temperature and vegetation dynamics across different climate zones at different timescales is still complex [29,30]. Most of these studies used statistical models to analyze the effect of water and heat conditions on vegetation. Recently, parametric methods have been reported to be unreliable; however, non-parametric methods such as Sen Slope and Mann–Kendall tests are widely used due to their reliability. These methods can be used to determine changes in vegetation dynamics. Some studies used partial correlation and cross-correlational analysis to quantify vegetation-driving factors. The partial correlation analysis is intended to explain the relationships between vegetation growth and the driving factors [31]. In addition, transient disturbances in the time series could be determined during changes in vegetation [29], and the Pettitt test is preferred to detect breakpoints or abrupt changes in vegetation and climate time series in many different regions [29,30].

The Equatorial Africa (EQA) region is located in the tropics, which are geographical zones that regulate hydrological and carbon cycles [32]. Previous studies on vegetation dynamics have been conducted in the Horn of Africa [33] and sub-Saharan Africa [34–37]. Moreover, recent studies have observed frequent changes in land use and land cover

(LULC) across sub-Saharan Africa. Thus, the effects of vegetation dynamics caused by climatic and non-climatic factors make this tropical region a hot spot for land–atmosphere interactions [32,38,39] and, by extension, a test bed for investigation [40]. Moreover, warming at night has been reported to be faster than that during the day in certain global lands (e.g., West Africa and Europe) [41]. However, our insight into vegetation responses to the influences of climate and LUCL remains unclear due to confounding factors, such as trends in minimum and maximum temperatures and land use changes. Therefore, it is essential to strengthen research to understand vegetation dynamics in a warming global climate. This study used the GIMMS NOAA/AVHRR NDVI dataset and precipitation and air surface temperatures (minimum and maximum) from the CRU (Climate Research Units) and AVHRR global LULC datasets, respectively. The objectives of this study are as follows:

(1) To investigate the spatiotemporal trends in vegetation and climate variables;
(2) To analyze the main climatic drivers of vegetation variability in the EQA region. All the analyses were conducted in the EQA region from 1982 to 2021.

To the best of our knowledge, further studies on the inter-relationship of vegetation and climate in the region have been recommended to include the effect of minimum and maximum temperatures on vegetation dynamics based on the updated NDVI.

2. Materials and Methods

2.1. Study Area

The study area is Equatorial Africa (EQA), which is a region in the tropical African continent with a geographical location of 18° W–55° E longitude and 2° N–20° N latitude. The tropical monsoon climate strongly influences EQA [22,42]. The West African monsoon [22] on the western side and the East African monsoon [42] on the eastern side are prevalent in the region. The average annual precipitation (P) is about 62.76 mm and varies depending on the seasons (summer = 115.48 mm, autumn = 73.01 mm, winter = 9.12 mm, and spring = 53.35 mm). The annual average temperature (TMEAN) is about 26.75 °C and varies depending on the seasons (summer = 27.75 °C, autumn = 26.59 °C, winter = 24.12 °C, and spring = 28.55 °C). The EQA climate system has a distinct wet and dry season [43]. The seasonal climatology values that were computed by the authors, based on the CRU dataset for the period 1982–2021 [44] and averaged over longitude 18° W–55° E and latitude 2° N–20° N, showed that the wettest season was summer and the driest season was winter; the warmest season was spring, and the coolest season was winter. The elevation ranges from 500 to 2000 m above mean sea level [45]. The highest elevations are found in the Ethiopian, Kenyan, and Cameroon highlands (Figure 1c). The ten (10) predominant land cover classification types [46] found in EQA are presented in Figure 1a.

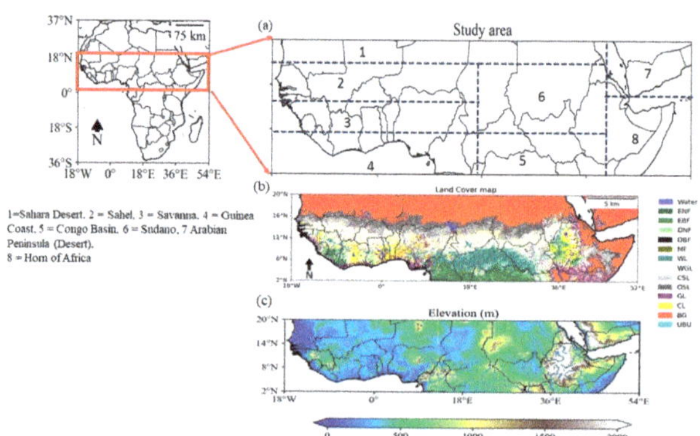

Figure 1. The location of the study area. (**a**) Study area with subregions: (1) Sahara Desert [21°–37° N], (2) Sahel [12° N–20° N, 18° W–18° E], (3) Savanna [8° N–12° N, 18° W–18° E], (4) Guinea Coast [4–8° N,

18° W–18° E], (5) Congo Basin [2°–4° N, 18°–25° E], (6) Sudano belt [12° N–20° N, 18°–25° E], (7) Arabian Peninsula (ARP) [12°–20° N, 25°–52° E], and (8) Horn of Africa (HOA) [8°–20° N, 25°–52° E]. (**b**) AVHRR land cover map: ENF—evergreen needleleaf forest (0.00%), EBF—evergreen broadleaf forest (3.89%), DNF—deciduous needleleaf forest (0.00%), DBF—deciduous broadleaf forest (0.13%), MF—mixed forest (0.00%), WL—woodland (7.68%), WGL—woodland grassland (20.20%), CSL—closed shrubland (5.52%), OSL—open Shrubland (9.80%), GL—grassland (4.38%), CL—cropland (3.67%), UBU—urban and built-up, BG—barren or sparsely vegetated (27.60%), and water (17.11%). (**c**) Elevation.

2.2. Data Sources

2.2.1. NDVI

The data used in this study include the Global Inventory Modeling and Mapping Studies (GIMMS) of the National Oceanic and Atmospheric Administration/Advanced Very High-Resolution Radiometer (AVHRR) Normalized Difference Vegetation Index third generation (NDVI3g) (i.e., GIMMS AVHRR NDVI3g) dataset, which includes precipitation, temperature, digital elevation model (DEM), and a land cover map. The GIMMS AVHRR NDVI3g data with a spatial resolution of 8 km and a temporal interval of 15-day compositing periods were obtained from the NASA website (https://daac.ornl.gov/VEGETATION/guides/Global_Veg_Greeness_GIMMS_3G.html, accessed on 10 May 2023). The National Oceanic and Atmospheric Administration/Advanced Very High-Resolution Radiometer Normalized Difference Vegetation Index third generation plus (AVHRR NDVI3g+) dataset covers January 1981 to December 2022 [18,19]. The AVHRR NDVI3g data acquisition process, which includes using various sensor platforms, advanced algorithms for preprocessing, and quality assurance procedures for atmospheric and radiometric correction, has been described in detail in previous studies [18,47]. The Africa region mask was defined to isolate the NDVI region from a gridded global dataset. To convert the bimonthly NDVI to the monthly NDVI, the maximum-value composite (MVC) method proposed by Holben et al. [48] was used. The NDVI grid values range from +1 to −1 and are computed as the ratio of (NIR − R)/(NIR + R) pixels, where NIR is the near-infrared wavelength and R is the red wavelength [18]. Positive NDVI values close to +1 indicate the presence of dense green foliage, while negative NDVI values near −1 indicate the presence of water bodies [49]. We removed negative and zero NDVI grid values that indicated non-vegetated surfaces and water bodies. The pixel grids were set to a monthly mean NDVI value of <0.1 over 39 years [19,50].

2.2.2. Climate Datasets (Precipitation and Temperature)

The Climatic Research Unit (CRU) data were used in this study to compute spatial variation, trends, and correlation between climate and vegetation. The CRU data consist of monthly spans from 1901 to the present with a resolution of 0.5° × 0.5°. The Climatic Research Unit (CRU) dataset was downloaded from the website (http://www.cru.uea.ac.uk, accessed on 10 May 2023). Gridded product data processing and validation were reported by Harris et al. [44]. In addition, the gridded data have been extensively used as they were obtained from over 4000 weather stations and interpolated based on spatial autocorrelation functions [51,52]. The temperature data comprise the monthly mean minimum temperature (TMIN) and the monthly mean maximum temperature (TMAX). The monthly mean temperature (TMEAN) was computed from the TMIN and TMAX, and the study period covered 1982–2021.

2.2.3. Land Use Land Cover

The spatial distribution of land use land cover (LULC) was obtained from the University of Maryland's Department of Geography through the website https://glad.umd.edu/ (accessed on 10 May 2023) with a spatial resolution of 1 km [46]. Out of the 14 separate land cover types, 10 were regrouped into four predominant land cover classes, including forest, grass, crop, and shrub, which were extracted for further analysis, and non-vegetated cover areas were excluded for consideration in the computation.

2.3. Methods

In this study, we applied various statistical approaches for data analyses.

2.3.1. Data Processing

All data were initially loaded, averaged, and selected from January 1982 to December 2021, and the climatology was calculated by averaging the data from 1982 to 2021 at different time scales. The seasonal scales were defined as a combined monthly average as follows: spring (i.e., March–April–May, MAM), summer (June–July–August, JJA), autumn (September–October–November, SON), and winter (December–January–February, DJF). All datasets were resampled to a spatial resolution of 0.5° × 0.5° using the bilinear interpolation method to match CRU dataset resolution. Figure 2 shows the methodological flow chart of the work.

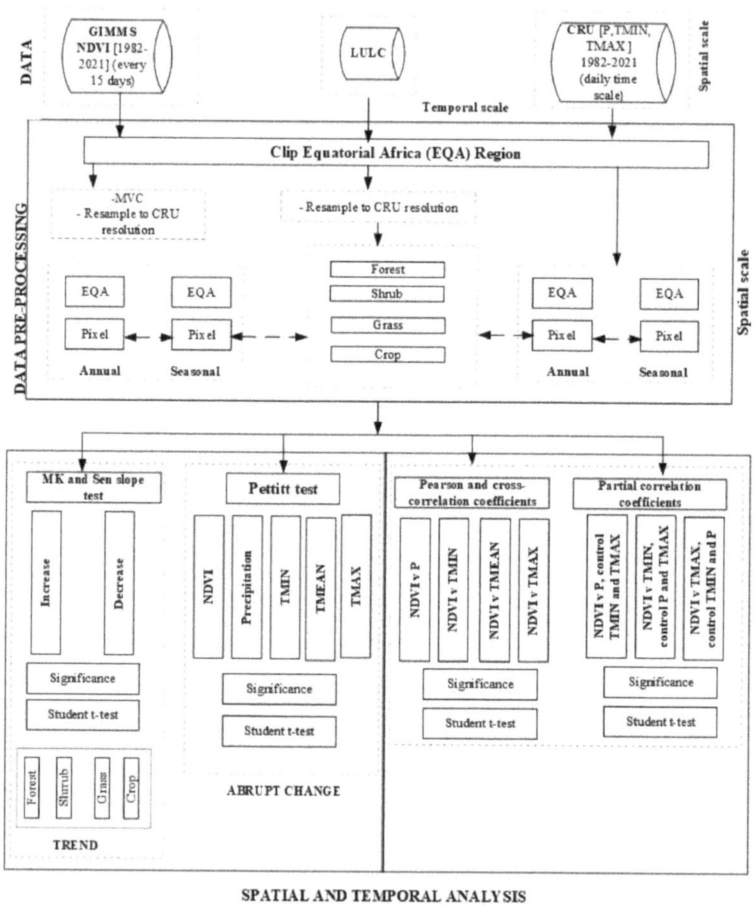

Figure 2. The work flowchart for this study.

2.3.2. Statistical Analysis

(a) Trend analysis and significance testcalculation

First, the linear trend was computed to understand the vegetation dynamics and its connections to climate in the EQA region. The linear trend analysis was computed using the non-parametric Mann–Kendall (MK) trend and Sen slope tests. The MK trend test was used to detect the significance of the time series trend, and the Sen slope test was used to compute the magnitude of the trend in the time series [53–56]. The MK trend test is a non-parametric test on data values (x) of a time series based on Equations (1)–(4) as follows:

$$\upsilon = \mathrm{f(t)} + \sum t \tag{1}$$

$$\mathrm{S} = \sum_{k-1}^{n-1} \sum_{j=k+1}^{n} sin\,(x_j - x_k) \tag{2}$$

where n is the length of the time series $x_i, \ldots x_n$, sgn () is the sign function, and x_j and x_k are values in years j and k, respectively. E[S] = 0 for series without a trend and the variance was calculated based on A3 as follows:

$$\delta^2(\mathrm{S}) = \frac{1}{18}\left[n(n-1)(2n+5) - \sum_{i=1}^{n} t_n(t_n - 1)(2t_n + 5)\right] \tag{3}$$

where n is the number of tied groups and t_n is the number of data values in the nth group. The test statistic Z is as defined in Equation (4):

$$\mathrm{Z} = \begin{cases} \frac{s-1}{\sqrt{\delta^2(s)}}\, if\ S > 0 \\ 0\, if\ S = 0 \\ \frac{s+1}{\sqrt{\delta^2(s)}} if\ S < 0 \end{cases} \tag{4}$$

The Z-statistics test the null hypothesis (Ho), which states that there is no trend, against the alternative hypothesis (H1), which states that there is a trend. H_1 signifies an increase or decrease trend in the data. The Sen slope is used to estimate the true slope as follows:

$$Y = \mathrm{m}x + \mathrm{c} \tag{5}$$

where Y and x are the dependent and independent variables, respectively; m is the gradient; and c is the intercept.

(b) Calculation of abrupt changes

Second, we used the Pettitt test [56] to detect abrupt changes in the time series of the NDVI and climate variables. The null hypothesis (Ho), of no change, was tested against the alternative hypothesis (Ha), which was changed. We implemented the function based on Equations (6)–(8), following Verstraeten et al. [57] as follows:

$$P_{ij} = \begin{cases} -1,\ x_i < x_j \\ x_i = x_j \\ 1\, x_i > x_j \end{cases} \tag{6}$$

where x_i and x_j denote the magnitude of climate variables; x_i precedes x_j.

$$\mathrm{Q_{t,T}} = \sum_{i=1}^{t} \sum_{j=t+1}^{T} P_{ij} \tag{7}$$

$Q_{t,T}$ is the Mann–Whitney statistic for samples, $x_1, \ldots x_t$ and $x_{t+1}, \ldots, x_T$, which denote the series of observed data.

The test statistic $Q_{,T}$ is computed based on expected values of t ranging from 1 to T. Based on the test statistic below, the change point is computed using a two-tailed test.

$$\mathrm{W_T} = \max|Q_{t,T}| \tag{8}$$

If the statistic W_T differs significantly from 0, then there is a change in year t that corresponds to the time for the largest absolute value of $Q_{t,T}$. The probability of a shift in one year is the maximum $|Q_{t,T}|$.

$$L = 2\exp\left(\frac{-6W_T^2}{T^3 + T^2}\right) \tag{9}$$

Based on the significance level (α), if we reject the Ho hypothesis (null hypothesis) when $L < \alpha$, we can conclude that X_t is a significant change point at the α level.

(c) Correlation Analysis Model

Third, the relationships between the NDVI and single climate variables were performed using Pearson correlation coefficients at annual and seasonal scales over 39 years using Equation (10). Furthermore, we computed the correlation between the trends in the NDVI response to LULC as follows:

$$r = \frac{\sum_{i=1}^{n}(x_i - \overline{x})(y_i - \overline{y})}{\sqrt{\sum_{i=1}^{n}(x_i - \overline{x})^2(y_i - \overline{y})^2}} \tag{10}$$

where r is the correlation coefficient (r), x_i and y_i denote the climate variables, i denotes the time, and n denotes the sample size.

In addition, a partial correlation coefficient (PCC) was also used to determine the main driving factors for vegetation growth [58,59]. The PCC determines the relationship between vegetation responses to the different climatic factors (i.e., precipitation and temperature). The PCC between the NDVI and each climate factor was computed (Equation (11)), with the other two as control variables. The climate variable with the greatest partial correlation coefficient (PCR) was interpreted as the main driver. The PCC analysis model is provided as follows:

$$\rho xyz = \frac{\rho xy - \rho xz \rho yx}{\sqrt{(1 - \rho^2 xz) * (1 - \rho^2 yz)}} \tag{11}$$

where ρxyz is the partial correlation of variables x and y conditional on z, ρxy is the correlation between variables x and y, and ρyz is the correlation between variables y and z. The correlation values range from -1 to +1 to denote negative and positive correlation, respectively.

Furthermore, we used cross-correlational analysis (CCA) to assess the time lag effects of the NDVI responses to precipitation or temperature at a specified time lag [60]. The CCA in Equations (12) and (13) was used to analyze the spatial patterns of time lag-correlation for the time lags for 1, 2, 3, 6, 9, and 12 months as follows:

$$ccf(\tau) = \sum x(t_i) * y(t_i + \tau) \tag{12}$$

$$ccc(\tau) = \frac{ccf(\tau)}{\sqrt{(Var(x)) * Var(y)}} \tag{13}$$

where $ccf(\tau)$ denotes the cross-correlation function, and $ccc(\tau)$ signifies cross-correlation. $Var(x)$ and $Var(y)$ are variations (standard deviations) of x and y, respectively. τ is a time lag x, and $x(t_i)$ and $y(t_i)$ are measured values of two variables at $t = t_i$. For example, when $\tau > 0$, it means A leads B; $\tau < 0$ means A lags B.

3. Results

3.1. Seasonal Analysis of the NDVI

Annually, the highest NDVI values of greater than 0.5 (>0.5) are in locations 2°–8° N, 18° W–8° E and 2°–8° N, 8°–35° E, with NDVI values of 0.20–0.39 in semi-arid regions along 2°–8° N, 36–52° E and 8°–14° N. Dense vegetation canopy tends to have positive NDVI values greater than 0.4 to 0.8, which is consistent with vegetation conditions in the humid regions of the Congo Basin. Moderate NDVI values (0.2–0.35) are located in the Sahel, Savanna, GC, and southern Sudano. The lowest values of 0.1 are located in arid regions (14°–20° N). Vegetation with NDVI values < 0.1 tends to be scattered vegetation consistent with conditions in the arid regions of the Sahara Desert, Arabian Peninsula, and Horn of Africa (Figure 3a). Seasonally, the NDVI values showed similar spatial variability in the NDVI distribution, albeit with differences in the NDVI values (Figure 3b–e). Winter and spring seasons showed a similar spatial pattern in the NDVI, albeit the values differed (Figure 3b,c). The summer season showed the lowest NDVI values in arid regions of the SD, northern Sudano, ARP, and HOA. Semi-arid areas of the Sahel, Savanna, GC, and southern Sudano showed moderate values between 0.2 and 0.3 and >0.4 (Congo Basin) (Figure 3d). The autumn season showed values of >0.4 in areas in the western Savanna, Guinea Coast, and Congo Basin (Figure 3e).

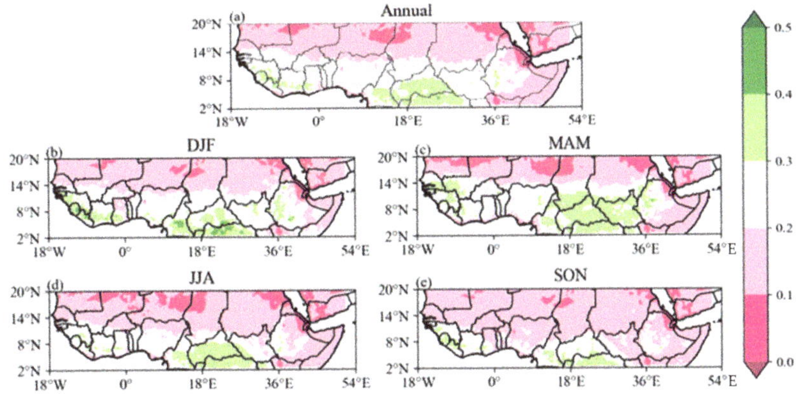

Figure 3. The spatial distribution of the maximum NDVI values across EQA from 1982 to 2021. (**a**) annual, (**b**) winter (DJF), (**c**) spring (MAM), (**d**) summer (JJA), and (**e**) autumn (SON).

The monthly NDVI values range from 0.04 to 0.22 (Figure 4, green bar). The monthly P ranges from 0 to 140 mm (Figure 4, blue line), and the TMEAN ranges from 22 to 30 °C (Figure 4, red line). All variables exhibited a clear periodic change, where the NDVI values peaked at 0.22 in October and reached their lowest value at 0.16 in July. P seasonality was highest (lowest) in August (January) at 138 (2 mm). The monthly TMEAN showed the highest (lowest) values in May (January).

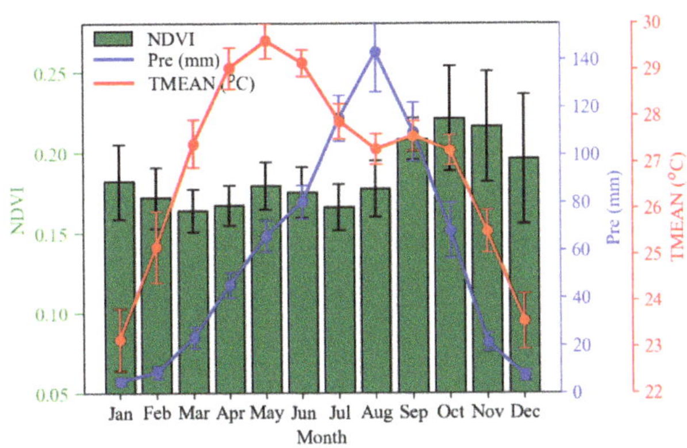

Figure 4. Monthly average changes in the NDVI (green bar) and climate factors in the EQA region. NDVI—green bar; precipitation—blue line; and mean temperature—red line.

3.2. Long-Term Changes in NDVI and Climate Drivers

3.2.1. Spatial Trends in NDVI and Climate Drivers

The NDVI trends varied at a decadal rate of −0.5 to 0.5. However, many regions showed no significant changes in NDVI trends across the EQA region (Figure 5). The annual NDVI values exhibited an increasing trend at a rate of 0.5 per decade in the western Savanna and Sahel and in a few patches in the eastern Sahel and southern Sudano. Similarly, the annual NDVI values exhibited patches with a significant negative trend scattered along the eastern Guinea Coast, Congo Basin, and the tip of the HOA at a rate of 0.1 per decade (Figure 5a). The seasonal NDVI trends range from 0.1 to 0.5 per decade and are presented in Figure 5b–e. The spring NDVI results showed mixed trends similar to the spatial pattern of annual NDVI trends, albeit with differences in trend values (Figure 5b). Summer showed widespread, significantly decreasing NDVI trends at 0.3 per decade in areas along the Guinea Coast, eastern Savanna, and Congo Basin (Figure 5c). Autumn (Figure 5d) and winter (Figure 5e) exhibited a similar increasing trend at 0.5 per decade along the Sahel. However, the winter season showed increasing NDVI trends extending to the Savanna, Guinea Coast, Congo Basin, southern Sudano, and patches scattered in the HOA (Figure 5e).

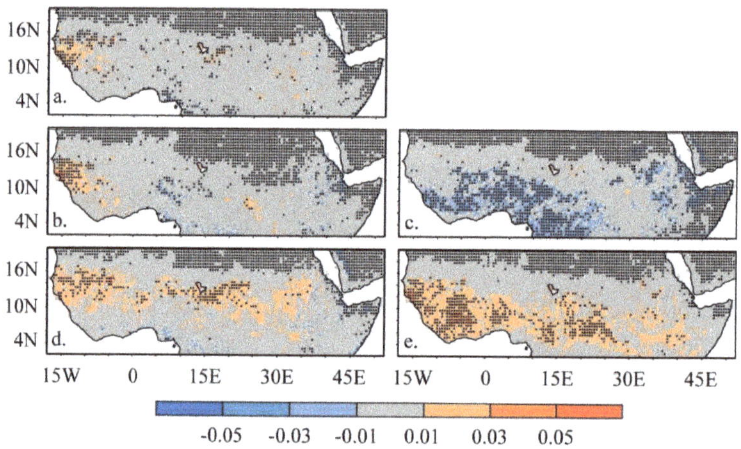

Figure 5. The spatial patterns of the NDVI trends in Equatorial Africa during 1982–2021. (**a**) Annual, (**b**) spring (MAM), (**c**) summer (JJA), (**d**) autumn (SON), and (**e**) winter (DJF). The blue–red legend denotes negative–positive trends. The dots in the maps denote the significant results at $p < 0.05$. Positive and negative changes in the NDVI are called greening and browning, respectively.

The linear trend in P ranges from −5 to +5 mm per decade, as shown in Figure 6. Annually, P showed significant increasing trends in the Sahel, Savanna, Sudano, and HOA. The central Guinea Coast areas showed positive trends at 1 mm per decade but were insignificant ($p < 0.05$). The spring season showed no significant P trends (Figure 6b). The P trends during the summer presented similar values to annual P trends, except the Guinea Coast showed significant decreasing trends (Figure 6c). Similarly, the spatial pattern of P trends in autumn was identical to the annual P trends (Figure 6d, SON). The winter season showed significantly increasing P trends along the Guinea Coast and patches in the Congo Basin (Figure 6e).

The annual TMEAN ranged from −0.5 to +0.5 °C and increased across the study area (Figure 7a). Areas in the Guinea Coast, central Sahel, Congo Basin, western Sahel, northern Sudano, and parts of the HOA increased at a rate of 0.3–0.5 °C (Figure 7a, year). Generally, the spatial patterns of TMEAN trends during the spring (Figure 7b, MAM) and winter (Figure 7e, DJF) seasons were identical to annual trends. Summer trends showed that the TMEAN increased at 0.1 °C per decade along the periphery of the EQA region (Figure 7c). During the autumn season, a significantly increasing trend at 0.3 °C is widespread across the area (Figure 7d). A similar trend analysis was shown for the TMIN (Figure S1) and TMAX (Figure S2). However, in the EQA region, the TMIN exhibited more pronounced warming trends than the TMAX.

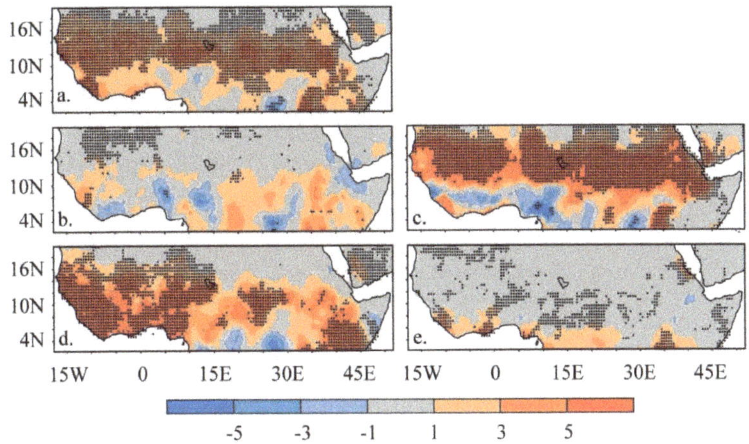

Figure 6. The spatial patterns of precipitation (P) trends. (**a**) Annual, (**b**) spring (MAM), (**c**) summer (JJA), (**d**) autumn (SON), and (**e**) winter (DJF). The blue–red legend denotes negative–positive trends. The dots in the maps denote the significant results at $p < 0.05$. Positive and negative changes in precipitation are called wetting and drying, respectively.

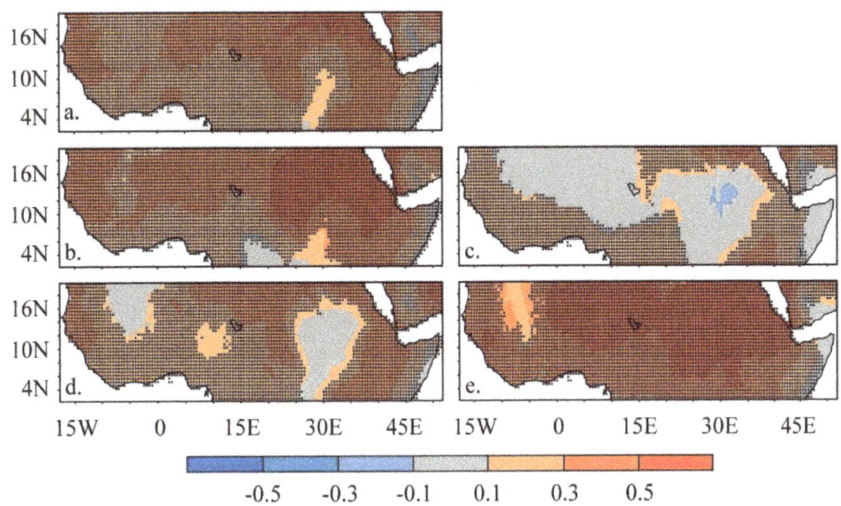

Figure 7. The spatial patterns of mean average temperature (TMEAN) trends. (**a**) Annual, (**b**) spring (MAM), (**c**) summer (JJA), (**d**) autumn (SON), and (**e**) winter (DJF). The blue–red legend denotes negative–positive trends. The dots in the maps denote the significant results at $p < 0.05$. Positive and negative temperature changes are called warming and cooling, respectively.

3.2.2. Temporal Trends in NDVI and Climate Drivers

Figure 8 presents the annual and seasonal variability in the NDVI and climate variables (i.e., precipitation and temperature) from 1982 to 2021. Table 1 illustrates the tabulated trend rate. The NDVI trends range from 0.05 to 0.30 (Figure 8a). Overall, the annual NDVI trends decreased at a decadal rate of -2.3×10^{-4} (Figure 8a, blue color). Moreover, spring (Figure 8a, orange color) and summer (Figure 8a, yellow color) showed that NDVI trends decreased at -3.9×10^{-4} and $-7.5 \times 10^{-4*}$ year^{-1}, while the NDVI showed increasing trends of 3.3×10^{-4} and 1.4×10^{-5} year^{-1} in autumn (Figure 8a, purple color) and winter (Figure 8a, green color), respectively. The annual P ranged from 0 to 120 mm (Figure 8b) and increased at 2.0* mm per decade from 1982 to 2021 (Table 1, P). On a seasonal scale, all four seasons showed a significant increasing trend. The summer season recorded the highest trend at 4.5 mm per decade, followed by autumn at 3.4* mm per decade. The increase in spring was 0.9 mm per decade, and winter was the lowest at 0.4 mm per decade. The average mean temperature (TMEAN) ranges from 22 °C to 30 °C (Figure 8c). The trend increased at an annual rate of 0.2* °C per decade (Table 1, TMEAN). Spring and autumn exhibited a trend rate of 0.2* °C per decade, respectively. However, the trend rate is slightly higher in winter and lowest in summer at 0.3* °C and 0.1 °C (α = 000) per decade, respectively. In addition, the TMIN and TMAX presented significantly increasing trends annually and in the four seasons, as presented in Figures S3 and S4, respectively. The TMIN ranges from 15 to 23 °C in Figure S3, and the TMAX ranges from 30 to 36 °C in Figure S4, respectively. Remarkably, all three temperature values (TMIN, TMEAN, and TMAX) exhibited an annual trend rate of 0.2* °C per decade, albeit with differences in values. At the seasonal level, the TMIN and TMEAN exhibited similar seasonal trend rates, except for the winter season, which presented rates of 0.2* °C and 0.3* °C per decade, respectively (Figures S3 and 8c). On the other hand, the results of the TMAX showed a similar value of trend rate with summer and winter at 0.1 °C and 0.3* °C per decade, respectively. In spring and autumn, temperatures increased slightly to 0.31* °C and 0.1* °C per decade, respectively (Figure S4, Table 1).

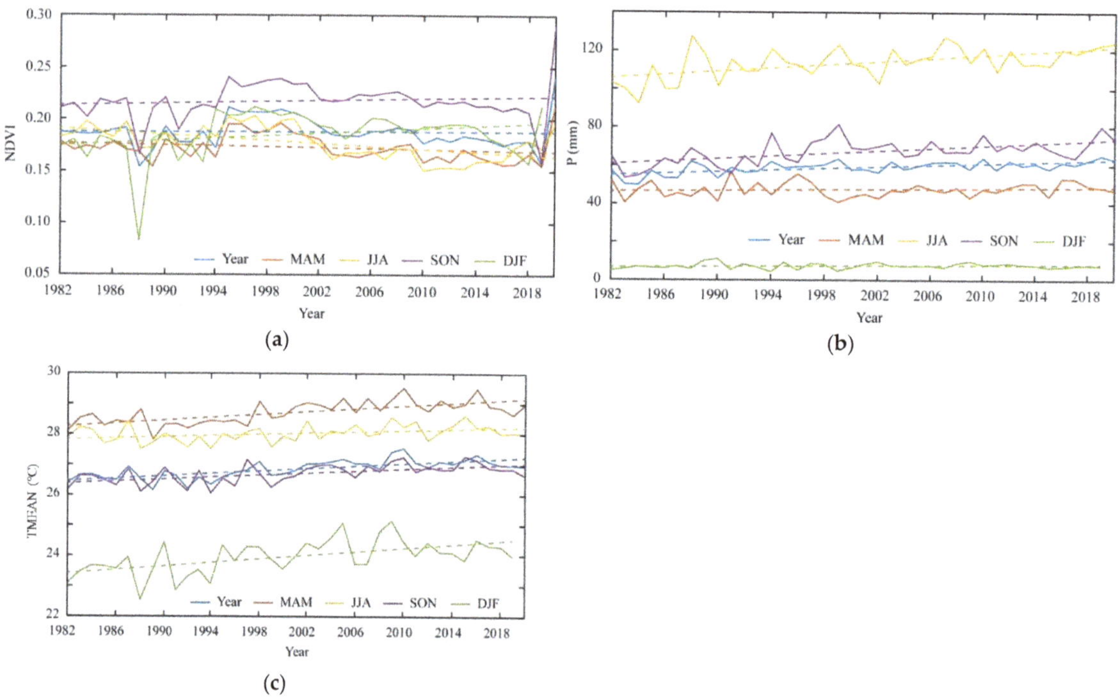

Figure 8. The inter-annual variations in (**a**) the mean NDVI, (**b**) P, and (**c**) TMEAN at annual and seasonal timescales from 1982 to 2021. The linear trend calculated using the linear regression (dashed lines) trend is calculated using the least squares linear trend fitting method over the period (at *p*-value < 0.05). Annual (blue color), spring (MAM, orange color), summer (JJA, yellow), autumn (SON, purple color), and winter (DJF, green color).

Table 1. The slope of the annual and seasonal NDVI and meteorological variables from 1982 to 2021 over the EQA region.

Variable	Time Scales				
	Annual	MAM	JJA	SON	DJF
NDVI	$m = -2.3 \times 10^{-4}$, $c = 0.19$, ($\alpha = 0.097$)	$m = -3.9 \times 10^{-4}$, $c = 0.18$, ($\alpha = 0.259$)	$m = -7.5 \times 10^{-4}$, $c = 0.19$, ($\alpha = 0.001$)	$m = 1.4 \times 10^{-5}$, $c = 0.22$, ($\alpha = 0.478$)	$m = 3.3 \times 10^{-4}$, $c = 0.18$, ($\alpha = 0.076$)
P	m = 0.20, c = 55.44, ($\alpha = 0.000$)	m = 0.09, c = 45.00, ($\alpha = 0.032$)	m = 0.45, c = 104.66, ($\alpha = 0.000$)	m = 0.34, c = 58.91, ($\alpha = 0.000$)	m = 0.04, c = 6.50, ($\alpha = 0.000$)
TMEAN	m = 0.02, c = 26.46, ($\alpha = 0.000$)	m = 0.02, c = 28.18, ($\alpha = 0.000$)	m = 0.01, c = 27.79, ($\alpha = 0.008$)	m = 0.02, c = 26.32, ($\alpha = 0.000$)	m = 0.03, c = 23.42, ($\alpha = 0.001$)
TMAX	m = 0.02, c = 33.14, ($\alpha = 0.000$)	m = 0.03, c = 35.17, ($\alpha = 0.000$)	m = 0.01, c = 33.40, ($\alpha = 0.043$)	m = 0.01, c = 32.89, ($\alpha = 0.001$)	m = 0.03, c = 31.08, ($\alpha = 0.001$)
TMIN	m = 0.02, c = 19.90 ($\alpha = 0.000$)	m = 0.02, c = 21.36, ($\alpha = 0.000$)	m = 0.01, c = 22.27, ($\alpha = 0.000$)	m = 0.02, c = 19.95, ($\alpha = 0.000$)	m = 0.02, c = 15.82. ($\alpha = 0.001$)

Minus (−) value indicates a decreasing trend. α is the *p*-value.

The annual NDVI declined in forests and croplands at a rate of -3.0×10^{-3} and -0.61×10^{-3} and increased in shrubs and grasslands by 2.2×10^{-3} and 4.0×10^{-3} across the region from 1982 to 2021 (Figure 9). Seasonally, the NDVI declined in the forest and cropland covers conservatively in spring, summer, and autumn, except in winter, where it increased but at different rates of magnitude (Figure 9). In contrast, the NDVI increased in grassland during the spring and autumn transition seasons, except for a decline in summer. In addition, the NDVI in shrubs decreased (increased) in spring and summer (autumn and winter), albeit with different trend values.

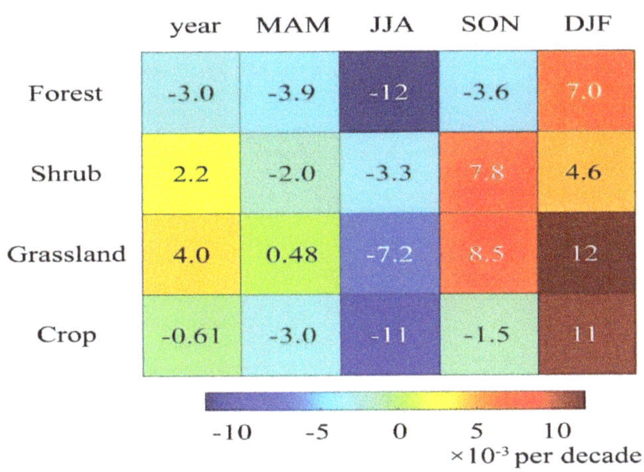

Figure 9. The temporal trends in the mean NDVI from 1982 to 2021 across EQA in four dominant vegetation types.

3.3. *Abrupt Change Analysis of NDVI and Climate Drivers*

Table 2 presents breakpoint changes for the NDVI and climate variables from 1982 to 2021. Overall, the breakpoint results showed considerable similarities and differences over the period. On the inter-annual scale, the NDVI, P, TMEAN, TMIN, and TMAX breakpoints varied and occurred at different years. The annual NDVI breakpoints were observed in 2009. For the different seasons, the spring and summer NDVI breakpoints occurred in spring and summer in 2002, while the autumn and winter breakpoints occurred in 1994 and 1993, respectively. The annual precipitation (P) breakpoint occurred in 2002 and seasonally, except in summer, which exhibited an identical breakpoint as the annual; the rest occurred in 2011 for spring, in 1996 for autumn, and winter in 1998. In the temperature values, the annual breakpoint for the TMEAN and TMIN was observed in 2001, while the TMAX occurred in 2000. Seasonally, the TMIN showed that summer and autumn had similar breakpoints in 2001, while the spring breakpoints occurred in 2000.

Table 2. The years of abrupt changes in the NDVI, P, TMEAN, TMIN, and TMAX from 1982 to 2021.

Variable	Years of Abrupt Changes				
	Annual	**MAM**	**JJA**	**SON**	**DJF**
NDVI	2009	2002	2002	1994	1993
P	2002	2011	2002	1996	1988
TMEAN	2001	1997	2001	2001	1994
TMIN	2001	2000	2001	2001	2001
TMAX	2000	1997	2001	2000	1994

An analysis of the trend slope before and after the breakpoint was performed (Table 3). The annual NDVI breakpoint displayed a positive (0.005) value before 2009 and a negative value after 2009, indicating that the NDVI trends were not monotonic across the EQA region from 1982 to 2021 as shown in Table 3. Seasonal analysis showed that the NDVI values exhibited positive trends in the spring before and after the breakpoint. The summer (JJA) NDVI values showed a positive trend before and after the breakpoints. The autumn and winter NDVI trends were negative before and positive after the breakpoints, respectively. For precipitation (P) analysis of trend estimates, we observed a significant positive annual trend before (0.3645) and after (0.4830) the breakpoint in 2009 ($p < 0.05$). Spring P exhibited positive values before and after the negative breakpoints. Summer, autumn, and winter P showed positive trends before and after the breakpoints. In addition, the TMEAN and TMIN showed no significant change in annual trends before and after the 2001 breakpoints. The TMIN, TMEAN, and TMAX exhibited positive trends before and after the spring, autumn, and winter breakpoints. In contrast, the TMIN, TMEAN, and TMAX showed negative trends before and after the breakpoints in the summer season.

Table 3. The trends in the abrupt changes before and after for the NDVI, P, TMEAN, TMIN, and TMAX from 1982 to 2021.

Trends before Abrupt Changes					
Variable	Annual	MAM	JJA	SON	DJF
NDVI	0.005	0011 *	0.0005	−0.0006	−0.0018
P	0.3645 *	−0.0339	0.7566 *	0.5653	0.4220
TMEAN	0.0114	−0.0021	−0.0060	0.0053	−0.0009
TMIN	0.0065	0.0058	−0.0063	0.0055	0.0193
TMAX	0.0151	0.0026	−0.0057	0.0029	−0.0033
Trends after Abrupt Changes					
Variable	Annual	MAM	JJA	SON	DJF
NDVI	−0.0013	0.0006	0.0006	0.0009 *	0.0015 *
P	0.4830 *	1.6933	0.9599 *	0.5658 *	0.0835 *
TMEAN	0.0114	0.0033	−0.0060	0.0055	0.0394 *
TMIN	0.0065	−0.0066	−0.0063	0.0053	0.0193
TMAX	0.0186	0.0044	−0.0057	0.0079	0.0476 *

* Asterisk significance level at $p < 0.05$.

3.4. Analysis of Factors That Drive NDVI Changes

Correlation Analysis of NDVI and Climate Drivers

The annual correlation coefficient (r) between the NDVI and climate factors shows R values ranging between −0.50 and +0.50 (Figure 10). Spatially, a strong positive correlation ($r > 0.5$, $p < 0.05$) was found in the Sahel and northern Sudano regions, and a negative correlation was observed in the SD and ARP ($R > 0.45$) between the NDVI and P (Figure 10a). However, mixed results of weak relationships ($r \leq 0.30$) were found in the Savanna, Guinea Coast, and Congo Basin. Figure 10b (NDVI-TMIN) shows that strong and significant positive $r \geq 0.5$ correlations mainly occurred in the western Sahel and Guinea Coast, while a weakly insignificant correlation was found in areas of southern Sudano ($r \leq 0.30$, $p > 0.05$). The spatial patterns of positive and negative correlation trends in Figure 10c (NDVI-TMEAN) and Figure 10d (NDVI-TMAX) present similar values, with differences in magnitudes. However, the spatial patterns of positive and negative correlation trends in Figure 10c (NDVI-TMEAN) and Figure 10d (NDVI-TMAX) are identical. Seasonally, a strong relationship between the NDVI and P was observed, and this spatial pattern is comparable to spring–summer–autumn in the Sahel, Savanna, and Guinea Coast. However, the spatial patterns of r values between the NDVI and the temperature values (TMIN, TMEAN, and TMAX) were comparable to the spring and summer seasons across the region (Figure S5).

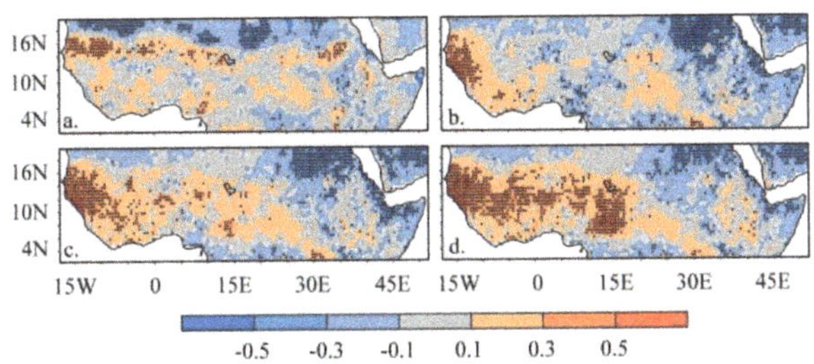

Figure 10. Spatial correlation analysis of the annual and seasonal NDVI with climatic variables: (**a**) NDVI vs. PRE, (**b**) NDVI vs. TMIN, (**c**) NDVI vs. TMEAN, and (**d**) NDVI vs. TMAX across EQA during 1982–2021. The blue–red legend denotes a negative–positive relationship. The dots in the maps denote the significant results at $p < 0.05$.

We analyzed the spatial patterns of time lag correlation to assess the time lag effects of the NDVI as responses to precipitation and temperature at the time lags of 1, 2, 3, 6, 9, and 12 months. Table 4 presents the time lag correlation coefficients between the NDVI and P and T from 1982 to 2021. Overall, the time lag of the NDVI negative correlation coefficients of the NDVI with responses to climatic factors differed with R values ranging from −0.6 to +0.6 over 1 to 3 months (Table 4), which signifies that the NDVI response to climate varies within 1 to 3 months, specifically the maximum and mean temperatures. Meanwhile, the TMIN leads the NDVI for up to 9 months, while the P leads the NDVI for up to 6 months. The response of the NDVI to these climate factors varied from 1 to 9 months. It can be seen that from the 12th month, the climate variables mostly lead to the NDVI. The spatial distribution of the time lag correlation is provided in Figures A1–A4 at the annual scale for the time lags for 1, 2, 3, 6, 9, and 12 months. Although significant positive correlations were found in the study area, negative correlations of the NDVI with climatic factors were observed in certain areas. This explains the heterogeneous response of the vegetation cover to climatic conditions (Figures A1–A4).

Table 4. The time lag correlation coefficients between the NDVI and P and T from 1982 to 2021.

Time Lag	Lag-Corr (NDVI, TMIN)	Lag-Corr (NDVI, TMEAN)	Lag-Corr (NDVI, TMAX)	Lag-Corr (NDVI, P)
1	−0.27	−0.27	−0.27	−0.21
2	−0.43	−0.42	−0.34	−0.33
3	−0.43	−0.39	−0.27	−0.35
6	−0.006	+0.11	0.28	−0.23
9	−0.32	+0.29	+0.20	+0.27
12	+0.14	+0.06	−0.07	+0.29

Note: A correlation coefficient (r) value ranges from −1 to +1. A minus sign indicates a negative correlation, and a plus sign indicates a positive correlation.

We performed a partial correlation analysis to clarify the factors driving temporal dynamics in the NDVI. Overall, the PCC results of the NDVI–climate relationship indicate that the TMIN and P results were statistically significant at $p < 0.05$ (Table 5). The TMIN showed moderate negative PCC with the NDVI (R = −0.45, $p < 0.05$), followed by P with low negative PCC (R = −0.19, $p < 0.05$). This result suggests that the TMIN was the main driver that moderately influenced the NDVI during the study period.

Table 5. The partial correlation coefficients between the NDVI and P and T (i.e., TMIN and TMAX) from 1981 to 2021.

Climate Factors	NDVI
P	−0.19 *
TMIN	−0.45 *
TMAX	+0.31

Note: A higher PCR value infers a greater effect and, hence, the main driving factor. A minus sign indicates the factor suppresses vegetation growth, and a plus sign encourages vegetation growth. The asterisks indicate significance at $p < 0.05$.

4. Discussion

This study investigated the long-term record of vegetation dynamics and main drivers across the Equatorial Africa (EQA) region based on the annual and seasonal scale. The spatiotemporal distribution of vegetation and climate were analyzed. Trend analysis was used to analyze the spatial heterogeneity and the dynamic variations in vegetation growth and climate. In addition, correlation (partial) analysis was used to analyze the relationship between vegetation growth and climate and the driving factors. The spatial patterns in the NDVI are highly seasonal and exhibit contrasting seasonal patterns. Forests dominate the Guinea Coast, and the Congo Basin region exhibits higher NDVI values. The Sahel, Sudano, and Savanna areas are semi-arid and dominated by shrub woodlands, grasslands, and crops exhibiting high values. In contrast, areas in the Sahara Desert, Arabian Peninsula, and parts of the Horn of Africa are arid and dominated by sparse shrubs, crops, and grasslands. Overall, the vegetation pattern and its seasonal distribution suggest that changes in vegetation productivity depend on season and location, which is related to climate (e.g., precipitation and temperature).

The distribution across the region has been confirmed in previous studies [15,31,61], suggesting that vegetation trends are increasing and decreasing in different areas worldwide [62]. Some studies found that vegetation increased while others declined [63,64]. Our study found that the vegetation trends displayed spatial heterogeneity, and the overall trend slightly declined from 1992 to 2021. Previous global studies have also reported that global average temperature and changing precipitation regimes are expected to alter moisture conditions in various global land regions [9]. Our results showed that the trend in the mean average temperature increased in agreement with past studies [9]. Increasing minimum and maximum temperatures contributed to increasing mean average temperatures, which is consistent with Meehl et al. [65]. Increasing minimum temperatures are related to vapor pressure feedback, while maximum temperatures are related to local conditions such as soil moisture [66,67]. In addition, our precipitation results exhibited an increasing trend in the semi-arid regions. In contrast, mixed trends are present in the humid regions, and there was no change in precipitation in the arid regions. Our analyses of the temporal trend changes to detect the timing and significance of changes in vegetation and climate drivers showed that the breakpoint years occurred and shifted at different years consistent with past studies [68] and the reasons that drive changes require further analysis. However, there seems to be a general agreement that in parts of the study area, such as the Sahel and Sudano regions, natural climate variability influences these changes more than human-related activities, as reported by Fensholt and Rasmussen [69].

Regarding the GIMMS NDVI observations, previous global studies have reported vegetation greening in the Northern Hemisphere (NH) [70]. Other studies have reported that greening has stalled or reversed [63,70–72]. Our results showed spatial vegetation heterogeneity as the different regions exhibited distinct vegetation variations in browning, greening, or stagnation. Our results showed that the vegetation trend increased in the western Savanna, Sahel, and parts of Sudano. These locations are dominated by shrubs, grasses, crops, and woodlands, and the climate zone is considered semi-arid [73,74]. Furthermore, this study showed that greening along the Sahel region is consistent with Zhao et al. [62] compared with global research that reported greening in the northern hemisphere [75]. In addition, stagnant or no change in vegetation activity suggests that vegetation transitioning stalled predominately forested areas along the Guinea Coast and Congo Basin, and the region's climate is considered a humid zone [73,74]. Likewise, vegetation transitioning stalled in the southern Sahara Desert, northern Sudano, Arabian Peninsula, and parts of the HOA. The region is arid, dominated by grasslands, crops, and sparse woodlands [73,74]. Our results of vegetation stalling are consistent with previous studies on arid and humid regions by Zhou et al. [10] and arid regions by Berdugo et al. [28]. Generally, changes in vegetation productivity are constrained by water and energy availability. Most global and regional studies have reported that wet regions are becoming wetter and drier areas are becoming drier [76,77]. Moreover, other global and regional studies have reported that watersheds can be water-limited or energy-limited to drive vegetation growth. In semi-arid regions, water-limited and energy-limited conditions dominate the constraints on woody foliage production [15,31,61]. Our results provide data on the semi-arid areas in the Sahel and Sudano regions, which show strong warming and wetting trends. Thus, it is unsurprising that the greening in these semi-arid regions is consistent with strong significant wetting, and the significantly weak warming trends in this study are consistent with previous research [15,31].

Previous global and regional studies have indicated that external climate factors influence vegetation changes [78]. Some studies reported that greening patterns are related to increasing temperature and precipitation [78]. Other studies have found that certain regions have stalled or even reversed due to changes in precipitation and temperature seasonality [63]. In this study, our correlation analysis showed spatial heterogeneity. Vegetation changes in the semi-arid region of the Sahel and major parts of the Savanna areas are strongly positively related to precipitation. Biasutti et al. [61] found that rainfall recovery in the Sahel supports foliage production. The analysis in the arid regions revealed a significant negative relationship between precipitation and vegetation, suggesting vegetation activity is significantly constrained by water conditions. The seasonal analysis of autumn precipitation correlated most significantly with vegetation change. Generally, the investigation showed that temperature values (minimum, average, and maximum) are significantly positive in the western Sahel and Savanna areas, whereas a significantly negative correlation is obvious in the northern Sudano and Arabian Peninsula. Semi-arid regions in the

Sahel and western Savanna demonstrate that the area is a hot spot for land–atmosphere interactions [39,79].

We further analyzed the time lag effects of the NDVI response to the different climate factors, considering monthly values of 1 to 3 months and 6, 9, and 12 months (Appendix A, Figures A1–A4). The subregions exhibited distinct inter-annual lags in vegetation response to temperature and precipitation variations. Overall, there was a significant time-lagged effect of climate factors on vegetation growth in general, with a lag of up to three months, after which the response decreased in correlation values in line with previous studies [80,81]. Similarly, minimum temperature and precipitation primarily affect vegetation growth responses. However, the response is negative based on the partial correlation coefficients. This indicates that precipitation inhibits vegetation growth because higher precipitation generally means lower temperatures. Understanding the mechanisms by which vegetation changes occur is challenging, and future studies should explore the combined effect of natural climate variability and human activities on changing vegetation dynamics. Our findings on climate drivers emphasize that regional vegetation changes can better capture specific patterns and dynamics.

5. Conclusions

This study investigated the spatiotemporal variations in the NDVI to examine the responses of vegetation to climate and environmental factors in the EQA region using trend analysis, abrupt change, correlation (partial) analysis, and MLR at annual and seasonal scales from 1982 to 2021. The following conclusions were drawn based on our findings:

1. The NDVI annual trends revealed a distinct spatial heterogeneity with obvious contrasting seasonal patterns in the Sahel, Savanna, Guinea Coast, Congo Basin, Sudano, Horn of Africa, Saharan Desert, and Arabian Peninsula at a rate of 0.5 per decade. Precipitation annual trends showed significant increasing trends in the Sahel, Savanna, Sudano, and western Guinea Coast at 0.1 mm per decade. Over the whole of the study area, the spatial patterns of the TMAX, TMIN, and TMEAN showed comparable positive trends at the annual rate of 0.2 °C per decade over the past 39 years;
2. The temporal NDVI trends decreased at an annual rate of -2.3 ($\times10^{-4}$) per decade, with trends decreasing in spring and summer and increasing in autumn and winter, i.e., -3.9 ($\times10^{-4}$) and -7.5 ($\times10^{-4}$); 3.3 ($\times10^{-4}$) and 1.4 ($\times10^{-4}$), respectively. Precipitation trends increased annually at a rate of 2.0 mm per decade and in all four seasons with rates of 4.5 mm10a^{-1}, 3.5 mm10a^{-1}, 0.9 mm10a^{-1}, and 0.4 mm10a^{-1}. The TMAX, TMIN, and TMEAN showed similar increasing annual trends at 0.2 °C (10a^{-1}) and in all four seasons;
3. The timing of the abrupt changes differed among the NDVI, P, and TMAX (i.e., 2009, 2002, and 2000), respectively, except for the TMIN and TMEAN in 2001. The NDVI breakpoints in spring and summer occurred in 2002 but differed in autumn (1994) and winter (1993). Seasonal P timing of abrupt changes differed in all four seasons (i.e., spring, summer, autumn, and spring), occurring in 2011, 2002, 1996, and 1998, respectively. The timing of abrupt changes between the TMAX and TMIN differed in spring (1997, 2000), summer and autumn (2000, 2001), and winter (1994, 2001), respectively, except in summer in 2001;
4. The annual trend showed that areas with significant greening were consistent with stronger wetter and weaker warming trends and vice versa. Spatially, summer and winter showed seasonal reversals in vegetation greening and browning trends, respectively. The spring and autumn transition seasons showed similar spatial trend patterns;
5. The relationship between the NDVI and precipitation is significantly positive in the Sahel, western Savanna, and Guinea Coast and negative in the Congo Basin, Sudano, Horn of Africa, Saharan Desert, and Arabian Peninsula. Similarly, the NDVI and temperature trends showed a significant positive relationship with temperature values (TMIN, TMEAN, and TMAX) in most of the Sahel, Savanna, and Guinea Coast areas

and a negative relationship with temperature in the Congo Basin, Sudano, Horn of Africa, Saharan Desert, and Arabian Peninsula. Across the study area, partial correlation analysis showed that vegetation growth response to climate variables was significant in precipitation and minimum temperature; however, the response was negative.

Supplementary Materials: The following supporting information can be downloaded at: https://www.mdpi.com/article/10.3390/f15071129/s1, Figure S1. Spatial patterns of the mean average minimum temperature (TMIN) trends: (a) annual, (b) spring (MAM), (c) summer (JJA), (d) autumn (SON), and (e) winter (DJF). Figure S2. Same as Figure S1 but for the average maximum temperature (TMAX). Figure S3. Inter-annual variations in the average minimum temperature (TMIN) at annual and seasonal timescales from 1982 to 2021. Figure S4. Same as Figure S3 but for the average maximum temperature (TMAX). Figure S5. Spatial correlation analysis of the seasonal NDVI with climatic variables: (a) NDVI vs. PRE, (b) NDVI vs. TMIN, (c) NDVI vs. TMEAN, and (d) NDVI vs. TMAX across EQA during 1982–2021. Top left: MAM, top right: JJA, bottom left: SON; and bottom right: DJF.

Author Contributions: Conceptualization, I.K.N., F.K.O. and J.L.; methodology, I.K.N. and F.K.O.; software, I.K.N. and F.K.O.; validation, I.K.N., F.K.O. and J.L.; formal analysis, I.K.N. and F.K.O.; investigation, I.K.N. and F.K.O.; data curation, I.K.N., F.K.O., N.A.P. and A.A.S.C.; writing—original draft preparation, I.K.N.; writing—review and editing, N.A.P., D.F.T.H., Z.J. and A.A.S.C.; visualization, N.A.P., D.F.T.H., Z.J. and A.A.S.C. All authors have read and agreed to the published version of the manuscript.

Funding: This research was funded by the Wuxi University Starting Project, 2021r010.

Data Availability Statement: The dataset used in this study is publicly available. The GLEAM algorithm is accessible via https://www.gleam.eu/ (accessed on 10 May 2033). The ERA5 variables are available at https://cds.climate.copernicus.eu/ (accessed on 10 May 2033). The GIMMS AVHRR NDVI3g dataset was obtained from the NASA website (https://daac.ornl.gov/VEGETATION/guides/Global_Veg_Greeness_GIMMS_3G.html (accessed on 10 May 2033)). The LULC from the University of Maryland was obtained from https://glad.umd.edu/ (accessed on 10 May 2033). The elevation data were obtained from NASA SRTM at 30 arc seconds (https://lpdaac.usgs.gov/products/srtmgl1v003 (accessed on 10 May 2033)).

Acknowledgments: We thank the developers, managers, and funding agencies of the data sources used in this study for granting access to these datasets in accordance with their specific data use and citation policies. We acknowledge the administrative and technical support provided by the School of Atmospheric Science and Remote Sensing, Wuxi University, as well as the use of a medical imaging platform with a CMOS image sensor (Wuxi Univ Starting Project, 2021r010) and the design of the minimally invasive surgery robot project by the School of Atmospheric Science and Remote Sensing, Wuxi University. Last, but not least, we thank all anonymous reviewers for their expertise, suggestions, and insights into this study.

Conflicts of Interest: The authors declare no conflicts of interest.

Appendix A

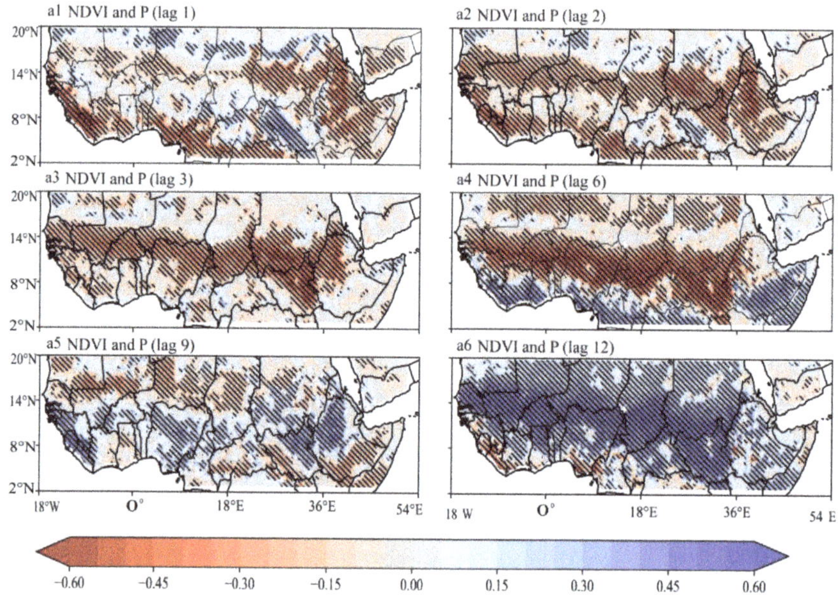

Figure A1. Spatial correlation analysis of time lag effects of NDVI with precipitation (P) at annual scale: (**a1**) NDVI vs. PRE 1 month; (**a2**) NDVI vs. PRE 2 months; (**a3**) NDVI vs. PRE 3 months; (**a4**) NDVI vs. PRE 6 months; (**a5**) NDVI vs. PRE 9 months; and (**a6**) NDVI vs. PRE 12 months across EAQ during 1982–2021. The blue–red legend denotes positive–negative relationships. The dots in the maps denote the significant results at $p < 0.05$.

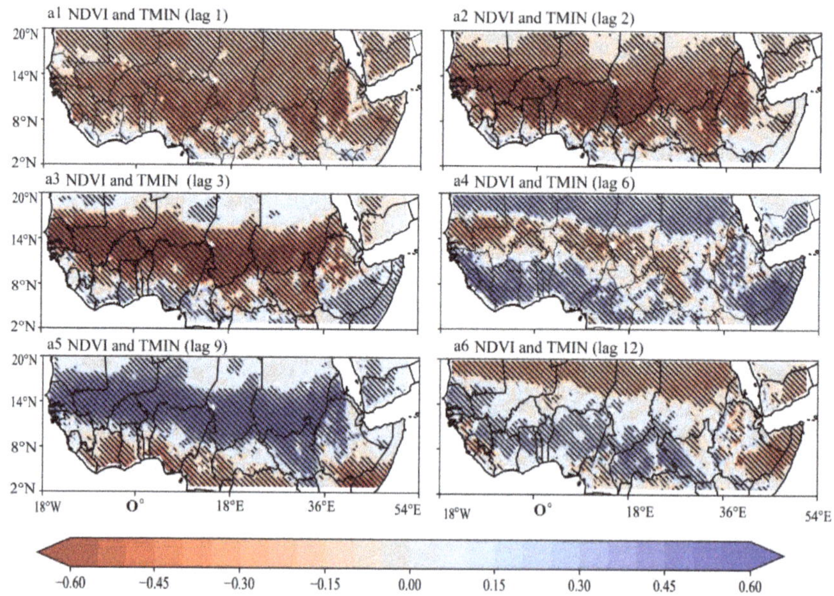

Figure A2. Same as Figure A1 but for the time lag effects of the NDVI with the mean minimum temperature (TMIN) at the annual scale.

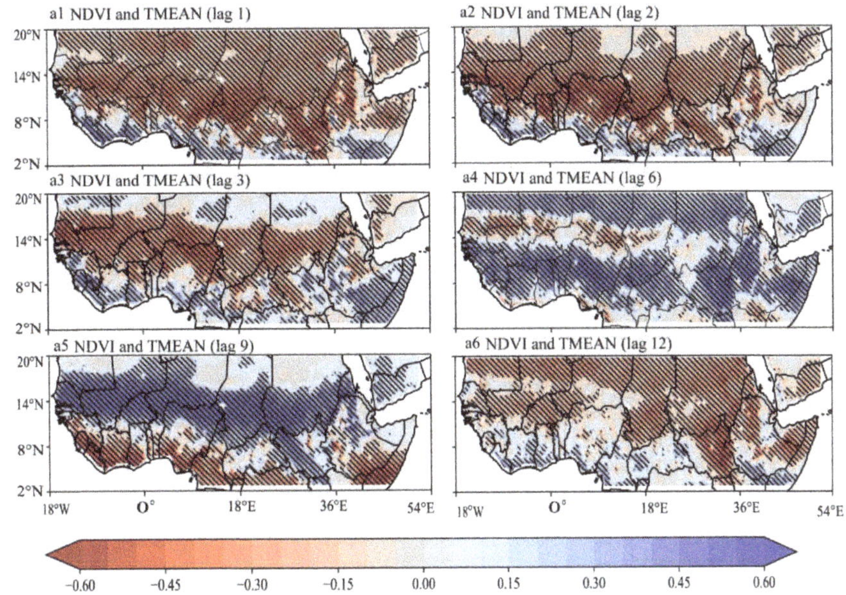

Figure A3. Same as Figure A1 but for the time lag effects of the NDVI with the mean average temperature (TMEAN) at the annual scale.

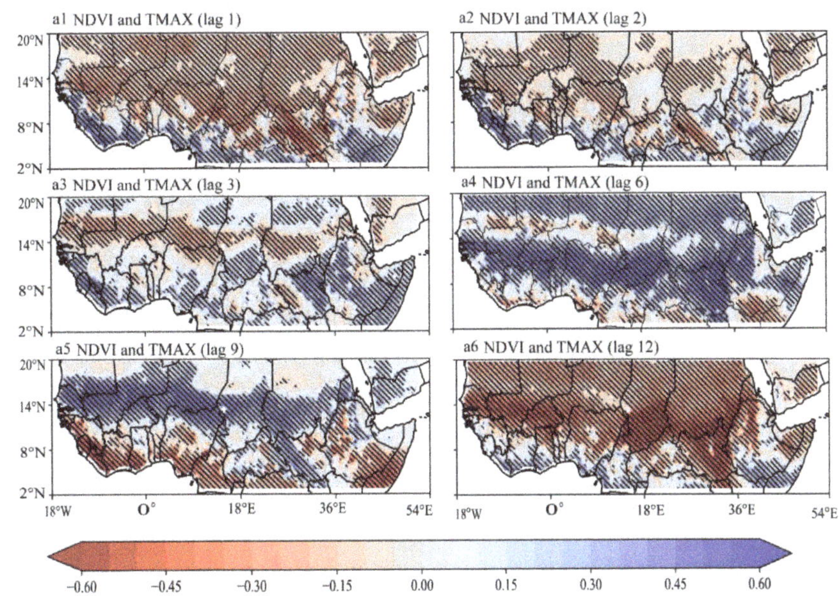

Figure A4. Same as Figure A1 but for the time lag effects of the NDVI with the mean maximum temperature (TMAX) at the annual scale.

References

1. Zeng, J.; Zhang, Q.; Zhang, Y.; Yue, P.; Yang, Z.S.; Wang, S.; Zhang, L.; Li, H.Y. Enhanced Impact of Vegetation on Evapotranspiration in the Northern Drought-Prone Belt of China. *Remote Sens.* **2023**, *15*, 221. [CrossRef]
2. Liu, X.; Sun, G.; Fu, Z.; Ciais, P.; Feng, X.; Li, J.; Fu, B. Compound droughts slow down the greening of the Earth. *Glob. Chang. Biol.* **2023**, *29*, 3072–3084. [CrossRef] [PubMed]

3. Zhang, J.; Wu, H.Q.; Zhang, Z.; Zhang, L.L.; Luo, Y.C.; Han, J.C.; Tao, F.L. Asian Rice Calendar Dynamics Detected by Remote Sensing and Their Climate Drivers. *Remote Sens.* **2022**, *14*, 4189. [CrossRef]
4. Terasaki Hart, D.E.; Yeo, S.; Almaraz, M.; Beillouin, D.; Cardinael, R.; Garcia, E.; Kay, S.; Lovell, S.T.; Rosenstock, T.S.; Sprenkle-Hyppolite, S.; et al. Priority science can accelerate agroforestry as a natural climate solution. *Nat. Clim. Chang.* **2023**, *13*, 1179–1190. [CrossRef]
5. Thackeray, C.W.; Hall, A.; Norris, J.; Chen, D. Constraining the increased freqency of global precipitation extremes under warming. *Nat. Clim. Chang.* **2022**, *12*, 441–448. [CrossRef]
6. Madakumbura, G.D.; Thackeray, C.W.; Norris, J.; Goldenson, N.; Hall, A. Anthropogenic influence on extreme precipitation over global land areas seen in multiple observational datasets. *Nat. Commun.* **2021**, *12*, 3944. [CrossRef]
7. Qiao, L.; Zuo, Z.; Zhang, R.; Piao, S.; Xiao, D.; Zhang, K. Soil mositure-atmosphere coupling accelerates global warming. *Nat. Commun.* **2023**, *14*, 4908. [CrossRef]
8. Fischer, E.; Sippel, S.; Knutti, R. Increasing probability of record634 shattering climate extremes. *Nat. Clim. Chang.* **2021**, *11*, 689–695. [CrossRef]
9. Intergovernmental Panel on Climate Change. *Technical Summary*; Cambridge University Press: Cambridge, UK, 2022.
10. Zhou, L.; Tian, Y.; Myneni, R.B.; Ciais, P.; Saatchi, S.; Liu, Y.Y.; Piao, S.; Chen, H.; vermote, E.F.; Song, C.; et al. Widespread decline of Congo rainforest greeness in the past decade. *Nature* **2014**, *509*, 86–90. [CrossRef]
11. Boulton, C.A.; Lenton, T.M.; Boers, N. Pronounced loss of Amazon rainforest resilence since the early 200s. *Nat. Clim. Chang.* **2022**, *12*, 271–278. [CrossRef]
12. Liang, L.; Wang, Q.; Guan, Q.; Du, Q.; Sun, Y.; Ni, F.; Lv, S.; Shan, Y. Assessing vegetation restoration prospects under different environmental elements in cold and arid mountainous region of China. *CATENA* **2023**, *226*, 107055. [CrossRef]
13. Chen, C.; Park, T.; Wang, X.; Piao, S.; Xu, B.; Chaturvedi, R.K.; Fuchs, R.; Brovkin, V.; Ciais, P.; Fensholt, R.; et al. China and India Lead in greening of the world through land-use management. *Nat. Sustain.* **2019**, *2*, 122–129. [CrossRef] [PubMed]
14. Yu, Y.; Notaro, M.; Wang, F.; Mao, J.; Shi, X.; Wei, Y. Observed positive vegetation-rainfall feedbacks in the Sahel dominated by a mositure recycling mechanism. *Nat. Commun.* **2017**, *8*, 1873. [CrossRef] [PubMed]
15. Brandt, M.; Hiernaux, P.; Rasmussen, K.; Tucker, C.J.; Wigneron, J.; Diouf, A.A.; Herrmann, S.M.; Zhang, W.; Kergoat, L.; Mbow, C.; et al. Changes in rainfall distribution promote woody fiologe production in the Sahel. *Commun. Biol.* **2019**, *2*, 133. [CrossRef] [PubMed]
16. Dardel, C.; Kergoat, L.; Hiernaux, P.; Mougin, E.; Grippa, M.; Tucker, C.J. Re-greening Sahel: 30 years of remote sensing data and field observations (Mali, Niger). *Remote Sens. Environ.* **2014**, *140*, 350–364. [CrossRef]
17. Beck, H.E.; McVicar, T.R.; van Dijk, A.I.; Schellekens, J.; de Jeu, R.A.; Bruijnzeel, L.A. Global evaluation of four AVHRR–NDVI data sets: Intercomparison and assessment against Landsat imagery. *Remote Sens. Environ.* **2011**, *115*, 2547–2563. [CrossRef]
18. Pinzon, J.E.; Pak, E.W.; Tucker, C.J.; Bhatt, U.S.; Frost, G.V.; Macander, M.J. Global Vegetation Greenness (NDVI) from AVHRR GIMMS-3G+, 1981–2022. Available online: https://daac.ornl.gov/VEGETATION/guides/Global_Veg_Greenness_GIMMS_3G.html (accessed on 10 June 2024).
19. Tucker, C.J.; Pinzon, J.E.; Brown, M.E.; Slayback, D.A.; Pak, E.W.; Mahoney, R.; Vermote, E.F.; El Saleous, N. An extended AVHRR 8-km NDVI dataset compatible with MODIS and SPOT vegetation NDVI data. *Int. J. Remote Sens.* **2005**, *26*, 4485–4498. [CrossRef]
20. Gessesse, A.A.; Melesse, A.M. Chapter 8—Temporal relationships between time series CHIRPS-rainfall estimation and eMODIS-NDVI satellite images in Amhara Region, Ethiopia. In *Extreme Hydrology and Climate Variability*; Melesse, A.M., Abtew, W., Senay, G., Eds.; Elsevier: Amsterdam, The Netherlands, 2019; pp. 81–92.
21. Nicholson, S. On the question of the "recovery" of the rains in the West African Sahel. *J. Arid Environ.* **2005**, *63*, 615–641. [CrossRef]
22. Sultan, B.; Janicot, S.; Drobinski, P. Characterization of the diurnal cycle of the West African Monsoon around the monsoon onset. *J. Clim.* **2007**, *20*, 4014–4032. [CrossRef]
23. Lian, X.; Jeong, S.; Park, C.-E.; Xu, H.; Li, L.Z.X.; Wang, T.; Gentine, P.; Peñuelas, J.; Piao, S. Biophysical impacts of northern vegetation changes on seasonal warming patterns. *Nat. Commun.* **2022**, *13*, 3925. [CrossRef]
24. Myneni, R.B.; Keeling, C.D.; Tucker, C.J.; Asrar, G.; Nemani, R.R. Increased plant growth in the northern high latitudes from 1981 to 1991. *Nature* **1997**, *386*, 698–702. [CrossRef]
25. Ichii, K.; Kawabata, A.; Yamaguchi, Y. Global correlation analysis for NDVI and climatic variables and NDVI trends: 1982–1990. *Int. J. Remote Sens.* **2002**, *23*, 3873–3878. [CrossRef]
26. Brown, M.; Pinzon, J.; Didan, K.; Morisette, J.; Tucker, C. Evaluation of the consistency of long-term NDVI time series derived from AVHRR, SPOT-vegetation, SeaWIFS, MODIS and LandSAT ETM+ sensors. *IEEE Trans. Geosci. Remote Sens.* **2006**, *44*, 1787–1793. [CrossRef]
27. Zhu, Z.; Piao, S.; Myneni, R.B.; Huang, M.; Zeng, Z.; Canadell, J.G.; Ciais, P.; Sitch, S.; Friedlingstein, P.; Arneth, A.; et al. Greening of the Earth and its drivers. *Nat. Clim. Chang.* **2016**, *6*, 791–795. [CrossRef]
28. Berdugo, M.; Delgado-Baquerizo, M.; Soliveres, S.; Hernandez-Clemente, R.; Zhao, Y.; Gaitan, J.J.; Gross, N.; Saiz, H.; Maire, V.; Lehman, A.; et al. Global ecosystem thresholds driven by aridity. *Science* **2020**, *367*, 787–790. [CrossRef] [PubMed]
29. Lamchin, M.; Lee, W.K.; Jeon, S.W.; Wang, S.W.; Lim, C.H.; Song, C.; Sung, M. Long-term trend and correlation between vegetation greenness and climate variables in Asia based on satellite data. *Sci. Total Environ.* **2018**, *618*, 1089–1095. [CrossRef]
30. Peteet, D. Sensitivity and rapidity of vegetational response to abrupt climate change. *Proc. Natl. Acad. Sci. USA* **2000**, *97*, 1359–1361. [CrossRef] [PubMed]

31. Brandt, M.; Mbow, C.; Diouf, A.A.; Verger, A.; Samimi, C.; Fensholt, R. Ground- and satellite-based evidence of the biophysical mechanisms behind the greening Sahel. *Glob. Chang. Biol.* **2015**, *21*, 1610–1620. [CrossRef]
32. Gentine, P.; Massmann, A.; Lintner, B.R.; Hamed Alemohammad, S.; Fu, R.; Green, J.K.; Kennedy, D.; Vilà-Guerau de Arellano, J. Land–atmosphere interactions in the tropics—A review. *Hydrol. Earth Syst. Sci.* **2019**, *23*, 4171–4197. [CrossRef]
33. Ghebrezgabher, M.G.; Yang, T.; Yang, X.; Eyassu Sereke, T. Assessment of NDVI variations in responses to climate change in the Horn of Africa. *Egypt. J. Remote Sens. Space Sci.* **2020**, *23*, 249–261. [CrossRef]
34. Hoscilo, A.; Balzter, H.; Bartholomé, E.; Boschetti, M.; Brivio, P.A.; Brink, A.; Clerici, M.; Pekel, J.F. A conceptual model for assessing rainfall and vegetation trends in sub-Saharan Africa from satellite data. *Int. J. Climatol.* **2015**, *35*, 3582–3592. [CrossRef]
35. Seaquist, J.W.; Hickler, T.; Eklundh, L.; Ardö, J.; Heumann, B.W. Disentangling the effects of climate and people on Sahel vegetation dynamics. *Biogeosciences* **2009**, *6*, 469–477. [CrossRef]
36. Hickler, T.; Eklundh, L.; Seaquist, J.W.; Smith, B.; Ardö, J.; Olsson, L.; Sykes, M.T.; Sjöström, M. Precipitation controls Sahel greening trend. *Geophys. Res. Lett.* **2005**, *32*, L21415. [CrossRef]
37. Ogou, F.K.; Igbawua, T. Investigation of changes in vegetation cover associated with changes in its hydro-climatic drivers in recent decades over North Sub-Saharan Africa. *Theor. Appl. Climatol.* **2022**, *149*, 1135–1152. [CrossRef]
38. Dirmeyer, P. The terrestrial segment of soil moisture-climate coupling. *Geophys. Res. Lett.* **2011**, *38*, L16702. [CrossRef]
39. Koster, R.D.; Dirmeyer, P.A.; Guo, Z.; Bonan, G.; Chan, E.; Cox, P.; Gordon, C.T.; Kanae, S.; Kowalczyk, E.; Lawrence, D.; et al. The Second Phase of the Global Land–Atmosphere Coupling Experiment: Soil Moisture Contributions to Subseasonal Forecast Skill. *J. Hydrometeorol.* **2011**, *12*, 805–822. [CrossRef]
40. Mahecha, M.; Bastos, A.; Bohn, F.; Eisenhauer, N.; Feilhauer, H.; Hickler, T.; Kalesse-Los, H.; Migliavacca, M.; Otto, F.; Peng, J.; et al. Biodiversity and climate extremes: Known interactions and research gaps. *Earth Future* **2023**, *12*, e2023EF003963. [CrossRef]
41. Cox, D.T.C.; Maclean, I.M.D.; Gardner, A.S.; Gaston, K.J. Global variation in diurnal asymmetry in temperature, cloud cover, specific humidity and precipitation and its association with leaf area index. *Glob. Chang. Biol.* **2020**, *26*, 7099–7111. [CrossRef] [PubMed]
42. Funk, C.; Hoell, A.; Shukla, S.; Husak, G.; Michaelsen, J. The East African Monsoon System: Seasonal Climatologies and Recent Variations. In *The Monsoons and Climate Change: Observations and Modeling*; de Carvalho, L.M.V., Jones, C., Eds.; Springer International Publishing: Cham, Switzerland, 2016; pp. 163–185.
43. Sultan, B.; Janicot, S. Abrupt shift of the ITCZ over West Africa and intra-seasonal variability. *Geophys. Res. Lett.* **2000**, *27*, 3353–3356. [CrossRef]
44. Harris, I.; Jones, P.; Osborn, T.; Lister, D. Updated high–resolution grids of monthly climatic observations–the CRU TS3.10 dataset. *Int. J. Climatol.* **2014**, *34*, 623–642. [CrossRef]
45. Nooni, I.K.; Ogou, F.K.; Hagan, D.F.; Saidou Chaibou, A.A.; Prempeh, N.A.; Nakoty, F.M.; Jin, Z.; Lu, J. The Relationship between Changes in Hydro-Climate Factors and Maize Crop Production in the Equatorial African Region from 1980 to 2021. *Atmosphere* **2024**, *15*, 542. [CrossRef]
46. Hansen, M.C.; Defries, R.S.; Townshend, J.R.G.; Sohlberg, R. Global land cover classification at 1 km spatial resolution using a classification tree approach. *Int. J. Remote Sens.* **2000**, *21*, 1331–1364. [CrossRef]
47. de Jong, R.; de Bruin, S.; de Wit, A.; Schaepman, M.E.; Dent, D.L. Analysis of monotonic greening and browning trends from global NDVI time-series. *Remote Sens. Environ.* **2011**, *115*, 692–702. [CrossRef]
48. Holben, B.N. Characteristics of maximum-value composite images from temporal AVHRR data. *Int. J. Remote Sens.* **1986**, *7*, 1417–1434. [CrossRef]
49. Basak, D.; Bose, A.; Roy, S.; Chowdhury, I.R. Chapter 17—Understanding the forest cover dynamics and its health status using GIS-based analytical hierarchy process: A study from Alipurduar district, West Bengal, India. In *Water, Land, and Forest Susceptibility and Sustainability*; Chatterjee, U., Pradhan, B., Kumar, S., Saha, S., Zakwan, M., Fath, B.D., Fiscus, D., Eds.; Elsevier: Amsterdam, The Netherlands, 2023; Volume 1, pp. 475–508.
50. Zeng, Y.; Hao, D.; Huete, A.; Dechant, B.; Berry, J.; Chen, J.M.; Joiner, J.; Frankenberg, C.; Bond-Lamberty, B.; Ryu, Y.; et al. Optical vegetation indices for monitoring terrestrial ecosystems globally. *Nat. Rev. Earth Environ.* **2022**, *3*, 477–493. [CrossRef]
51. Mitchell, T.D.; Jones, P.D. An improved method of constructing a database of monthly climate observations and associated high-resolution grids. *Int. J. Climatol.* **2005**, *25*, 693–712. [CrossRef]
52. New, M.; Hulme, M.; Jones, P.D. Representing twentieth century spaceetime climate variability. Part 2: Development of 1901e96 monthly grids of terrestrial surface climate. *J. Clim.* **2000**, *13*, 2217–2238. [CrossRef]
53. Mann, H.B. Non-parametric tests against trend. *Econometrica* **1945**, *13*, 245–259. [CrossRef]
54. Kendall, M. *Rank Correlation Measures*; Charles Griffin: London, UK, 1975; Volume 202.
55. Sen, P.K. Estimates of the Regression Coefficient Based on Kendall's Tau. *J. Am. Stat. Assoc.* **1968**, *63*, 1379–1389. [CrossRef]
56. Pettitt, A.N. A non-parametric approach to the change-point problem. *J. R. Stat. Soc.* **1979**, *28*, 126–135. [CrossRef]
57. Verstraeten, G.; Poesen, J.; Demaree, G.; Salles, C. Long-term (105 years) variability in rain erosivity as derived from 10-min rainfall depth data for Ukkel (Brussels, Belgium): Implications for assessing soil erosion rates. *J. Geophys. Res.* **2006**, *111*, D22109. [CrossRef]
58. Adeyeri, O.E.; Ishola, K.A. Variability and Trends of Actual Evapotranspiration over West Africa: The Role of Environmental Drivers. *Agric. For. Meteorol.* **2021**, *308–309*, 108574. [CrossRef]

59. Nooni, I.K.; Wang, G.; Hagan, D.F.T.; Lu, J.; Ullah, W.; Li, S. Evapotranspiration and its Components in the Nile River Basin Based on Long-Term Satellite Assimilation Product. *Water* **2019**, *11*, 1400. [CrossRef]
60. Akimoto, F.; Matsunami, A.; Kamata, Y.; Kodama, I.; Kitagawa, K.; Arai, N.; Higuchi, T.; Itoh, A.; Haraguchi, H. Cross-Correlation Analysis of Atmospheric Trace Concentrations of N_2O, CH_4 and CO_2 Determined by Continuous Gas-Chromatographic Monitoring. *Energy* **2005**, *30*, 299–311. [CrossRef]
61. Biasutti, M. Rainfall trends in the African Sahel: Characteristics, processes, and causes. *WIREs Clim. Chang.* **2019**, *10*, e591. [CrossRef]
62. Zhao, L.; Dai, A.; Dong, B. Changes in global vegetation activity and its driving factors during 1982–2013. *Agric. For. Meteorol.* **2018**, *249*, 198–209. [CrossRef]
63. Wang, X.; Piao, S.; Ciais, P.; Li, J.; Friedlingstein, P.; Koven, C.; Chen, A. Spring temperature change and its implication in the change of vegetation growth in North America from 1982 to 2006. *Proc. Natl. Acad. Sci. USA* **2011**, *108*, 1240–1245. [CrossRef] [PubMed]
64. Piao, S.L.; Wang, X.H.; Ciais, P.; Zhu, B.; Wang, T.; Liu, J. Changes in satellite-derived vegetation growth trend in temperate and boreal Eurasia from 1982 to 2006. *Glob. Chang. Biol.* **2011**, *17*, 3228–3239. [CrossRef]
65. Meehl, G.A.; Stocker, T.F.; Collins, W.D.; Gaye, A.J.; Gregory, J.M.; Kitoh, A.; Knutti, R.; Murphy, J.M.; Noda, A.; Raper, S.C.B.; et al. *Global Climate Projection*; Cambridge University Press: Cambridge, UK; New York, NY, USA, 2007.
66. Hatfield, J.L.; Prueger, J.H. Temperature extremes: Effect on plant growth and development. *Weather Clim. Extrem.* **2015**, *10*, 4–10. [CrossRef]
67. Alfaro, E.T.; Gershunov, D.; Cayan, D. Prediction of summer maximum and minimum temperature over the central and western United States: The roles of soil moisture and sea surface temperature. *J. Clim.* **2006**, *19*, 1407–1421. [CrossRef]
68. Verbesselt, J.; Hyndman, R.; Zeileis, A.; Culvenor, D. Phenological change detection while accounting for abrupt and gradual trends in satellite image time series. *Remote Sens. Environ.* **2010**, *114*, 2970–2980. [CrossRef]
69. Fensholt, R.; Rasmussen, K. Analysis of trends in the Sahelian 'rain-use efficiency' using GIMMS NDVI, RFE and GPCP rainfall data. *Remote Sens. Environ.* **2011**, *115*, 438–451. [CrossRef]
70. de Jong, R.; Verbesselt, J.; Schaepman, M.E.; de Bruin, S. Trend changes in global greening and browning: Contribution of short-term trends to longer-term change. *Glob. Chang. Biol.* **2012**, *18*, 642–655. [CrossRef]
71. De Jong, R.; Verbesselt, J.; Zeileis, A.; Schaepman, M.E. Shifts in Global Vegetation Activity Trends. *Remote Sens.* **2013**, *5*, 1117–1133. [CrossRef]
72. Zhao, M.; Running, S.W. Drought-induced reduction in global terrestrial net primary production from 2000 through 2009. *Science* **2010**, *329*, 940–943. [CrossRef] [PubMed]
73. Mukherjee, S.; Mishra, A.K.; Zscheischler, J.; Entekhabi, D. Interaction between dry and hot extremes at a global scale using a cascade modeling framework. *Nat. Commun.* **2023**, *14*, 277. [CrossRef]
74. Adepoju, K.; Adelabu, S.; Fashae, O. Vegetation Response to Recent Trends in Climate and Landuse Dynamics in a Typical Humid and Dry Tropical Region under Global Change. *Adv. Meteorol.* **2019**, *2019*, 4946127. [CrossRef]
75. Piao, S.; Wang, X.; Park, T.; Chen, C.; Lian, X.; He, Y.; Bjerke, J.W.; Chen, A.; Ciais, P.; Tømmervik, H.; et al. Characteristics, drivers and feedbacks of global greening. *Nat. Rev. Earth Environ.* **2020**, *1*, 14–27. [CrossRef]
76. Zaitchik, B.F.; Rodell, M.; Biasutti, M.; Seneviratne, S.I. Wetting and drying trends under climate change. *Nat. Water* **2023**, *1*, 502–513. [CrossRef]
77. Xiong, J.; Guo, S.; Abhishek; Chen, J.; Yin, J. Global evaluation of the "dry gets drier, and wet gets wetter" paradigm from a terrestrial water storage change perspective. *Hydrol. Earth Syst. Sci.* **2022**, *26*, 6457–6476. [CrossRef]
78. Zhou, L.; Tucker, C.J.; Kaufmann, R.K.; Slayback, D.; Shabanov, N.V.; Myneni, R.B. Variations in northern vegetation activity inferred from satellite data of vegetation index during 1981 to 1999. *J. Geophys. Res. Atmos.* **2001**, *106*, 20069–20083. [CrossRef]
79. Seneviratne, S.I.; Lüthi, D.; Litschi, M.; Schär, C. Land–atmosphere coupling and climate change in Europe. *Nature* **2006**, *443*, 205–209. [CrossRef] [PubMed]
80. Vicente-Serrano, S.M.; Gouveia, C.; Camarero, J.J. Response of vegetation to drought time-scales across global land biomes. *Proc. Natl. Acad. Sci. USA* **2012**, *110*, 52–57. [CrossRef] [PubMed]
81. Anderson, L.O.; Malhi, Y.; Aragão, L.E.O.C.; Ladle, R.; Arai, E.; Barbier, N.; Phillips, O.L. Remote sensing detection of droughts in Amazonian forest canopies. *New Phytol.* **2010**, *187*, 733–750. [CrossRef] [PubMed]

forests

Article

Predicted Future Changes in the Mean Seasonal Carbon Cycle Due to Climate Change

Mauro Morichetti [1,*], Elia Vangi [1,2] and Alessio Collalti [1,3]

[1] National Research Council of Italy, Forest Modelling Laboratory, Institute for Agriculture and Forestry Systems in the Mediterranean (CNR-ISAFOM), Via Madonna Alta 128, 06128 Perugia, Italy; elia.vangi@isafom.cnr.it (E.V.); alessio.collalti@cnr.it (A.C.)
[2] geoLAB—Laboratory of Forest Geomatics, Department of Agriculture, Food, Environment and Forestry, Università degli Studi di Firenze, Via San Bonaventura 13, 50145 Firenze, Italy
[3] National Biodiversity Future Center (NBFC), Piazza Marina 61, 90133 Palermo, Italy
* Correspondence: mauro.morichetti@cnr.it; Tel.: +39-0755014541

Abstract: Through photosynthesis, forests absorb annually large amounts of atmospheric CO_2. However, they also release CO_2 back through respiration. These two, opposite in sign, large fluxes determine how much of the carbon is stored or released back into the atmosphere. The mean seasonal cycle (MSC) is an interesting metric that associates phenology and carbon (C) partitioning/allocation analysis within forest stands. Here, we applied the 3D-CMCC-FEM model and analyzed its capability to represent the main C-fluxes, by validating the model against observed data, questioning if the sink/source mean seasonality is influenced under two scenarios of climate change, in five contrasting European forest sites. We found the model has, under current climate conditions, robust predictive abilities in estimating NEE. Model results also predict a consistent reduction in the forest's capabilities to act as a C-sink under climate change and stand-aging at all sites. Such a reduction is predicted despite the number of annual days as a C-sink in evergreen forests increasing over the years, indicating a consistent downward trend. Similarly, deciduous forests, despite maintaining a relatively stable number of C-sink days throughout the year and over the century, show a reduction in their overall annual C-sink capacity. Overall, both types of forests at all sites show a consistent reduction in their future mitigating potential.

Keywords: carbon cycle; climate change; process-based model; mean seasonal cycle; forest ecosystems

Citation: Morichetti, M.; Vangi, E.; Collalti, A. Predicted Future Changes in the Mean Seasonal Carbon Cycle Due to Climate Change. *Forests* **2024**, *15*, 1124. https://doi.org/10.3390/f15071124

Academic Editor: Giacomo Alessandro Gerosa

Received: 21 May 2024
Revised: 14 June 2024
Accepted: 26 June 2024
Published: 28 June 2024

1. Introduction

Forests play a pivotal role in the biosphere–atmosphere feedback by annually absorbing large amounts of atmospheric CO_2 through photosynthesis (GPP; ~150 PgC year^{-1}) and releasing it back because of, e.g., ecosystem respiration (R_{eco}), a relatively close amount yet not necessarily equal, which varies year by year [1–3]. The net ecosystem exchange (NEE) of CO_2 between ecosystems and the atmosphere is the net balance between these two gross fluxes opposite in sign, and it governs much of the overall terrestrial annual net carbon (C) budget. Imbalances between CO_2 sources (even including carbon lost by fires and other processes) and sinks directly increase or decrease atmospheric CO_2 levels [4]. Terrestrial ecosystems—and forests in particular—are contributing substantially to climate change mitigation, provided that they are C-sinks and not C-sources [5]. Forests that might absorb more than they emit are commonly considered carbon sinks (with NEE-negative in sign), while if they emit more than they absorb are considered as carbon sources (with NEE-positive in sign). The magnitude of this exchange of CO_2, however, is subject to substantial variability and trends, in large part as a response to variations and trends in climate [6]. Indeed, forests have been shown to be extremely sensitive to changes in environmental conditions (e.g., climate, seasonality, atmospheric CO_2 concentration, nitrogen deposition), to aging [7,8], and to disturbances [9], including management practices [10,11], which can control both photosynthesis and respiration. Therefore, estimating NEE, GPP, and R_{eco} is a key step for better understanding the underlying mechanisms constraining ecosystem functioning [12].

Europe and the Mediterranean are expected to become in the near future a 'Hot Spot' of climate change [13–16]. The literature reports that under climate change scenarios forests are expected to grow faster, to mature earlier but also to die younger, curtailing their life span [17], because of, mainly, warming and increased atmospheric CO_2 concentration (the so-called 'CO_2-fertilization effect') [18]. Conversely, there is a general lack of evidence and knowledge on how, overall, forest ecosystems will, on the whole, react to climate change. Forest carbon balance will be impacted by climate change because various main processes are impacted, which, in turn, may react and respond differently to climate change, also because they are vulnerable and sensitive to separate environmental drivers. Indeed, while an extensive line of evidence shows that the increased availability of CO_2 may amplify the photosynthetic rate and assimilation capacity [19], such an increase is largely debated since there is no evidence that such positive changes will generally continue indefinitely [20,21]. Similarly, the effects of warming are largely discussed because, while it is documented that some species may take advantage of a longer vegetation season (e.g., deciduous species), there are also negative effects linked to warming such as heat waves and the often-associated drought events [22,23], including late frost [24] and disturbances [25], which can be detrimental to growth till tree survival. In addition, there is a general concern that the changing temperature response of respiration turns boreal forests from carbon sinks into carbon sources [26]. Indeed, warming has also been found to accelerate both autotrophic as well as heterotrophic respiration (the two components of ecosystem respiration), meaning that increased temperature may lead forests to release more carbon, potentially more than absorbed annually [27]. Conversely, drought has been shown to reduce microbial respiration and then heterotrophic respiration [28]. How these processes (i.e., photosynthesis and ecosystem respiration) will be impacted by climate change annually will determine much of the future forest annual C-budget.

The Mean Seasonal Cycle (MSC) metric, which reflects the average distribution of flux (i.e., NEE, GPP, and R_{eco}) throughout the days of a year, is an insightful measure linking phenology with carbon partitioning and allocation within seasonal climatic variability. By capturing the typical fluctuations in a specific region due to changing seasons, the MSC highlights the expected seasonal changes in climate data, averaged over many years, to smooth out anomalies and emphasize the regular, cyclical nature of these changes. Many studies [29–32] have indeed shown that climate change will impact both the phenology by changing the date for the beginning and the end of the growing season, as well as by changing the shape of the Leaf Area Index (LAI) distribution over the year, which, at the same time, will influence the way, among the other things, when photosynthesis can start and how recent and old photosynthates are partitioned and used to build new tissues and to replenish the reserves used for the metabolism, as well as carbon allocation [33] and the C-dynamic.

Process-based models are valuable tools to understand how and to what extent future climate change will impact these two fluxes (GPP and R_{eco}) in the MSC, both processes being controlled by warming and changes in precipitation regime and atmospheric CO_2 concentration [34,35]. Here, we at first applied and validated, under current observed climate conditions, the 'Three Dimensional–Coupled Model Carbon Cycle–Forest Ecosystem Module' (3D-CMCC-FEM), a biogeochemical, biophysical process-based forest ecosystem model designed to simulate carbon, nitrogen, and water cycle in forest ecosystems and, secondly, under climate change conditions.

Specifically, we question and analyze: (1) the capability of the 3D-CMCC-FEM to represent under the current climate the main C-fluxes governing the C-cycle in terms of net ecosystem exchange (NEE), gross primary production (GPP), and ecosystem respiration (R_{eco}), by validating the model against independent data from the Fluxnet network; and, provided that the model is close to the observed data; (2) how, and if, the sink/source mean seasonality will be influenced under two locally bias-corrected scenarios of warming (RCP 2.6 and 6.0) and atmospheric CO_2 enrichment from three CMIP5 Earth System Models, within ISIMIP-PROFOUND initiative, in five well-studied and long-monitored contrasting forest sites (three evergreens and two deciduous) on a longitudinal transect through Europe up to the end of the century.

2. Materials and Methods

2.1. Model Description (3D-CMCC-FEM 'v.5.6')

The 'Three Dimensional–Couple Model Carbon Cycle–Forest Ecosystem Module' (hereafter '3D-CMCC-FEM') is a biogeochemical, biophysical, process-based forest ecosystem model (see [10,11,36–44] and reference therein). The model is designed to simulate carbon, nitrogen, and water cycles in forest ecosystems at commonly 1-hectare spatial resolution and the main eco-physiological processes (e.g., photosynthesis) at daily temporal resolution. The most recent code versions since Collalti et al. [17] adopt the biogeochemical photosynthesis model of Farquhar, von Caemmerer, and Berry [45] to compute gross primary productivity (GPP). The biogeochemical photosynthesis model is parameterized as in Bernacchi et al. [46,47] and temperature acclimation for leaves as in Kattge and Knorr [48]. The 3D-CMCC-FEM considers, as in De Pury and Farquhar [49], light interception, reflection, transmission, and assimilation (and leaf respiration) for both sun and shaded leaves. Autotrophic respiration (R_A) is computed mechanistically following the 'Growth and Maintenance Respiration Paradigm' (GMRP; [50]), which is divided into the metabolic costs for synthesizing new tissues (growth respiration, R_G) and the metabolic costs for maintaining the existing ones (maintenance respiration, R_M). In 3-D-CMCC-FEM, the maintenance respiration is based on Nitrogen amount (a fixed fraction of carbon mass varying between the six tree compartments) and is temperature-controlled by a standard Arrhenius relationship [36]. 'Type I' and 'Type II' acclimation of respiration to temperature (i.e., short- and long-term acclimation; [10,39,51]) are also accounted for. Any imbalance between carbon assimilation and carbon losses because of plants' respiration is buffered by a seventh pool, the Non-Structural Carbon pool (NSC; starch and sugars undistinguished), which has priority in the carbon allocation all over the year. The net primary production (NPP) is the GPP minus R_A. Biomass production (BP) is the amount of NPP not used for replenishing the NSC pool. Indeed, other forms of non-structural carbon losses (e.g., biogenic volatile organic compounds, BVOCs, or root exudates to mycorrhizas) are not accounted for by the model. The phenological scheme, as well as the carbon partitioning/allocation scheme, distinguished for deciduous and evergreen tree species, is in-depth described in Collalti et al. [10,36,39,52]. Heterotrophic respiration follows a BIOME-BGC-like approach (which follows the CENTURY model; [53,54]) distinguishing decomposition for litter and soil pools with each of the four different conceptual pools characterized by different decomposability degrees (i.e., fast, medium, slow, and a recalcitrant carbon pool) [39,55]. Altogether, litter and soil decomposition emissions form heterotrophic respiration (R_H), which, summed up to R_A, constitutes ecosystem respiration (R_{eco}). Net ecosystem exchange (NEE) is equal to R_{eco} minus GPP. Therefore, negative values indicate carbon uptake from the atmosphere (i.e., the system acts as a C-sink, NEE < 0), and positive values indicate carbon release (i.e., the system acts as a C-source, NEE > 0). The 3D-CMCC-FEM's sensitivity to its 54 species-specific parameters and how it varies along the forest development and under climate change is described in Collalti et al. [17].

In the present study, we used version 5.6 [41], which slightly differs from v.5.5, described in Collalti et al. [8] and Dalmonech et al. [11], for a new scheme (and relative parameterization) for sapwood and live wood turnover and dynamics and some additional new forest management schemes (not used here) described in Testolin et al. [43].

2.2. Case Study Areas

Five case studies were selected as representative of the main European forest species (and climate) and at the same time because of long-monitored sites and part of the Fluxnet network [56], the ISIMIP (Inter-Sectoral Impact Model Intercomparison Project, [57]; https://www.isimip.org/, accessed on 1 January 2020) initiative and the PROFOUND database [42,58,59]: the temperate European beech (*Fagus sylvatica* L.) forest of Collelongo, Italy (IT-Col), and of Sorø, Denmark (DK-Sor), the maritime pine (*Pinus pinaster* Ait.) forest of Le Bray, France (FR-Lbr), the boreal Scots pine (*Pinus sylvestris* L.) forest of Hyytiälä, Finland (FI-Hyy), and the temperate Norway spruce (*Picea abies* (L.) H. Karst) forest of Bílý Kříž, Czech Republic (CZ-Bk1) (Figure 1). Stand characteristics are described in Table 1.

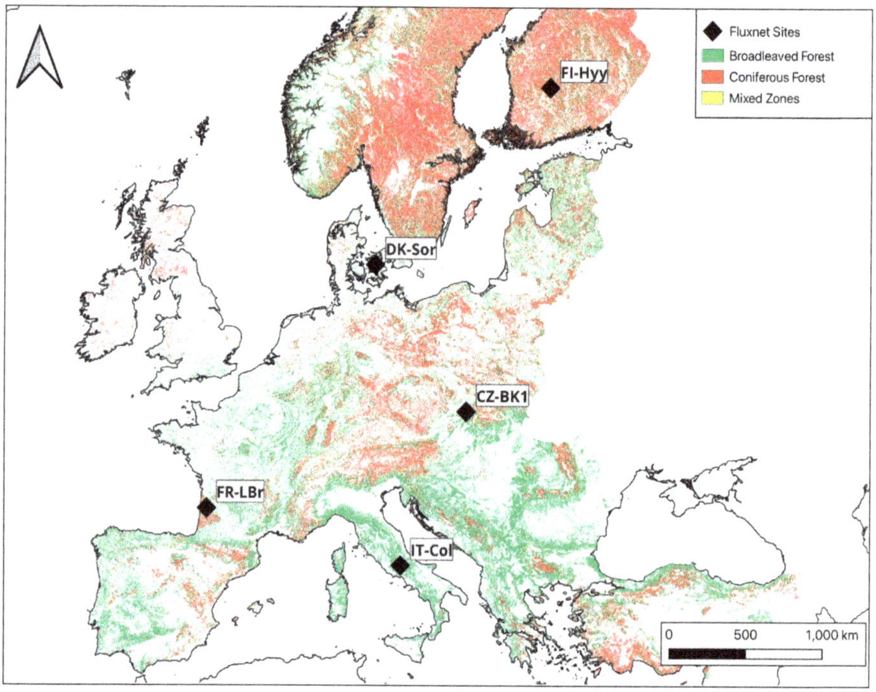

Figure 1. Forest classification for broadleaved and coniferous forests at the European level in the spatial resolution of 100 m [60]. Location of the Fluxnet sites considered.

Table 1. Site description and stand initialization data are used in simulations with data from the PROFOUND database [58]. Model initialization data (e.g., diameter at breast height—DBH, tree height, stand age, and stand density) correspond to the stand conditions of the first year in the historical simulations.

Fluxnet Code	Nation	Site Name	Coordinate			Climate	Species Composition	Period of Simulation and Climate Data		Mean DBH	Tree Height	Mean Stand Age	Stand Density
			Lat (°)	Lon (°)	Elevation (m)			Historical (Observed)	Historical + RCP (Modeled)	(cm)	(m)	(years)	(n. trees ha^{-1})
IT-Col	Italy	Collelongo	41.8	13.6	1560	Temperate	*Fagus sylvatica*	1997–2014	1997–2099	20.2	17.3	105	900
FR-LBr	France	Le Bray	44.7	−0.8	61	Temperate	*Pinus pinaster*	1996–2008	1997–2099	26.7	17.8	26	614
FI-Hyy	Finland	Hyytiälä	61.9	24.3	181	Boreal	*Pinus sylvestris*	1996–2014	1997–2099	13	11.3	35	870
CZ-Bk1	Czech Republic	Bílý Kříž	49.3	18.3	875	Temperate	*Picea abies*	2000–2008	1997–2099	9.1	7.5	19	2388
DK-Sor	Denmark	Sorø	55.3	11.4	40	Boreal	*Pinus sylvestris*	1996–2008	1997–2099	24.3	21	75	353

2.3. Input, Meteorological Data, and Climate Change Scenarios

To run, 3D-CMCC-FEM needs a set of input data from state variables representing the stand at the beginning of simulations and that account for structural characteristics (e.g., tree height, average age, diameter at breast height; see Table 1) as well as carbon and nitrogen pools (e.g., stem carbon and nitrogen); meteorological forcing data (e.g., daily maximum and minimum temperature, daily precipitation); annual atmospheric CO_2 concentrations; and species-specific parameters (e.g., maximum stomatal conductance). All the data used to initialize the model in the present study for the five stands come from the ISIMIP initiative and the PROFOUND database [58]. More specifically, daily observed meteorological data for model validation come from the Fluxnet2015 Dataset [56], while daily modeled historical (1997–2005) and future climate scenarios (2006–2099) are those from the 'ISIMIP 2bLBC' experiments ('2b experiments Locally Bias Corrected') coming from three different Earth System Models (ESMs; GFDL-ESM2M, IPSL-CM5A-LR, and MIROC5, respectively) based on the Climate Model Intercomparison Project 5 (CMIP5) driven by two Representative Concentration Pathways (RCPs) of atmospheric greenhouse gas concentration trajectories, namely, RCP 2.6 and RCP 6.0. [61,62]. The ISIMIP 2bLBC have the same structure as those in the 2b experiments but have been corrected by improving the method described in Hempel et al. [63] and subsequently by the methods described in Frieler et al. [57] and Lange [64] using the observed meteorology at the local level [58]. Therefore, the 2bLBC climate data represent the more consistent and closer modeled climate data with the observational data. The annual atmospheric CO_2 concentrations for the historical period are based on Meinshausen et al. [65] and have been extended for the period from 2006 to 2015 with data from Dlugokencky and Tans [66]. Values specific for each RCP for the period 2016 to 2099 are also based on Meinshausen et al. [65] and were used within the Farquhar, von Caemmerer, and Berry [45] photosynthesis model with values varying, at the end of this century, from 421.4 μmol mol^{-1} (RCP 2.6) to 666.4 μmol mol^{-1} (RCP 6.0), respectively. NEE, GPP, and R_{eco} data, for model validation, come from the Fluxnet2015 Dataset [56]. Other variables have been validated at these forest stands in the past (although using slightly different model versions) and described in Collalti et al. [8,10,37], Marconi et al. [40], Mahnken et al. [42], and Dalmonech et al. [11].

2.4. Model Runs, Validation, and Mean Seasonal Cycle Analyses under Climate Change

The model simulations for model validation under measured forcing climate ran for CZ-Bk1 from 2000 to 2008, for IT-Col from 1997 to 2014, for FI-Hyy from 1996 to 2014, for FR-Lbr from 1996 to 2008, and for DK-Sor from 1996 to 2008 (see Table 1). For all sites model simulations under climate change scenarios began in 1997 and finished in December 2099. Model validation was performed by comparing modeled NEE, GPP, and R_{eco} against measured eddy covariance estimates (for GPP and R_{eco} using the night-time method with constant USTAR [67], as reported in the Fluxnet2015 Dataset [56].

To analyze 3D-CMCC-FEM's capabilities to simulate NEE, GPP, and R_{eco} for daily and monthly time series, a set of commonly used statistical metrics have been applied to compare measured vs. modeled data (under observed climate forcing). To avoid considering bad quality data, a filtering procedure for quality-check has been applied; days with less than 60% of valid data were not considered and excluded both in the model and in the observed datasets. Therefore, daily NEE, GPP, and R_{eco} eddy covariance data with low-quality values (i.e., less than 0.6; [68]) were removed. Consequently, the corresponding daily modeled data were removed too. The monthly NEE, GPP, and R_{eco} values (both from eddy covariance and the model) were aggregated from the daily ones. The common statistic we applied includes Pearson's correlation coefficient (r), Relative Mean Bias (RMB), Normalized Root Mean Square Error (NRMSE), and Modeling Efficiency (ME)

In climate change projections, we considered the potential modifications in the ability of forest stands to absorb or emit carbon throughout the season and across the years under two different locally bias-corrected climate change scenarios, each coming from three Earth System Models. This involves averaging the daily values of the MSC of the three fluxes considered. It is noteworthy that NEE represents the equilibrium between carbon absorption by vegetation during photosynthesis and carbon release through vegetation and microbial respiration. Not only the length of the growing season but also the balance between the yearly amount of photosynthesis and R_{eco} has been shown to control much of the variability across the sites and the decades analyzed [47,49]. This calculation is derived from the variance between GPP and R_{eco} encompassing both autotrophic respiration (R_A, including ground components) and heterotrophic respiration (R_H) [69].

Under climate change scenarios, we account for potential changes in the sink/source capacity of the stands during the season by averaging NEE, GPP, and R_{eco} values every ten years up to 2100 [70] and accounting for changes in the sink/source and source/sink length during the year, computed as the number of total days of the year (DoY) where a forest stand behaves as a C-sink (GPP > R_{eco} with NEE < 0) or a C-source (GPP < R_{eco} with NEE > 0), as described by the NEE. In addition, we also account for the changes in the DoY where NEE changes its sign at least for ten consecutive days to avoid artifact effects of pulsing, e.g., the 'Birch effect' on R_{eco} [71], and to account for unstable conditions and no clear source/sink and sink/source seasonal transition during the year. Therefore, we discuss changes in the MSC under the RCP 2.6 and RCP 6.0 for NEE, being the net result of opposite fluxes (i.e., GPP and R_{eco}), through its changes in negative values, i.e., days in the year where NEE < 0 and describing CO_2 uptake from the atmosphere, and positive, i.e., days in the year where NEE > 0 and describing CO_2 release to the atmosphere. In this way, we account for potential changes that may lead to anticipations or delays in the switch from source/sink and sink/source capacity, which often happens during spring and autumn during the year. The analyses under climate change scenarios (2006–2099) also include the changes in the shape of the Mean Seasonal Cycle (MSC) for NEE, GPP, and R_{eco} values and for the changes (both in the absolute and the percentage values) in the annual value. Changes in MSC were estimated on the ten-year average values using the 1998–2008 decade as a benchmarking reference. Furthermore, we account for changes in the annual values of NEE, GPP, and R_{eco} due to climate change.

3. Results

3.1. Model Validation vs. Fluxnet Data

The NEE, GPP, and R_{eco}, as modeled by 3D-CMCC-FEM, exhibit strong correlations with the observed daily and monthly eddy covariance data and for the overall MSC across all five sites (Figures A1 and 2 and Tables A1–A3 in Appendix A). Some slight differences are observed for the daily values for temperate European beech forests of IT-Col and DK-Sor when representing the NEE and R_{eco} between the 100 and 200 DoY (Figure 2(1.a–1.c) and Figure 2(5.a–5.c)). Modeled overestimations for GPP of about 5 gC m^{-2} day^{-1} at the peak of production (~180 DoY) for both the European beech forest at IT-Col (Figure 2(1.b)) and the maritime pine forest at FR-Lbr (Figure 2(2.b)) was observed. The highest correlation coefficients between modeled and observed data were observed for DK-Sor and IT-Col (r = 0.97 and 0.96, respectively) for the daily NEE and GPP (Table A1, Appendix A), while FI-Hyy shows the best correlation for daily R_{eco} (r = 0.94) and monthly NEE (r = 0.99) (Tables A2 and A3).

At the IT-Col and CZ-Bk1, the model shows the best performances in simulating NEE (RMB = 0.07 and 0.14, respectively—Table A1), while the lowest values were reached by DK-Sor and IT-Col for the GPP (RMB ranging between −0.06 and 0.55, Table A2). Overall, RMB values for GPP are relatively low across all case studies and time scales. Regarding R_{eco} at FI-Hyy and DK-Sor, the model tended to underestimate both daily (RMB of −0.3 and −0.42, respectively) and monthly (−0.29 and −0.42, respectively). Conversely, at IT-Col, the model exhibited a slight overestimation (RMB daily 0.82 and monthly 0.77, Table A3).

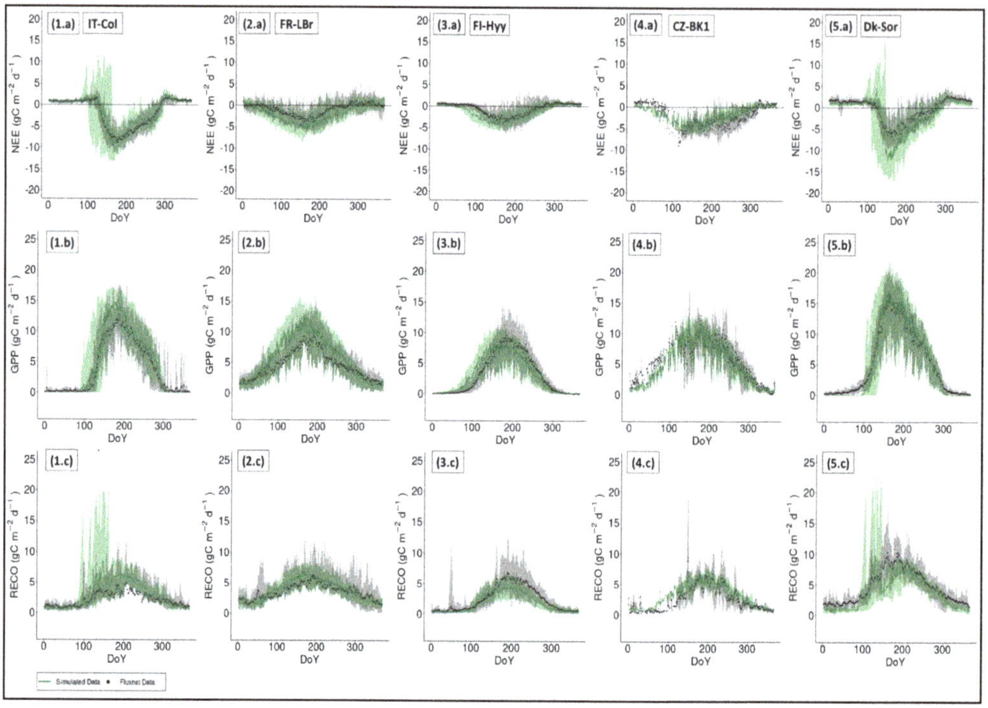

Figure 2. The green lines represent the modeled (**a**) NEE, (**b**) GPP, and (**c**) R_{eco} amounts (gC m^{-2} day^{-1}) per DoY (Day of Year) for the five selected case studies, i.e., (1) Collelongo—IT-Col, (2) Le Bray—FR-Lbr, (3) Hyytiälä—FI-Hyy, (4) Bílý Kříž—CZ-Bk1, and (5) Sorø—DK-Sor compared to relative observed data (depicted as black dots) from the Fluxnet2015 Dataset [56]. The lower and upper lines of the shaded area represent, respectively, the minimum and maximum values of the observed and modeled datasets considered.

The model reports negative NRMSE values for NEE across all time scales, indicating a slight overestimation (values ranging from −0.57 to −6.59) with the exception of the CZ-Bk1 site for the daily NEE value (1.39), and at the IT-Col site for the monthly NEE value (0.99). Regarding the GPP, (Table A2) DK-Sor showed the lowest NRMSE for daily and monthly values (0.19 and 0.16, respectively). Conversely, at CZ-Bk1 and FR-Lbr forests, the model displays the highest NRMSE for daily (1.21) and monthly (0.40) values, respectively. Last, Table A3 shows the validation results for R_{eco}. At FR-LBR, the highest accuracy and precision were reported with the lowest NRMSE for daily and monthly values (0.24 and 0.20 respectively). For the other sites, we found almost the same model capability described for the above fluxes (i.e., NEE and GPP) with a lower correlation and a slightly higher error in terms of daily R_{eco} at CZ-Bk1 (NRMSE = 1.16).

ME exhibits values close to one across all time scales and sites for NEE, with values ranging from −0.10 (FR-LBr) to 0.92 (IT-Col) for daily NEE values and −0.15 (FR-LBr) to 0.95 (IT-Col) for monthly values. The modeled GPP has a similar trend of the observed NEE with values ranging from 0.28 for FR-LBr to 0.97 for DK-Sor for daily values and FR-LBr = 0.34 to DK-Sor = 0.92 for monthly values (Table A2). The R_{eco} achieved lower performances than GPP and NEE in terms of ME; Table A3 displays values from 1.17 for IT-Col to 0.75 for FI-Hyy and IT-Col = −0.64 to FI-Hyy = 0.77, for daily and monthly time scale, respectively.

Finally, the lowest MAE was found at IT-Col, for NEE (0.67 gC m^{-2} day^{-1} and 11.42 gC m^{-2} $month^{-1}$) and GPP (0.91 gC m^{-2} day^{-1} and 20.68 gC m^{-2} $month^{-1}$), while the lowest MAE for R_{eco} was for FR-LBr (0.71 gC m^{-2} day^{-1} and 18.01 gC m^{-2} month^{-1}) (Tables A1–A3 in Appendix A).

3.2. Mean Seasonal NEE Cycle under Climate Change Scenario

Figures 3 and 4 display the 10-year average NEE seasonal cycle under climate scenarios (i.e., RCP 2.6 and RCP 6.0) for the five case studies. Overall, across all the sites and scenarios considered, there is a consistent reduction in the absolute NEE over time (i.e., NEE is 'less negative' showing a reduction in the sink capacity) with changes in the source/sink (NEE becomes negative and the site turns into a C-sink) and sink/source (NEE becomes positive and the site turns into a C-source) switch over the year(s). However, RCP 2.6 generally exhibits lower reductions in annual and mean seasonal NEE when compared to RCP 6.0 across most study sites and time intervals. The loss in the modeled sink capacity is because R_{eco} increases more than GPP, and the differences between the scenarios are related to an increase in R_{eco} higher than that for GPP under RCP 6.0.

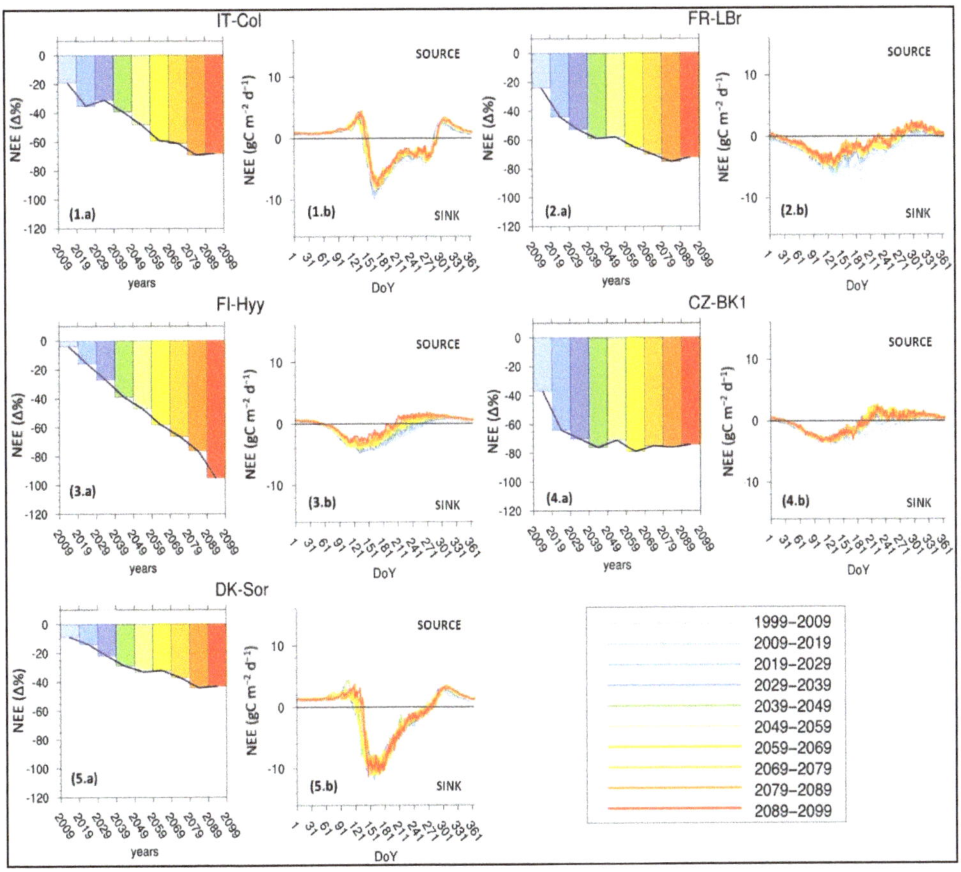

Figure 3. Ten-year average NEE seasonal cycle under the RCP 2.6 climate scenario for 5 case studies selected, i.e., (1) Collelongo—IT-Col, (2) Le Bray—FR-LBr, (3) Hyytiälä—FI-Hyy, (4) Bílý Kříž—CZ-Bk1, and (5) Sorø—DK-Sor. The histograms (**a**) represent the annual NEE variation (%) from the first decade taken as a benchmark of simulation (1999–2009). The xy plots (**b**) show the Mean Seasonal NEE Cycle of daily values (gC m^{-2} day^{-1}).

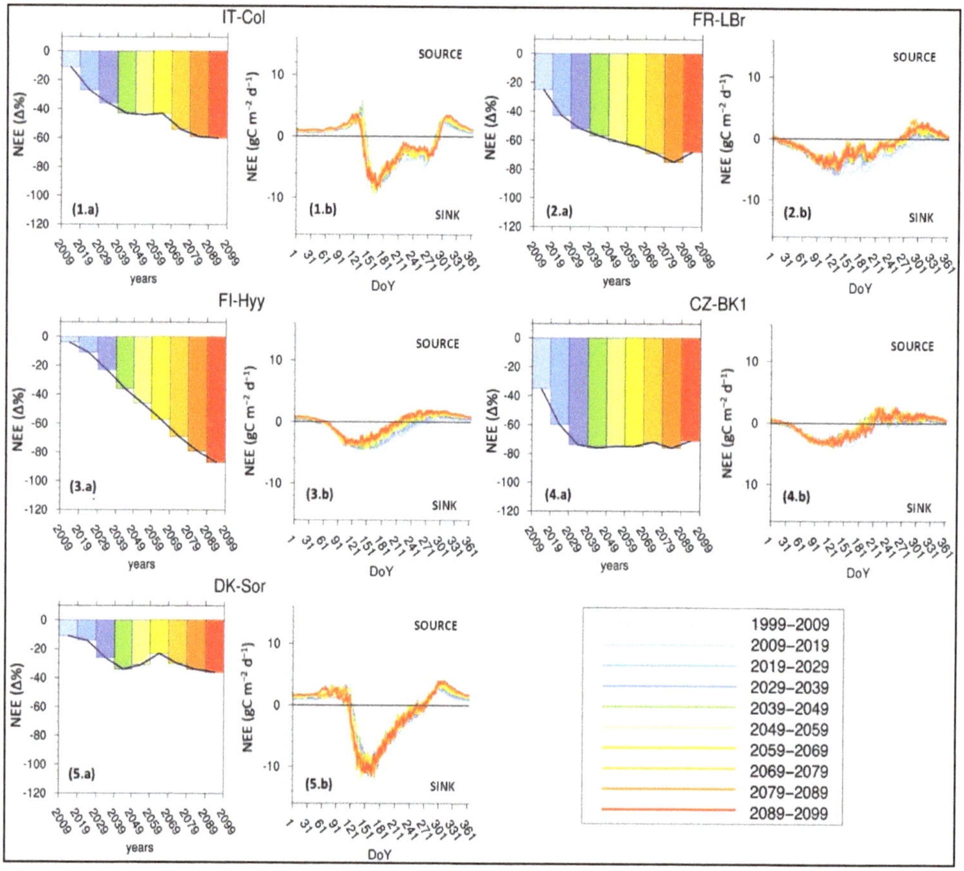

Figure 4. Ten-year average NEE seasonal cycle under the RCP 6.0 climate scenario for 5 case studies selected, i.e., (1) Collelongo—IT-Col, (2) Le Bray—FR-LBr, (3) Hyytiälä—FI-Hyy, (4) Bílý Kříž—CZ-Bk1, and (5) Sorø—DK-Sor. The histograms (**a**) represent the annual NEE variation (%) from the first decade taken as a benchmark of simulation (1999–2009). The xy plots (**b**) show the Mean Seasonal NEE Cycle of daily values (gC m^{-2} day^{-1}).

The rate of decrease in the annual NEE varies among locations and forest species. For example, the beech forests at IT-Col and DK-Sor show a reduction in the NEE (i.e., the site is less C-sink) of about 68%, and 43% (Figures 3 and 4—panel 1 and panel 5) at the end of the century, generally exhibiting a more moderate decrease compared to evergreen sites. The Scots pine forest shows, at the end of the century, a reduction in the sink capacity of 95%, standing out as the most significant decrease in NEE (Figures 3 and 4—panel 3). Over time, the rate of decline shows a tendency to speed up, hinting at a reduction in the capacity of the carbon sink because of, despite an overall increase in the GPP (Figures A2 and A3), the increased ecosystem respiration due to increased temperatures (Figures A4 and A5). By the end of the century, substantial reductions in NEE, i.e., the sites become less C-sink, across all locations and scenarios were simulated, with some locations experiencing over 70–90% reduction compared to the 1999–2009 decade (Table A4).

The GPP, similar to NEE, generally increases across forest types and scenarios over time, albeit with varying degrees (Figures A2 and A3). Specifically, RCP 6.0 shows higher growth rates, particularly in later years (2059–2099), with increases ranging from 28 to 51% across all studied forests, compared to 10–26% for RCP 2.6 (Table A5). Analysis of long-term trends suggests saturation and subsequent slight decreases in GPP growth rates under RCP 2.6, with this phenomenon being most noticeable in the forests of FI-Hyy. RCP 6.0 consistently shows a higher percentage of change in R_{eco} compared to RCP 2.6. By the end of the century (2089–2099), R_{eco} changes from 37 to 106% for RCP 2.6 and 60 to 121% for RCP 6.0 across all forests examined (Table A6). Boreal and maritime pine forests experienced higher changes compared to deciduous forests. For example, by 2089–2099, under RCP 6.0, changes reach 142% and 121% at FI-Hyy and FR-Lbr forests, while deciduous forests like European beech in IT-Col and DK-Sor experience lower changes at 67% and 60% respectively.

3.3. Changes in NEE Dynamics under Different Climate Scenarios

We considered fluctuations in the length of sink/source and source/sink forest stand behaviors throughout the year. This calculation involved determining the total number of days in a year ('N. days year^{-1}') where a forest stand exhibited either C-sink (NEE < 0) or C-source behavior (NEE > 0). The 10-year average number of days as a C-sink and as a C-source under the RCP 2.6 and 6.0 for the five case studies selected is presented in Figure 5a. The number of days identified as a C-sink in the evergreen forests for the scenario RCP 2.6 (i.e., FR-LBr, CZ-Bk1, and FI-Hyy) starts relatively low in the first decade (1999–2009) but increases significantly over time, showing a consistent upward trend (Figure 5(2.a,3.a,4.a). At the CZ-Bk1 site, the number of days considered as a C-sink starts at 47 N. days year^{-1} in the 1999–2009 decade and increases steadily over the decades, peaking at 199 N. days year^{-1} in 2059–2069 and remaining relatively high thereafter. For FI-Hyy forests, the number of days as a C-sink starts at 107 N. days year^{-1} in 1999–2009 and rises to 217 days in 2089–2099, with a peak of 225 in 2079–2089. Lastly, at the forests of FR-LBr, there were no days as a C-sink in 1999–2009, but they increased steadily over time, reaching 114 days by 2079–2089. The number of days as a C-source varies inversely to those as a C-sink across decades, demonstrating a general reduction over time for evergreen forests. The number of days functioning as a C-source decreased, from ~300 in 1999–2009 to ~200 N. days year^{-1} in 2089–2099 (Table A1), for evergreen forests. Deciduous forests on the contrary almost maintain a relatively stable number of days as a C-sink through the century (Figure 5(1.a,5.a). The number of days as a C-sink at IT-Col forests ranges from 212 to 224 N. days year^{-1}, while at DK-Sor from 214 to 222 N. days year^{-1}, with slight fluctuations observed across decades and no clear overall trend. The same trend for the capacity of the stand to act as a C-source, but with slightly different days in the range from 137 to 159 N. days year^{-1} recorded in the forest of Sorø, and from 140 to 149 N. days year^{-1} for the IT-Col site. The discrepancies between the already described RCP 2.6 and RCP 6.0 scenarios are minimal as they exhibit a similar trend for evergreen and deciduous, respectively, with only a slight change in the number of days, mainly in the last decade, with a magnitude of ~10 N. days year^{-1}.

To assess the changes in MSC due to climate change scenarios, we focused on the shifting patterns of the DoY over decades, particularly examining when forest stands transitioned from a C-source to a C-sink and vice versa (see Figure 5b). Regardless of the scenario considered, the European beech forests revealed constant transition periods across decades. In IT-Col, the shift from being a C-source to a C-sink occurred between DoY 136 and 150, with the opposite transition from sink to source around DoY 285. Meanwhile, in DK-Sor, the same transitions happened between DoY 115 and 127, and reversed between DoY 272 and 280 (Figure 5(1.b,5.b)). Even for evergreen forests, there are no noticeable differences in the shift corresponding to RCP 2.6 and RCP 6.0 scenarios. At CK-BZ1, the shift to a C-sink occurs relatively early in the year, spanning from DoY 2 to 31 across the decades. At FI-Hyy, the transition timings vary widely, ranging from DoY 30 to 50 over the decades. Similarly, at FR-LBr, the C-sink transitions occur from DoY 1 to 16 across the decades, with some decades exhibiting earlier shifts. On the other hand, transitions to C-sources at CZ-BK1 occur from late November to early December (DoY 201–329), displaying a decreasing trend over the years. In contrast, at FI-Hyy, transitions take place from mid-September to late September (DoY 197–285), while at FR-LBr, they occur from late September to early October (DoY 255–285) (Figure 5(2.a,3.a,4.a)).

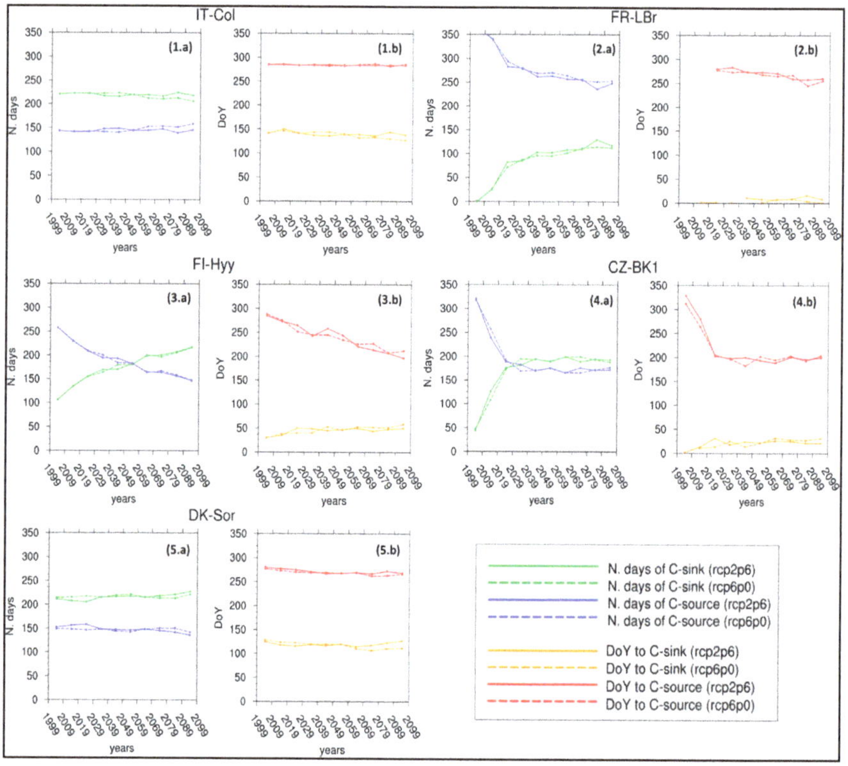

Figure 5. Ten-year average number of days (N. days year^{-1}) as C-sink and as C-source (**a**) and DoY (Day of Year) in which C-source switch to C-sink and C-source to C-sink (**b**) under the RCP 2.6 and 6.0 for the 5 case studies selected, i.e., (1) Collelongo—IT-Col, (2) Le Bray—FR-LBr, (3) Hyytiälä—FI-Hyy, (4) Bílý Kříž—CZ-Bk1, and (5) Sorø—DK-Sor. Data missing for some intervals are because of filtering and data removals to avoid pulsing artifacts, e.g., the 'Birch effect' and unstable conditions (see Section 2).

4. Discussion

4.1. Model Validation

To sum up statistical metrics, the 3D-CMCC-FEM performs best in replicating the mean seasonal patterns of the three fluxes in European beech forests DK-Sor and IT-Col. R_{eco} also reaches a satisfactory performance for the boreal Scots pine forest of FI-Hyy, and the Scots pine forest of CZ-Bk1. The robust predictive ability of 3D-CMCC-FEM in estimating NEE across different timeframes, forest species, and climates, as proved by its alignment with the eddy covariance Fluxnet2015 Dataset [56], underscores its effectiveness in capturing the complex dynamics of carbon fluxes within forest ecosystems, as documented by previous works (see, e.g., [10,11,36,40,42]). Nevertheless, there are slight inconsistencies, particularly during peak photosynthesis periods (around the 200th day of the year) in both deciduous and evergreen forests, which have been acknowledged in the existing literature. Studies indicate that estimates of ecosystem respiration derived from eddy covariance often underestimate actual values for forest ecosystems [72–75].

4.2. Mean Seasonal NEE Cycle under Climate Change Scenario

The sensitivity of forest ecosystems to changes in environmental factors such as climate change, seasonal variations, and atmospheric CO_2 levels has been thoroughly evidenced in the literature. Indeed, the interaction among these variables plays a crucial role in shaping the carbon exchange dynamics [6,12,17,31,76].

The analysis of the mean NEE seasonal cycle under different climate change scenarios presents several fascinating findings. The first key observation is the consistent reduction in forest stand capabilities to act as carbon sinks from the atmosphere in the coming years, across all study sites and climate change scenarios. Despite the overall lack of understanding regarding how forest ecosystems will respond to climate change, several new studies concur regarding the diminishing forest carbon uptake capabilities, thus confirming our statement [26,31,77–79].

In broader terms, climate change affects forest carbon balance by influencing key processes, which can respond differently due to their sensitivity to various environmental drivers [80]. Indeed, the reduction in NEE, within the forests observed in this study, has a disparity in the intensity of climate impacts across these scenarios: RCP 2.6 exhibits less intense NEE reduction compared to RCP 6.0 scenarios. The warming accelerates both autotrophic and heterotrophic respiration meaning that increased temperature may lead forests to an increase in R_{eco} [27,58,81,82]. On the other hand, the increasing atmospheric CO_2 concentration intensifies the GPP through the 'carbon fertilization effect' (i.e., reported to be the cause of 44% of the GPP increase since the 2000s) [18,19]. Under the RCP 2.6 scenario, R_{eco} exhibits a steady linear increase until the century's end, while GPP follows a bell-shaped curve, reaching saturation around the mid-century. Consequently, as GPP saturates, its compensatory capacity reduces, while R_{eco} continues to rise due to further temperature increases in the latter half of the century resulting in a further decline in NEE. Conversely, for the RCP 6.0 scenario, GPP saturation occurs only towards the end of the century, and while one might anticipate a reversal in NEE trends considering this factor, it does not materialize. This is because, under the RCP 6.0 scenario, a higher temperature increase is predicted compared to RCP 2.6 (i.e., [58]), leading to a more pronounced rise in ecosystem respiration relative to photosynthesis. As a consequence, there is a greater decrease in the stand forest's carbon sink capacity. The result refers to all forests studied in this work, but in the forest of FI-Hyy, the phenomena look more pronounced (Figures A2 and A3 in Appendix A).

The fate of ecosystem carbon flux depends not only on atmospheric and climate conditions but also on the type of forests analyzed. The reduction in C-sink capabilities is particularly notable in evergreen forests, which exhibit a higher decrease in NEE compared to evergreen sites. The boreal Scots pine forest of FI-Hyy stands out with the most significant reduction in NEE, indicating a heightened vulnerability to climate change effects. This finding is supported by the study of Hadden and Grelle [26] who found that, over a 17-year period, the forest ecosystem in a boreal forest stand in northern Sweden transitioned from being a carbon sink to a carbon source. This could mean that past efforts to validate the neutrality hypothesis [83] with climate change impact show limitations, and we need new research directions and new perspectives to better capture changes in the carbon fluxes between the ecosystem and atmosphere [84]. Indeed, a long-standing debate around the C-neutrality of old-growth forests (and some of the sites become old-growth at the end of simulations) raises concerns, and debates will increase about the reduction of the sink capacity of aging forests, as assumed by Odum's theory. As found in [7] (but see [85]), we also found that even a >200-year-old stand (as IT-Col in 2099) still has sink capacity. The annual NEE decrease (and much less the changes in the MSC), as shown by the model results, is certainly a function not only of climate but also of the inherent effects related to, e.g., biomass, both live and dead, accumulation (which led to increases in respiratory costs), and changes in, e.g., the forest structure (which led to decreases in the carbon assimilation), as stands age. Nevertheless, such effects on annual NEE are expected to be greatly exacerbated by climate change.

4.3. Changes in NEE Dynamics under Different Climate Scenarios

Climate change impacts plant phenology by altering the start and end dates of the growing season, which influences when photosynthesis can begin and consequently affects C-fluxes [20,29,30,32,86].

A primary finding is that, regardless of the scenario analyzed, the number of days identified as a C-sink in evergreen forests increases significantly over time, indicating a consistent upward trend. Similarly, the number of days classified as a C-source decreases over the decades, showing a general reduction. The second finding is that for evergreen forests, the DoY to C-sink tends to increase (indicating a forward shift in the year when the system becomes a sink), and the DoY to C-source decreases (indicating a backward shift in the year when the system becomes a source), aligning with the overall trend of fewer C-sink days and more C-source days over time. In contrast, deciduous forests maintain a relatively stable number of C-sink (and C-source) days throughout the century, reflecting a steady DoY when the system becomes a C-sink (or a C-source), despite an anticipated beginning of the growing season but compensated by higher respiration rates. Indeed, for the deciduous, the 3D-CMCC-FEM simulates the bud-brake through a thermic sum function and leaf and fine root development (and the relative growth respiration peak) in a well-defined and short time during spring [36]. Conversely, for the evergreens, leaf and fine root growth development is spread all over the spring. At the same time, leaf fall in the deciduous starts under certain hours of solar radiation, and thus, this is not under the control of climate, while in the evergreen, it happens all over the year, and under the control of climate, balanced by incoming photosynthates for new leaves and fine roots. Ultimately, deciduous spring C-sink capacity is counterbalanced by high C-emissions mainly because of growth respiration in spring. Such behavior is different for evergreens, which lengthen their C-sink capacity during spring. However, the lengthening of the growing season does not automatically mirror an increase in the net sink capacity because R_{eco} shows an increase much more than GPP. It is generally acknowledged that the changing temperature response of respiration transforms forests from C-sinks to C-sources [26,76,87–89], while the stability of the carbon sink/source dynamics over the decades for deciduous ecosystems is a relatively recent finding. The DoY for deciduous forests remains unchanged because the earlier start of the growing season, triggered by rising temperatures, is balanced by an earlier increase in respiration. This compensates for the earlier rise in GPP at the level of NEE. Overall, the lack of change in the number of C-sink (and C-source) days across decades and the reduction in the NEE suggest that, over the long term, deciduous forests are more efficient in using photosynthates compared to evergreen forests [90,91].

5. Limitations

The modeling framework presented has certain limitations that must be acknowledged. First, we deliberately decided to not simulate the effects of anthropogenic disturbances, e.g., forest management, nor the ones from natural disturbances caused by climate change, such as windstorms, forest fires, and insect outbreaks, so as to concentrate on the effects of climate change alone and avoid these potentially confounding effects. Climate extreme events are presumed to be incorporated into the climate scenarios used to drive the model and are, therefore, somewhat already accounted for. Moreover, indirect changes due to climate change in key factors like nitrogen deposition, phosphorus, or ozone—which could potentially amplify or mitigate our findings—were not evaluated. Nonetheless, some research (e.g., [92]) indicates that this issue might not be universally relevant. These studies highlight the strong response of various tree species to CO_2 fertilization across different levels of nutrient availability. Lastly, the potential for species migration to and from the study areas was not considered. However, such dynamics might require longer timescales than those covered in this study and it is unlikely (although still possible in theory) that species composition may completely change throughout the simulations.

6. Conclusions

The MSC metric is an interesting and descriptive metric that associates phenology and carbon partitioning allocation within forest stands. Climate change impacts both the phenology, by changing the date for the beginning and the end of the growing season, and the ecosystem carbon allocation.

We applied the process-based forest model 3D-CMCC-FEM to evaluate the potential modifications on the ability of different forest stands to absorb or emit carbon throughout the season and across the years up to 2100. Before that, we validated the model under current climate conditions and found a robust predictive ability of 3D-CMCC-FEM in estimating NEE, GPP, and R_{eco} across different timeframes, forest species, and climates.

The analysis of the mean NEE seasonal cycle under different climate change scenarios presents a consistent reduction in the forest stand capabilities to act as a carbon sink in the coming years, across all study sites, and climate change scenarios. The reduction in NEE ability has different intensities of climate impacts across these scenarios. The RCP 2.6 scenario demonstrates a less pronounced reduction in NEE compared to the RCP 6.0 scenario. This disparity primarily stems from variations in key variables, such as the differing rates of temperature increase between the two scenarios, as well as the CO_2 fertilization effect, while, in all sites, age effects depend on the age at the beginning of the simulations. The reduction in C-sink capabilities is mainly notable in evergreen forests, which exhibit a higher decrease in NEE compared to deciduous forest sites.

Finally, we found that the number of days as a C-sink in evergreen forests increases over the years, indicating a consistent upward trend. Conversely, the number of days as a C-source decreases over the decades, showing a general reduction. This statement aligns with the forward shift of DoY to C-sink, and the backward shift of DoY to C-source. In contrast, deciduous forests maintain a relatively stable number of C-sink (and C-source) days throughout the century, reflecting a fixed DoY when the system becomes a C-sink (or a C-source). The DoY for deciduous forests remains constant as the earlier onset of the growing season, driven by warming temperatures, is offset by an earlier uptick in respiration. Decades pass with little change in the number of days as a C-sink (and a C-source), alongside a decrease in NEE. This indicates that deciduous forests, over the long haul, demonstrate greater efficiency in utilizing photosynthates when compared to evergreen forests.

Author Contributions: M.M.: Data curation, Formal analysis, Investigation, Writing—original draft, Writing—review and editing; E.V.: Data curation, Writing—original draft, Writing—review and editing; A.C.: Conceptualization, Formal analysis, Investigation, Writing—original draft, Writing—review and editing. All authors have read and agreed to the published version of the manuscript.

Funding: OptForEU Horizon Europe research and innovation programme under grant agreement No. 101060554; National Recovery and Resilience Plan (NRRP), Mission 4 Component 2 Investment 1.4—Call for tender No. 3138 of 16 December 2021, rectified by Decree n.3175 of 18 December 2021 of Italian Ministry of University and Research funded by the European Union—NextGenerationEU under award number: Project code CN_00000033, Concession Decree No. 1034 of 17 June 2022 adopted by the Italian Ministry of University and Research, CUP B83C22002930006; Project title "National Biodiversity Future Centre—NBFC"; A.C. also acknowledges funding from the MIUR Project (PRIN 2020)—Research Projects of National Relevance funded by the Italian Ministry of University and Research entitled: "Unraveling interactions between WATER and carbon cycles during drought and their impact on water resources and forest and grassland ecosySTEMs in the Mediterranean climate" (WATERSTEM, project number: 20202WF53Z), and "WAFER" at CNR (Consiglio Nazionale delle Ricerche); A.C. and E.V. also acknowledge funding from the MIUR Project (PRIN 2020)—Research Projects of National Relevance funded by the Italian Ministry of University and Research entitled: "Multi-scale observations to predict Forest response to pollution and climate change" (MULTIFOR, project number: 2020E52THS).

Data Availability Statement: The 3D-CMCC-FEM model code is publicly available and can be found on the GitHub platform at: https://github.com/Forest-Modelling-Lab/3D-CMCC-FEM (accessed on 24 June 2024). The 3D-CMCC-FEM output data used in this work can be downloaded at: https://zenodo.org/records/11124413 (accessed on 7 May 2024). Correspondence and requests for additional materials should be addressed to the corresponding author.

Acknowledgments: We are thankful to D. Dalmonech for supporting data preparation and analysis. We also thank the ISIMIP project and the COST Action FP1304 PROFOUND (Towards Robust Projections of European Forests under Climate Change), supported by COST (European Cooperation in Science and Technology) for providing us the climate historical and future scenarios and site data used in this work. This work used eddy covariance data acquired and shared by the FLUXNET community, including these networks: AmeriFlux, AfriFlux, AsiaFlux, CarboAfrica, CarboEurope-IP, CarboItaly, CarboMont, ChinaFlux, Fluxnet-Canada, GreenGrass, ICOS, KoFlux, LBA, NECC, OzFlux-TERN, TCOS-Siberia, and USCCC. The ERA-Interim reanalysis data are provided by ECMWF and processed by LSCE. The FLUXNET eddy covariance data processing and harmonization was carried out by the European Fluxes Database Cluster, AmeriFlux Management Project, and Fluxdata Project of FLUXNET, with the support of CDIAC and ICOS Ecosystem Thematic Center, and the OzFlux, ChinaFlux, and AsiaFlux offices. We acknowledge the World Climate Research Programme's Working Group on Coupled Modelling, which is responsible for CMIP, and we thank the respective climate modeling groups for producing and making available their model output. The U.S. Department of Energy's Program for Climate Model Diagnosis and Intercomparison at Lawrence Livermore National Laboratory provides coordinating support for CMIP and led the development of software infrastructure in partnership with the Global Organization for Earth System Science Portals.

Conflicts of Interest: The authors declare no conflicts of interest.

Appendix A

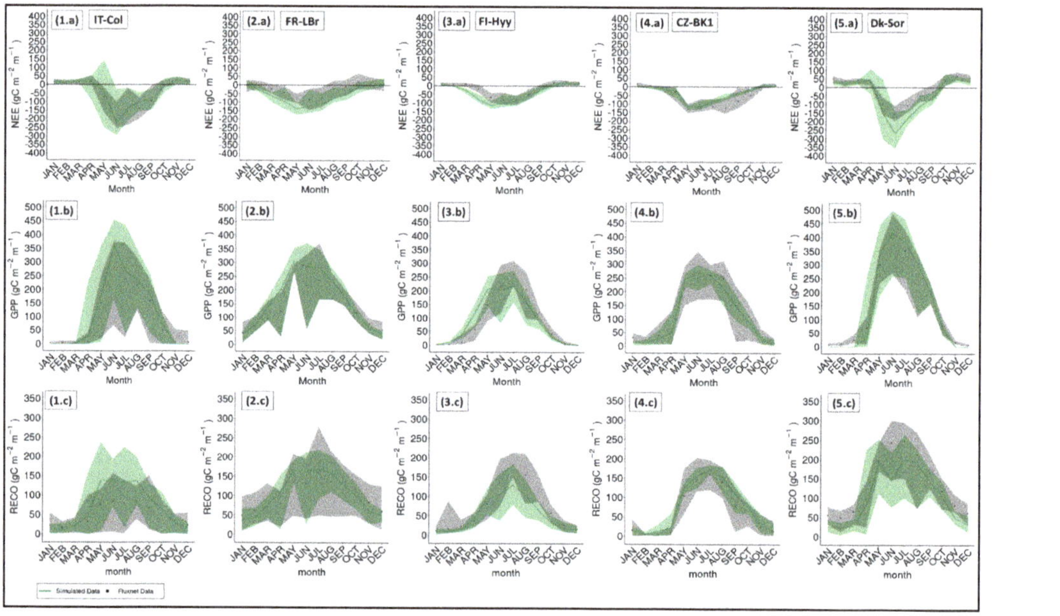

Figure A1. The green lines represent the modeled (**a**) NEE, (**b**) GPP, and (**c**) R_{eco} amounts (gC m^{-2} $month^{-1}$) per month for the five selected case studies (i.e., (1) Collelongo—IT-Col, (2) Le Bray—FR-Lbr, (3) Hyytiälä—FI-Hyy, (4) Bílý Křiž—CZ-Bk1, and (5) Sorø—DK-Sor) compared to relative observed data (depicted as black dots) from the Fluxnet2015 Dataset (Pastorello et al., 2020 [56]). The lower and upper lines of the shaded area represent, respectively, the minimum and maximum values of the observed and modeled datasets considered.

Table A1. Summary of the statistics between simulated and measured NEE from the Fluxnet2015 Dataset (Pastorello et al., 2020 [56]), calculated on the 5 cases studies selected (i.e., Collelongo—IT-Col, Le Bray—FR-LBr, Hyytiälä—FI-Hyy, Bílý Křiž—CZ-Bk1, and Sorø—DK-Sor). The table shows the daily and monthly values for Person's Coefficient (r—dimensionless), Relative Mean Bias (RMB—dimensionless), Normalized Root Mean Square Error (NRMSE—dimensionless), Modeling Efficiency (ME—dimensionless), and Mean Absolute Error (MAE, gC m^{-2} $time^{-1}$).

	IT-Col	**FR-LBr**	**FI-Hyy**	**CZ-Bk1**	**DK-Sor**
			Daily NEE		
r	0.96	0.93	0.93	0.85	0.97
RMB	0.07	−0.86	−0.48	0.14	−0.46
NRMSE	−0.57	−1.56	−1.54	1.39	−6.59
ME	0.92	−0.10	0.58	0.69	0.46
MAE	0.67	1.20	0.67	1.15	1.42
			Monthly NEE		
r	0.98	0.95	0.93	0.92	0.99
RMB	0.07	−0.88	−0.49	0.21	−0.46
NRMSE	0.99	−1.55	−1.52	−0.50	−6.14
ME	0.95	−0.15	0.55	0.79	0.55
MAE	11.42	33.39	19.63	19.12	37.78

Table A2. Summary of the statistics between simulated and measured GPP from the Fluxnet2015 Dataset (Pastorello et al., 2020 [56]), calculated on the 5 cases studies selected (i.e., Collelongo—IT-Col, Le Bray—FR-Lbr, Hyytiälä—FI-Hyy, Bílý Křiž—CZ-Bk1, and Sorø—DK-Sor). The table shows the daily and monthly values for Person's Coefficient (r—dimensionless), Relative Mean Bias (RMB—dimensionless), Normalized Root Mean Square Error (NRMSE—dimensionless), Modeling Efficiency (ME—dimensionless), and Mean Absolute Error (MAE, gC m^{-2} $time^{-1}$).

	IT-Col	**FR-LBr**	**FI-Hyy**	**CZ-Bk1**	**DK-Sor**
			Daily GPP		
r	0.99	0.95	0.93	0.95	0.99
RMB	0.15	0.55	0.01	−0.14	0.00
NRMSE	0.34	0.42	0.39	1.21	0.19
ME	0.91	0.28	0.86	0.85	0.97
MAE	0.91	1.51	0.91	0.97	0.75
			Monthly GPP		
r	0.99	0.96	0.93	0.99	1.00
RMB	0.15	0.52	0.01	−0.06	0.00
NRMSE	0.31	0.40	0.38	0.15	0.16
ME	0.92	0.34	0.86	0.97	0.97
MAE	20.68	41.69	26.05	14.25	20.26

Table A3. Summary of the statistics between simulated and measured R_{eco} from the Fluxnet2015 Dataset (Pastorello et al., 2020 [56]), calculated on the 5 cases studies selected (i.e., Collelongo—IT-Col, Le Bray—FR-LBr, Hyytiälä—FI-Hyy, Bílý Křiž—CZ-Bk1, and Sorø—DK-Sor). The table shows the daily and monthly values for Person's Coefficient (r—dimensionless), Relative Mean Bias (RMB—dimensionless), Normalized Root Mean Square Error (NRMSE—dimensionless), Modeling Efficiency (ME—dimensionless), and Mean Absolute Error (MAE, gC m^{-2} $time^{-1}$).

	IT-Col	**FR-LBr**	**FI-Hyy**	**CZ-Bk1**	**DK-Sor**
			Daily R_{eco}		
r	0.90	0.89	0.94	0.86	0.89
RMB	0.82	0.16	−0.30	0.30	−0.42
NRMSE	0.72	0.24	0.41	1.16	0.35
ME	−1.17	0.50	0.75	0.63	0.62
MAE	1.14	0.71	0.74	0.88	1.29

Table A3. *Cont.*

	IT-Col	FR-LBr	FI-Hyy	CZ-Bk1	DK-Sor
			Monthly R_{eco}		
r	0.99	0.94	0.95	0.99	0.95
RMB	0.77	0.14	−0.29	0.19	−0.42
NRMSE	0.83	0.20	0.39	0.24	0.29
ME	−0.64	0.68	0.77	0.93	0.71
MAE	25.42	18.01	21.02	11.61	31.82

Table A4. Summary of the NEE variation (%) from the first decade of simulation (1999–2009), considering both climate change scenarios (RCP 2.6 and 6.0) for 5 case studies selected (i.e., Collelongo—IT-Col, Le Bray—FR-LBr, Hyytiälä—FI-Hyy, Bílý Kříž—CZ-Bk1, and Sorø—DK-Sor). In bold values where changes were the highest between the decades while underlined the lowest ones. Note that negative values indicate that NEE becomes less negative, e.g., −100% indicates a reduction in the negative values of NEE.

			NEE		
			RCP 2.6		
year	CZ-Bk1	IT-Col	FI-Hyy	FR-LBr	DK-Sor
2009–2019	**−37%**	−19%	−4%	−24%	−9%
2019–2029	**−64%**	−35%	−16%	−44%	−14%
2029–2039	**−70%**	−31%	−27%	−53%	−22%
2039–2049	**−76%**	−39%	−39%	−59%	−29%
2049–2059	**−71%**	−48%	−47%	−58%	−33%
2059–2069	**−79%**	−59%	−58%	−65%	−32%
2069–2079	**−75%**	−61%	−66%	−70%	−37%
2079–2089	−76%	−69%	**−76%**	−75%	−44%
2089–2099	−74%	−68%	**−95%**	−72%	−43%
			RCP 6.0		
year	CZ-Bk1	IT-Col	FI-Hyy	FR-LBr	DK-Sor
2009–2019	**−35%**	−11%	0%	−25%	−11%
2019–2029	**−60%**	−27%	−11%	−43%	−14%
2029–2039	**−74%**	−36%	−23%	−52%	−26%
2039–2049	**−76%**	−43%	−36%	−57%	−34%
2049–2059	**−75%**	−44%	−46%	−61%	−31%
2059–2069	**−75%**	−43%	−57%	−64%	−23%
2069–2079	**−72%**	−54%	−69%	−69%	−30%
2079–2089	−76%	−59%	**−79%**	−75%	−34%
2089–2099	−71%	−60%	−87%	−68%	−36%

Table A5. Summary of the GPP variation (%) from the first decade of simulation (1999–2009), considering both climate change scenarios (RCP 2.6 and 6.0) for 5 case studies selected (i.e., Collelongo—IT-Col, Le Bray—FR-LBr, Hyytiälä—FI-Hyy, Bílý Kříž—CZ-Bk1, and Sorø—DK-Sor). In bold values where changes were the highest between the decades while underlined the lowest ones.

			GPP		
			RCP 2.6		
year	CZ-Bk1	IT-Col	FI-Hyy	FR-LBr	DK-Sor
2009–2019	7%	6%	**22%**	12%	8%
2019–2029	10%	5%	**34%**	15%	15%
2029–2039	16%	11%	**39%**	18%	18%
2039–2049	17%	13%	**42%**	19%	19%
2049–2059	18%	12%	**43%**	20%	20%
2059–2069	18%	11%	**41%**	18%	23%
2069–2079	17%	12%	**36%**	18%	18%
2079–2089	16%	9%	**32%**	18%	15%
2089–2099	17%	10%	**26%**	16%	15%

Table A5. *Cont.*

			GPP		
			RCP 6.0		
year	CZ-BK1	IT-Col	FI-Hyy	FR-LBr	DK-Sor
2009–2019	7%	7%	**23%**	11%	7%
2019–2029	10%	9%	**36%**	14%	13%
2029–2039	13%	10%	**43%**	18%	14%
2039–2049	15%	12%	**44%**	20%	15%
2049–2059	20%	17%	**49%**	20%	20%
2059–2069	23%	23%	**50%**	23%	28%
2069–2079	27%	26%	**51%**	26%	31%
2079–2089	28%	29%	**50%**	26%	31%
2089–2099	31%	30%	**51%**	28%	34%

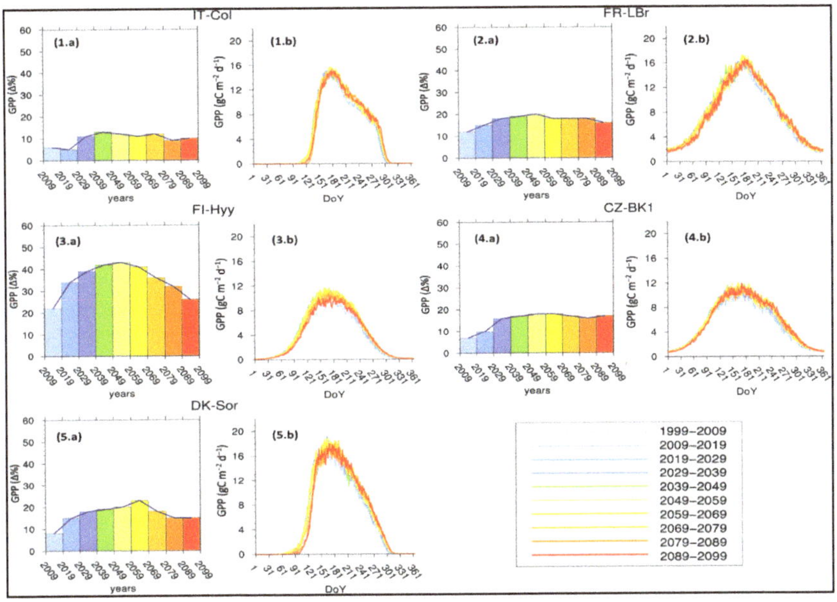

Figure A2. Ten-year average GPP seasonal cycle under the RCP 2.6 climate scenario for 5 case studies selected, i.e., (1) Collelongo—IT-Col, (2) Le Bray—FR-LBr, (3) Hyytiälä—FI-Hyy, (4) Bílý Kříž—CZ-Bk1, and (5) Sorø—DK-Sor. The histograms (**a**) represent the annual GPP variation (%) from the first decade taken as a benchmark of simulation (1999–2009). The xy plots (**b**) show the Mean Seasonal GPP Cycle of monthly values (gC m^{-2} day^{-1}).

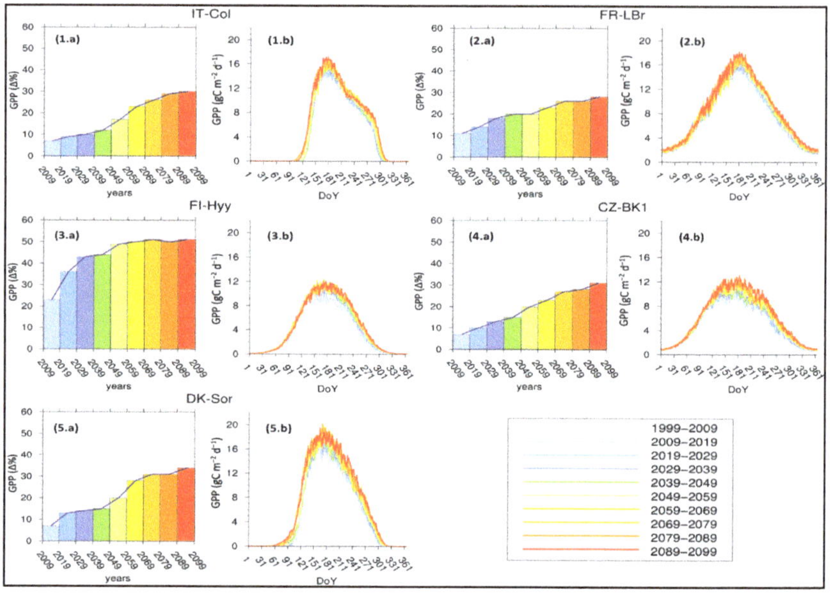

Figure A3. Ten-year average GPP seasonal cycle under the RCP 6.0 climate scenario for 5 case studies selected, i.e., (1) Collelongo—IT-Col, (2) Le Bray—FR-LBr, (3) Hyytiälä—FI-Hyy, (4) Bílý Kříž—CZ-Bk1, and (5) Sorø—DK-Sor. The histograms (**a**) represent the annual GPP variation (%) from the first decade taken as a benchmark of simulation (1999–2009). The xy plots (**b**) show the Mean Seasonal GPP Cycle of monthly values (gC m^{-2} day^{-1}).

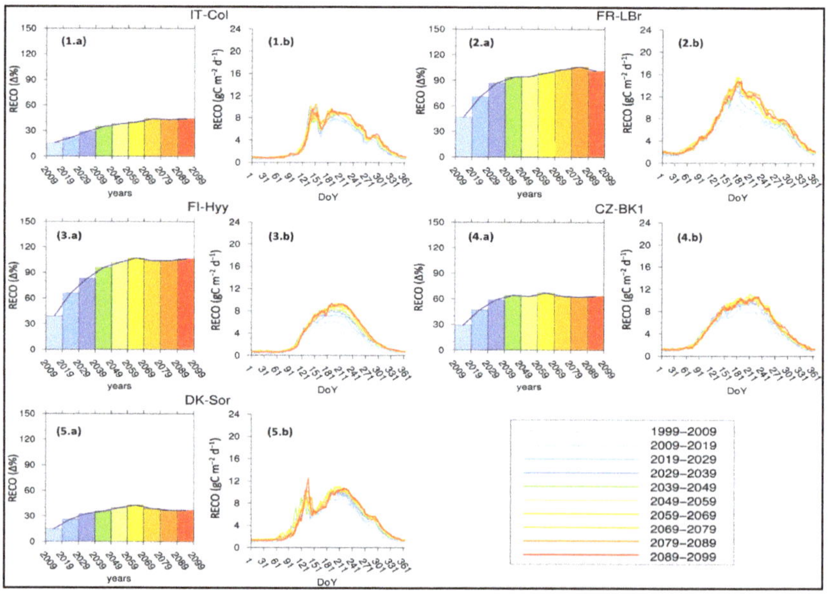

Figure A4. Ten-year average R_{eco} seasonal cycle under the RCP 2.6 climate scenario for 5 case studies selected, i.e., (1) Collelongo—IT-Col, (2) Le Bray—FR-LBr, (3) Hyytiälä—FI-Hyy, (4) Bílý Kříž—CZ-Bk1, and (5) Sorø—DK-Sor. The histograms (**a**) represent the annual R_{eco} variation (%) from the first decade taken as a benchmark of simulation (1999–2009). The xy plots (**b**) show the Mean Seasonal R_{eco} Cycle of monthly values (gC m^{-2} day^{-1}).

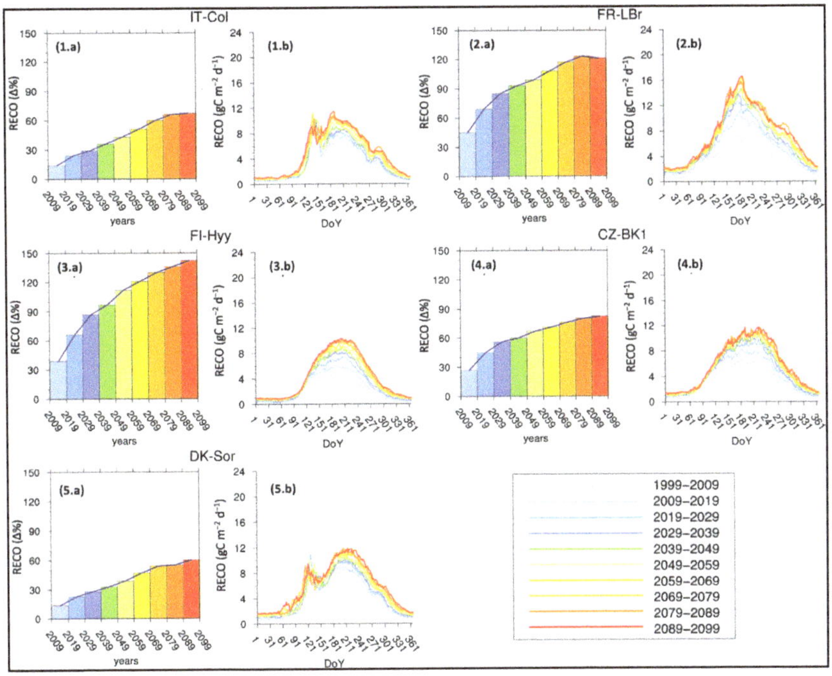

Figure A5. Ten-year average R_{eco} seasonal cycle under the RCP 6.0 climate scenario for 5 case studies selected, i.e., (1) Collelongo—IT-Col, (2) Le Bray—FR-LBr, (3) Hyytiälä—FI-Hyy, (4) Bílý Kříž—CZ-Bk1, and (5) Sorø—DK-Sor. The histograms (**a**) represent the annual R_{eco} variation (%) from the first decade taken as a benchmark of simulation (1999–2009). The xy plots (**b**) show the Mean Seasonal R_{eco} Cycle of monthly values (gC m^{-2} day^{-1}).

Table A6. Summary of the R_{eco} changes (%) from the first decade of simulation (1999–2009), considering both climate change scenarios (RCP 2.6 and 6.0) for 5 case studies selected (i.e., Collelongo—IT-Col, Le Bray—FR-LBr, Hyytiälä—FI-Hyy, Bílý Kříž—CZ-Bk1, and Sorø—DK-Sor). In bold values where changes were the highest between the decades while underlined the lowest ones.

		R_{eco}			
		RCP 2.6			
year	CZ-Bk1	IT-Col	FI-Hyy	FR-LBr	DK-Sor
2009–2019	29%	16%	39%	**47%**	15%
2019–2029	47%	22%	66%	**71%**	26%
2029–2039	59%	29%	83%	**87%**	33%
2039–2049	64%	35%	**96%**	94%	36%
2049–2059	63%	38%	**102%**	95%	40%
2059–2069	67%	40%	**107%**	99%	43%
2069–2079	63%	44%	**104%**	103%	39%
2079–2089	62%	43%	104%	**106%**	37%
2089–2099	63%	44%	**106%**	101%	37%
		RCP 6.0			
year	CZ-Bk1	IT-Col	FI-Hyy	FR-LBr	DK-Sor
2009–2019	27%	14%	39%	**45%**	14%
2019–2029	45%	24%	66%	**69%**	23%
2029–2039	56%	29%	**87%**	85%	28%
2039–2049	60%	36%	**97%**	93%	33%
2049–2059	67%	43%	**112%**	99%	39%
2059–2069	71%	51%	**121%**	108%	47%
2069–2079	76%	60%	**130%**	117%	54%
2079–2089	80%	66%	**136%**	123%	55%
2089–2099	82%	67%	**142%**	121%	60%

Table A7. Summary of the number of the days in a year (N. days year^{-1}) as C-sink, considering both climate change scenarios (RCP 2.6 and 6.0) for 5 case studies selected (i.e., Collelongo—IT-Col, Le Bray—FR-LBr, Hyytiälä—FI-Hyy, Bílý Kříž—CZ-Bk1, and Sorø—DK-Sor). In bold values where changes were the highest between the decades while underlined the lowest ones within each forest stand.

Days as C-Sink					
			RCP 2.6		
year	CZ-Bk1	IT-Col	FI-Hyy	FR-Lbr	DK-Sor
1999–2009	<u>44</u>	221	<u>107</u>	0	212
2009–2019	126	223	135	<u>25</u>	208
2019–2029	176	223	156	82	<u>206</u>
2029–2039	182	217	170	85	216
2039–2049	195	<u>216</u>	171	103	217
2049–2059	189	220	182	102	218
2059–2069	**199**	220	199	108	216
2069–2079	189	217	201	109	219
2079–2089	194	**225**	208	**129**	222
2089–2099	193	219	**218**	117	**228**
			RCP 6.0		
year	CZ-Bk1	IT-Col	FI-Hyy	FR-Lbr	DK-Sor
1999–2009	<u>47</u>	221	<u>107</u>	0	215
2009–2019	108	223	134	<u>26</u>	216
2019–2029	172	222	155	71	218
2029–2039	195	223	164	87	216
2039–2049	194	**224**	180	96	220
2049–2059	190	220	181	95	**222**
2059–2069	**199**	212	201	101	216
2069–2079	**199**	211	197	111	<u>214</u>
2079–2089	193	213	206	**114**	<u>214</u>
2089–2099	188	<u>206</u>	**217**	112	**222**

Table A8. Summary of number of the days in a year (n. days year^{-1}) source, considering both climate change scenarios (RCP 2.6 and 6.0) for 5 case studies selected (i.e., Collelongo—IT-Col, Le Bray—FR-LBr, Hyytiälä—FI-Hyy, Bílý Kříž—CZ-Bk1, and Sorø—DK-Sor). In bold values where changes were the highest between the decades while underlined the lowest ones within each forest stand.

Days as C-Source					
			RCP 2.6		
year	CZ-Bk1	IT-Col	FI-Hyy	FR-LBr	DK-Sor
1999–2009	**321**	144	**258**	**365**	153
2009–2019	239	142	230	340	157
2019–2029	189	142	209	283	**159**
2029–2039	183	148	195	280	149
2039–2049	<u>170</u>	**149**	194	262	148
2049–2059	176	145	183	263	147
2059–2069	166	145	166	257	149
2069–2079	176	148	164	256	146
2079–2089	171	<u>140</u>	157	<u>236</u>	143
2089–2099	172	146	<u>147</u>	248	<u>137</u>

Table A8. *Cont.*

Days as C-Source					
			RCP 6.0		
year	CZ-Bk1	IT-Col	FI-Hyy	FR-LBr	DK-Sor
1999–2009	**318**	144	**258**	**365**	150
2009–2019	257	142	231	339	149
2019–2029	193	143	210	294	147
2029–2039	170	142	201	278	149
2039–2049	171	<u>141</u>	185	269	145
2049–2059	175	145	184	270	<u>143</u>
2059–2069	<u>166</u>	153	164	264	149
2069–2079	<u>166</u>	154	168	254	**151**
2079–2089	172	152	159	<u>251</u>	**151**
2089–2099	177	**159**	<u>148</u>	253	<u>143</u>

Table A9. Summary of the changes in the source/sink DoY (Day of Year), considering both climate change scenarios (RCP 2.6 and 6.0) for 5 case studies selected (i.e., Collelongo—IT-Col, Le Bray—FR-LBr, Hyytiälä—FI-Hyy, Bílý Kříž—CZ-Bk1, and Sorø—DK-Sor). In bold values where changes were the highest between the decades while underlined the lowest ones within each forest stand. Data are missing for some intervals because of filtering and data removal to avoid pulsing artifacts, e.g., the 'Birch effect' (see Section 2).

DoY to C-Sink					
			RCP 2.6		
year	CZ-Bk1	IT-Col	FI-Hyy	FR-LBr	DK-Sor
1999–2009	<u>2</u>	141	<u>30</u>	-	125
2009–2019	13	**150**	35	<u>1</u>	118
2019–2029	**31**	142	**50**	2	116
2029–2039	18	137	-	-	119
2039–2049	24	<u>136</u>	45	11	117
2049–2059	22	139	47	8	120
2059–2069	26	139	**50**	7	<u>115</u>
2069–2079	25	<u>136</u>	44	9	118
2079–2089	21	144	48	**16**	123
2089–2099	21	138	**50**	8	**127**
			RCP 6.0		
year	CZ-Bk1	IT-Col	FI-Hyy	FR-LBr	DK-Sor
1999–2009	-	142	<u>30</u>	-	**129**
2009–2019	<u>11</u>	**146**	38	<u>1</u>	124
2019–2029	13	141	40	-	123
2029–2039	25	144	40	<u>1</u>	120
2039–2049	14	144	53	-	120
2049–2059	21	140	46	<u>1</u>	120
2059–2069	**32**	132	53	**8**	111
2069–2079	28	133	52	**8**	<u>107</u>
2079–2089	27	130	51	4	111
2089–2099	31	<u>127</u>	**59**	<u>1</u>	112

Table A10. Summary of the changes in the sink/source DoY (Day of Year), considering both climate change scenarios (RCP 2.6 and 6.0) for 5 case studies selected (i.e., Collelongo—IT-Col, Le Bray—FR-LBr, Hyytiälä—FI-Hyy, Bílý Kříž—CZ-Bk1, and Sorø—DK-Sor). In bold values where changes were the highest between the decades while underlined the lowest ones within each forest stand. Data are missing for some intervals because of filtering and data removal to avoid pulsing artifacts, e.g., the 'Birch effect' (see Section 2).

DoY to C-Source					
		RCP 2.6			
year	CZ-Bk1	IT-Col	FI-Hyy	FR-LBr	DK-Sor
1999–2009	**329**	285	**285**	-	**280**
2009–2019	280	**286**	273	-	277
2019–2029	203	284	265	**277**	275
2029–2039	198	285	243	273	271
2039–2049	200	285	258	274	267
2049–2059	194	284	244	268	268
2059–2069	189	284	221	265	269
2069–2079	201	284	214	267	267
2079–2089	196	284	207	245	272
2089–2099	200	284	197	255	268
		RCP 6.0			
year	CZ-Bk1	IT-Col	FI-Hyy	FR-LBr	DK-Sor
1999–2009	**312**	285	**288**	-	**277**
2009–2019	264	285	275	-	273
2019–2029	205	284	252	280	270
2029–2039	197	284	245	**283**	269
2039–2049	183	283	245	273	269
2049–2059	202	283	235	273	268
2059–2069	195	285	226	271	269
2069–2079	203	**287**	227	259	262
2079–2089	193	282	208	258	263
2089–2099	204	286	212	260	266

References

1. Ciais, P.; Sabine, C.; Bala, G.; Peters, W. Carbon and other biogeochemical cycles. In *Climate Change 2013 the Physical Science Basis: Working Group I Contribution to the Fifth Assessment Report of the Intergovernmental Panel on Climate Change*; Cambridge University Press: Cambridge, UK, 2013; Volume 9781107057, pp. 465–570. ISBN 9781107415324.
2. Anav, A.; Friedlingstein, P.; Beer, C.; Ciais, P.; Harper, A.; Jones, C.; Murray-Tortarolo, G.; Papale, D.; Parazoo, N.C.; Peylin, P.; et al. Spatiotemporal patterns of terrestrial gross primary production: A review. *Rev. Geophys.* **2015**, *53*, 785–818. [CrossRef]
3. Liang, M.C.; Laskar, A.H.; Barkan, E.; Newman, S.; Thiemens, M.H.; Rangarajan, R. New constraints of terrestrial and oceanic global gross primary productions from the triple oxygen isotopic composition of atmospheric CO_2 and O_2. *Sci. Rep.* **2023**, *13*, 1–10. [CrossRef] [PubMed]
4. Intergovernmental Panel on Climate Change. *Climate Change 2021—The Physical Science Basis*; Cambridge University Press: Cambridge, UK, 2023. [CrossRef]
5. Quéré, C.; Andrew, R.; Friedlingstein, P.; Sitch, S.; Hauck, J.; Pongratz, J.; Pickers, P.; Ivar Korsbakken, J.; Peters, G.; Canadell, J.; et al. Global Carbon Budget 2018. *Earth Syst. Sci. Data* **2018**, *10*, 2141–2194. [CrossRef]
6. Rödenbeck, C.; Zaehle, S.; Keeling, R.; Heimann, M. How does the terrestrial carbon exchange respond to inter-Annual climatic variations? A quantification based on atmospheric CO_2 data. *Biogeosciences* **2018**, *15*, 2481–2498. [CrossRef]
7. Luyssaert, S.; Schulze, E.D.; Börner, A.; Knohl, A.; Hessenmöller, D.; Law, B.E.; Ciais, P.; Grace, J. Old-growth forests as global carbon sinks. *Nature* **2008**, *455*, 213–215. [CrossRef] [PubMed]
8. Collalti, A.; Ibrom, A.; Stockmarr, A.; Cescatti, A.; Alkama, R.; Fernández-Martínez, M.; Matteucci, G.; Sitch, S.; Friedlingstein, P.; Ciais, P.; et al. Forest production efficiency increases with growth temperature. *Nat. Commun.* **2020**, *11*, 5322. [CrossRef] [PubMed]
9. Amiro, B.D.; Barr, A.G.; Barr, J.G.; Black, T.A.; Bracho, R.; Brown, M.; Chen, J.; Clark, K.L.; Davis, K.J.; Desai, A.R.; et al. Ecosystem carbon dioxide fluxes after disturbance in forests of North America. *J. Geophys. Res. Biogeosci.* **2010**, *115*, 237–251. [CrossRef]
10. Collalti, A.; Trotta, C.; Keenan, T.F.; Ibrom, A.; Bond-Lamberty, B.; Grote, R.; Vicca, S.; Reyer, C.P.O.; Migliavacca, M.; Veroustraete, F.; et al. Thinning Can Reduce Losses in Carbon Use Efficiency and Carbon Stocks in Managed Forests Under Warmer Climate. *J. Adv. Model. Earth Syst.* **2018**, *10*, 2427–2452. [CrossRef]

11. Dalmonech, D.; Marano, G.; Amthor, J.S.; Cescatti, A.; Lindner, M.; Trotta, C.; Collalti, A. Feasibility of enhancing carbon sequestration and stock capacity in temperate and boreal European forests via changes to management regimes. *Agric. For. Meteorol.* **2022**, *327*, 109203. [CrossRef]
12. Tramontana, G.; Migliavacca, M.; Jung, M.; Reichstein, M.; Keenan, T.F.; Camps-Valls, G.; Ogee, J.; Verrelst, J.; Papale, D. Partitioning net carbon dioxide fluxes into photosynthesis and respiration using neural networks. *Glob. Chang. Biol.* **2020**, *26*, 5235–5253. [CrossRef]
13. Santini, M.; Collalti, A.; Valentini, R. Climate change impacts on vegetation and water cycle in the Euro-Mediterranean region, studied by a likelihood approach. *Reg. Environ. Change* **2014**, *14*, 1405–1418. [CrossRef]
14. Noce, S.; Collalti, A.; Valentini, R.; Santini, M. Hot spot maps of forest presence in the Mediterranean basin. *iForest* **2016**, *9*, 766–774. [CrossRef]
15. Lionello, P.; Scarascia, L. The relation between climate change in the Mediterranean region and global warming. *Reg. Environ. Change* **2018**, *18*, 1481–1493. [CrossRef]
16. FAO. State of Mediterranean Forests (SoFMF), Concept Paper; Marseille, 2011. Available online: http://www.fao.org/docrep/013/ma723e/ma723e00.pdf (accessed on 10 May 2024).
17. Collalti, A.; Thornton, P.E.; Cescatti, A.; Rita, A.; Borghetti, M.; Nolè, A.; Trotta, C.; Ciais, P.; Matteucci, G. The sensitivity of the forest carbon budget shifts across processes along with stand development and climate change. *Ecol. Appl.* **2019**, *29*, e01837. [CrossRef] [PubMed]
18. Friedlingstein, P.; Bopp, L.; Ciais, P.; Dufresne, J.L.; Fairhead, L.; LeTreut, H.; Monfray, P.; Orr, J. Positive feedback between future climate change and the carbon cycle. *Geophys. Res. Lett.* **2001**, *28*, 1543–1546. [CrossRef]
19. Chen, C.; Riley, W.J.; Prentice, I.C.; Keenan, T.F. CO_2 fertilization of terrestrial photosynthesis inferred from site to global scales. *Proc. Natl. Acad. Sci. USA* **2022**, *119*, e2115627119. [CrossRef]
20. Keenan, T.F.; Prentice, I.C.; Canadell, J.G.; Williams, C.A.; Wang, H.; Raupach, M.; Collatz, G.J. Recent pause in the growth rate of atmospheric CO_2 due to enhanced terrestrial carbon uptake. *Nat. Commun.* **2016**, *7*, 13428. [CrossRef]
21. Wang, S.; Zhang, Y.; Ju, W.; Chen, J.M.; Ciais, P.; Cescatti, A.; Sardans, J.; Janssens, I.A.; Wu, M.; Berry, J.A.; et al. Recent global decline of CO_2 fertilization effects on vegetation photosynthesis. *Science* **2020**, *370*, 1295–1300. [CrossRef]
22. Yuan, Z.; Ali, A.; Jucker, T.; Ruiz-Benito, P.; Wang, S.; Jiang, L.; Wang, X.; Lin, F.; Ye, J.; Hao, Z.; et al. Multiple abiotic and biotic pathways shape biomass demographic processes in temperate forests. *Ecology* **2019**, *100*, e02650. [CrossRef]
23. Grossiord, C.; Buckley, T.N.; Cernusak, L.A.; Novick, K.A.; Poulter, B.; Siegwolf, R.T.W.; Sperry, J.S.; McDowell, N.G. Plant responses to rising vapor pressure deficit. *New Phytol.* **2020**, *226*, 1550–1566. [CrossRef]
24. D'Andrea, E.; Rezaie, N.; Prislan, P.; Gričar, J.; Collalti, A.; Muhr, J.; Matteucci, G. Frost and drought: Effects of extreme weather events on stem carbon dynamics in a Mediterranean beech forest. *Plant Cell Environ.* **2020**, *43*, 2365–2379. [CrossRef] [PubMed]
25. McDowell, N.G.; Allen, C.D.; Anderson-Teixeira, K.; Aukema, B.H.; Bond-Lamberty, B.; Chini, L.; Clark, J.S.; Dietze, M.; Grossiord, C.; Hanbury-Brown, A.; et al. Pervasive shifts in forest dynamics in a changing world. *Science* **2020**, *368*, eaaz9463. [CrossRef] [PubMed]
26. Hadden, D.; Grelle, A. Changing temperature response of respiration turns boreal forest from carbon sink into carbon source. *Agric. For. Meteorol.* **2016**, *223*, 30–38. [CrossRef]
27. Mitchard, E.T.A. The tropical forest carbon cycle and climate change. *Nature* **2018**, *559*, 527–534. [CrossRef] [PubMed]
28. Bastida, F.; Torres, I.F.; Andrés-Abellán, M.; Baldrian, P.; López-Mondéjar, R.; Větrovský, T.; Richnow, H.H.; Starke, R.; Ondoño, S.; García, C.; et al. Differential sensitivity of total and active soil microbial communities to drought and forest management. *Glob. Chang. Biol.* **2017**, *23*, 4185–4203. [CrossRef] [PubMed]
29. Gunderson, C.A.; Edwards, N.T.; Walker, A.V.; O'Hara, K.H.; Campion, C.M.; Hanson, P.J. Forest phenology and a warmer climate—Growing season extension in relation to climatic provenance. *Glob. Chang. Biol.* **2012**, *18*, 2008–2025. [CrossRef]
30. Richardson, A.D.; Keenan, T.F.; Migliavacca, M.; Ryu, Y.; Sonnentag, O.; Toomey, M. Climate change, phenology, and phenological control of vegetation feedbacks to the climate system. *Agric. For. Meteorol.* **2013**, *169*, 156–173. [CrossRef]
31. Keenan, T.F.; Gray, J.; Friedl, M.A.; Toomey, M.; Bohrer, G.; Hollinger, D.Y.; Munger, J.W.; O'Keefe, J.; Schmid, H.P.; Wing, I.S.; et al. Net carbon uptake has increased through warming-induced changes in temperate forest phenology. *Nat. Clim. Change* **2014**, *4*, 598–604. [CrossRef]
32. Peano, D.; Materia, S.; Collalti, A.; Alessandri, A.; Anav, A.; Bombelli, A.; Gualdi, S. Global Variability of Simulated and Observed Vegetation Growing Season. *J. Geophys. Res. Biogeosci.* **2019**, *124*, 3569–3587. [CrossRef]
33. De Kauwe, M.G.; Medlyn, B.E.; Zaehle, S.; Walker, A.P.; Dietze, M.C.; Wang, Y.P.; Luo, Y.; Jain, A.K.; El-Masri, B.; Hickler, T.; et al. Where does the carbon go? A model–data intercomparison of vegetation carbon allocation and turnover processes at two temperate forest free-air CO2 enrichment sites. *New Phytol.* **2014**, *203*, 883–899. [CrossRef]
34. Vacchiano, G.; Magnani, F.; Collalti, A. Modeling Italian forests: State of the art and future challenges. *Iforest-Biogeosci. For.* **2012**, *5*, 113. [CrossRef]
35. Maréchaux, I.; Langerwisch, F.; Huth, A.; Bugmann, H.; Morin, X.; Reyer, C.P.O.; Seidl, R.; Collalti, A.; Dantas de Paula, M.; Fischer, R.; et al. Tackling unresolved questions in forest ecology: The past and future role of simulation models. *Ecol. Evol.* **2021**, *11*, 3746–3770. [CrossRef] [PubMed]

36. Collalti, A.; Marconi, S.; Ibrom, A.; Trotta, C.; Anav, A.; D'andrea, E.; Matteucci, G.; Montagnani, L.; Gielen, B.; Mammarella, I.; et al. Validation of 3D-CMCC Forest Ecosystem Model (v.5.1) against eddy covariance data for 10 European forest sites. *Geosci. Model Dev.* **2016**, *9*, 479–504. [CrossRef]
37. Collalti, A.; Biondo, C.; Buttafuoco, G.; Maesano, M.; Caloiero, T.; Lucà, F.; Pellicone, G.; Ricca, N.; Salvati, R.; Veltri, A.; et al. Simulation, calibration and validation protocols for the model 3D-CMCC-CNR-FEM: A case study in the Bonis' watershed (Calabria, Italy). *For.-Riv. Selvic. Ecol. For.* **2017**, *14*, 247–256. [CrossRef]
38. Collalti, A.; Tjoelker, M.G.; Hoch, G.; Mäkelä, A.; Guidolotti, G.; Heskel, M.; Petit, G.; Ryan, M.G.; Battipaglia, G.; Matteucci, G.; et al. Plant respiration: Controlled by photosynthesis or biomass? *Glob. Chang. Biol.* **2020**, *26*, 1739–1753. [CrossRef]
39. Collalti, A.; Dalmonech, D.; Vangi, E.; Marano, G.; Puchi, P.F.; Morichetti, M.; Saponaro, V.; Orrico, M.R.; Grieco, E. *Monitoring and Predicting Forest Growth and Dynamics*; CNR Edizioni: Rome, Italy, 2024; ISBN 978-88-8080-655-4. [CrossRef]
40. Marconi, S.; Chiti, T.; Nolè, A.; Valentini, R.; Collalti, A. The role of respiration in estimation of net carbon cycle: Coupling soil carbon dynamics and canopy turnover in a novel version of 3D-CMCC forest ecosystem model. *Forests* **2017**, *8*, 220. [CrossRef]
41. Dalmonech, D.; Vangi, E.; Chiesi, M.; Chirici, G.; Fibbi, L.; Giannetti, F.; Marano, G.; Massari, C.; Nolè, A.; Xiao, J.; et al. Regional estimates of gross primary production applying the Process-Based Model 3D-CMCC-FEM vs. Remote-Sensing multiple datasets. *Eur. J. Remote Sens.* **2024**, *57*, 2301657. [CrossRef]
42. Mahnken, M.; Cailleret, M.; Collalti, A.; Trotta, C.; Biondo, C.; D'Andrea, E.; Dalmonech, D.; Marano, G.; Mäkelä, A.; Minunno, F.; et al. Accuracy, realism and general applicability of European forest models. *Glob. Chang. Biol.* **2022**, *28*, 6921–6943. [CrossRef] [PubMed]
43. Testolin, R.; Dalmonech, D.; Marano, G.; Bagnara, M.; D'Andrea, E.; Matteucci, G.; Noce, S.; Collalti, A. Simulating diverse forest management options in a changing climate on a Pinus nigra subsp. laricio plantation in Southern Italy. *Sci. Total Environ.* **2023**, *857*, 159361. [CrossRef] [PubMed]
44. Collalti, A.; Dalmonech, D.; Grieco, E.; Marano, G.; Vangi, E.; Puchi, P.; Orrico, M.R. 3D-CMCC-FEM(Coupled Model Carbon Cycle)BioGeoChemical and BiophysicalForest Ecosystem Model. 2023. Available online: http://www.forest-modelling-lab.com (accessed on 20 May 2024).
45. Farquhar, G.D.; von Caemmerer, S.; Berry, J.A. A biochemical model of photosynthetic CO_2 assimilation in leaves of C3 species. *Planta* **1980**, *149*, 78–90. [CrossRef]
46. Bernacchi, C.J.; Singsaas, E.L.; Pimentel, C.; Portis, A.R.; Long, S.P. Improved temperature response functions for models of Rubisco-limited photosynthesis. *Plant. Cell Environ.* **2001**, *24*, 253–259. [CrossRef]
47. Bernacchi, C.J.; Calfapietra, C.; Davey, P.A.; Wittig, V.E.; Scarascia-Mugnozza, G.E.; Raines, C.A.; Long, S.P. Photosynthesis and stomatal conductance responses of poplars to free-air CO_2 enrichment (PopFACE) during the first growth cycle and immediately following coppice. *New Phytol.* **2003**, *159*, 609–621. [CrossRef] [PubMed]
48. Kattge, J.; Knorr, W. Temperature acclimation in a biochemical model of photosynthesis: A reanalysis of data from 36 species. *Plant. Cell Environ.* **2007**, *30*, 1176–1190. [CrossRef] [PubMed]
49. De Pury, D.G.G.; Farquhar, G.D. Simple scaling of photosynthesis from leaves to canopies without the errors of big-leaf models. *Plant. Cell Environ.* **1997**, *20*, 537–557. [CrossRef]
50. Amthor, J.S. The McCree-de Wit-Penning de Vries-Thornley respiration paradigms: 30 Years later. *Ann. Bot.* **2000**, *86*, 1–20. [CrossRef]
51. Smith, N.G.; Dukes, J.S. Plant respiration and photosynthesis in global-scale models: Incorporating acclimation to temperature and CO_2. *Glob. Chang. Biol.* **2013**, *19*, 45–63. [CrossRef] [PubMed]
52. Merganičová, K.; Merganič, J.; Lehtonen, A.; Vacchiano, G.; Sever, M.Z.O.; Augustynczik, A.L.D.; Grote, R.; Kyselová, I.; Mäkelä, A.; Yousefpour, R.; et al. Forest carbon allocation modelling under climate change. *Tree Physiol.* **2019**, *39*, 1937–1960. [CrossRef] [PubMed]
53. Parton, W.J.; Scurlock, J.M.O.; Ojima, D.S.; Gilmanov, T.G.; Scholes, R.J.; Schimel, D.S.; Kirchner, T.; Menaut, J.-C.; Seastedt, T.; Garcia Moya, E.; et al. Observations and modeling of biomass and soil organic matter dynamics for the grassland biome worldwide. *Glob. Biogeochem. Cycles* **1993**, *7*, 785–809. [CrossRef]
54. Parton, W.J.; Schimel, D.S.; Cole, C.V.; Ojima, D.S. Analysis of Factors Controlling Soil Organic Matter Levels in Great Plains Grasslands. *Soil Sci. Soc. Am. J.* **1987**, *51*, 1173–1179. [CrossRef]
55. Thornton, P.E.; Rosenbloom, N.A. Ecosystem model spin-up: Estimating steady state conditions in a coupled terrestrial carbon and nitrogen cycle model. *Ecol. Modell.* **2005**, *189*, 25–48. [CrossRef]
56. Pastorello, G.; Trotta, C.; Canfora, E.; Chu, H.; Christianson, D.; Cheah, Y.W.; Poindexter, C.; Chen, J.; Elbashandy, A.; Humphrey, M.; et al. The FLUXNET2015 dataset and the ONEFlux processing pipeline for eddy covariance data. *Sci. Data* **2020**, *7*, 225. [CrossRef]
57. Frieler, K.; Lange, S.; Piontek, F.; Reyer, C.P.O.; Schewe, J.; Warszawski, L.; Zhao, F.; Chini, L.; Denvil, S.; Emanuel, K.; et al. Assessing the impacts of 1.5 °C global warming—Simulation protocol of the Inter-Sectoral Impact Model Intercomparison Project (ISIMIP2b). *Geosci. Model Dev.* **2017**, *10*, 4321–4345. [CrossRef]
58. Reyer, C.P.O.; Silveyra Gonzalez, R.; Dolos, K.; Hartig, F.; Hauf, Y.; Noack, M.; Lasch-Born, P.; Rötzer, T.; Pretzsch, H.; Meesenburg, H.; et al. The PROFOUND Database for evaluating vegetation models and simulating climate impacts on European forests. *Earth Syst. Sci. Data* **2020**, *12*, 1295–1320. [CrossRef]

59. Grünig, M.; Rammer, W.; Albrich, K.; André, F.; Augustynczik, A.L.D.; Bohn, F.; Bouwman, M.; Bugmann, H.; Collalti, A.; Cristal, I.; et al. A harmonized database of European forest simulations under climate change. *Data Br.* **2024**, *54*, 110384. [CrossRef]
60. ESA. Forest Type 2018 (Raster 10 m), Europe, 3-Yearly, Oct. 2020. Available online: https://sdi.eea.europa.eu/catalogue/copernicus/api/records/db1af59f-f01f-4bd4-830c-f0eb652500c1?language=all (accessed on 21 March 2024).
61. Moss, R.H.; Edmonds, J.A.; Hibbard, K.A.; Manning, M.R.; Rose, S.K.; Van Vuuren, D.P.; Carter, T.R.; Emori, S.; Kainuma, M.; Kram, T.; et al. The next generation of scenarios for climate change research and assessment. *Nature* **2010**, *463*, 747–756. [CrossRef] [PubMed]
62. van Vuuren, D.P.; Edmonds, J.; Kainuma, M.; Riahi, K.; Thomson, A.; Hibbard, K.; Hurtt, G.C.; Kram, T.; Krey, V.; Lamarque, J.F.; et al. The representative concentration pathways: An overview. *Clim. Change* **2011**, *109*, 5–31. [CrossRef]
63. Hempel, S.; Frieler, K.; Warszawski, L.; Schewe, J.; Piontek, F. A trend-preserving bias correction—The ISI-MIP approach. *Earth Syst. Dyn.* **2013**, *4*, 219–236. [CrossRef]
64. Lange, S. Bias correction of surface downwelling longwave and shortwave radiation for the EWEMBI dataset. *Earth Syst. Dyn.* **2018**, *9*, 627–645. [CrossRef]
65. Meinshausen, M.; Smith, S.J.; Calvin, K.; Daniel, J.S.; Kainuma, M.L.T.; Lamarque, J.; Matsumoto, K.; Montzka, S.A.; Raper, S.C.B.; Riahi, K.; et al. The RCP greenhouse gas concentrations and their extensions from 1765 to 2300. *Clim. Change* **2011**, *109*, 213–241. [CrossRef]
66. Dlugokencky, E.; Tans, P. *Trends in Atmospheric Carbon Dioxide*; National Oceanic & Atmospheric Administration, Earth System Research Laboratory (NOAA/ESRL): Washington, DC, USA, 2014.
67. Reichstein, M.; Falge, E.; Baldocchi, D.; Papale, D.; Aubinet, M.; Berbigier, P.; Bernhofer, C.; Buchmann, N.; Gilmanov, T.; Granier, A.; et al. On the separation of net ecosystem exchange into assimilation and ecosystem respiration: Review and improved algorithm. *Glob. Chang. Biol.* **2005**, *11*, 1424–1439. [CrossRef]
68. Papale, D.; Reichstein, M.; Aubinet, M.; Canfora, E.; Bernhofer, C.; Kutsch, W.; Longdoz, B.; Rambal, S.; Valentini, R.; Vesala, T.; et al. Towards a standardized processing of Net Ecosystem Exchange measured with eddy covariance technique: Algorithms and uncertainty estimation. *Biogeosciences* **2006**, *3*, 571–583. [CrossRef]
69. Nolè, A.; Collalti, A.; Magnani, F.; Duce, P.; Ferrara, A.; Mancino, G.; Marras, S.; Sirca, C.; Spano, D.; Borghetti, M. Assessing temporal variation of primary and ecosystem production in two Mediterranean forests using a modified 3-PG model. *Ann. For. Sci.* **2013**, *70*, 729–741. [CrossRef]
70. Jiang, F.; Ju, W.; He, W.; Wu, M.; Wang, H.; Wang, J.; Jia, M.; Feng, S.; Zhang, L.; Chen, J.M. A 10-year global monthly averaged terrestrial net ecosystem exchange dataset inferred from the ACOS GOSAT v9 XCO2retrievals (GCAS2021). *Earth Syst. Sci. Data* **2022**, *14*, 3013–3037. [CrossRef]
71. Jarvis, P.; Rey, A.; Petsikos, C.; Wingate, L.; Rayment, M.; Pereira, J.; Banza, J.; David, J.; Miglietta, F.; Borghetti, M.; et al. Drying and wetting of Mediterranean soils stimulates decomposition and carbon dioxide emission: The "Birch effect". *Tree Physiol.* **2007**, *27*, 929–940. [CrossRef] [PubMed]
72. Lavigne, M.B.; Ryan, M.G.; Anderson, D.E.; Baldocchi, D.D.; Crill, P.M.; Fitzjarrald, D.R.; Goulden, M.L.; Gower, S.T.; Massheder, J.M.; McCaughey, J.H.; et al. Comparing nocturnal eddy covariance measurements to estimates of ecosystem respiration made by scaling chamber measurements at six coniferous boreal sites. *J. Geophys. Res. Atmos.* **1997**, *102*, 28977–28985. [CrossRef]
73. Speckman, H.N.; Frank, J.M.; Bradford, J.B.; Miles, B.L.; Massman, W.J.; Parton, W.J.; Ryan, M.G. Forest ecosystem respiration estimated from eddy covariance and chamber measurements under high turbulence and substantial tree mortality from bark beetles. *Glob. Chang. Biol.* **2015**, *21*, 708–721. [CrossRef] [PubMed]
74. Campioli, M.; Malhi, Y.; Vicca, S.; Luyssaert, S.; Papale, D.; Peñuelas, J.; Reichstein, M.; Migliavacca, M.; Arain, M.A.; Janssens, I.A. Evaluating the convergence between eddy-covariance and biometric methods for assessing carbon budgets of forests. *Nat. Commun.* **2016**, *7*, 13717. [CrossRef]
75. Ryan, M.G. The enduring mystery of differences between eddy covariance and biometric measurements for ecosystem respiration and net carbon storage in forests. *New Phytol.* **2023**, *239*, 2060–2063. [CrossRef]
76. Cheng, W.; Huang, L.; Liu, Z.; Dong, J.; Moore, J.C.; MacMartin, D.G.; Deng, X. Seasonal and regional changes in terrestrial carbon uptake under an overshoot scenario. *Resour. Conserv. Recycl.* **2023**, *195*, 106997. [CrossRef]
77. van der Woude, A.M.; Peters, W.; Joetzjer, E.; Lafont, S.; Koren, G.; Ciais, P.; Ramonet, M.; Xu, Y.; Bastos, A.; Botía, S.; et al. Temperature extremes of 2022 reduced carbon uptake by forests in Europe. *Nat. Commun.* **2023**, *14*, 6218. [CrossRef]
78. Wolf, S.; Paul-Limoges, E. Drought and heat reduce forest carbon uptake. *Nat. Commun.* **2023**, *14*, 14–17. [CrossRef] [PubMed]
79. Chen, Z.; Wang, W.; Forzieri, G.; Cescatti, A. Transition from positive to negative indirect CO_2 effects on the vegetation carbon uptake. *Nat. Commun.* **2024**, *15*, 1500. [CrossRef] [PubMed]
80. De Marco, A.; Sicard, P.; Feng, Z.; Agathokleous, E.; Alonso, R.; Araminiene, V.; Augustatis, A.; Badea, O.; Beasley, J.C.; Branquinho, C.; et al. Strategic roadmap to assess forest vulnerability under air pollution and climate change. *Glob. Chang. Biol.* **2022**, *28*, 5062–5085. [CrossRef] [PubMed]
81. Nissan, A.; Alcolombri, U.; Peleg, N.; Galili, N.; Jimenez-Martinez, J.; Molnar, P.; Holzner, M. Global warming accelerates soil heterotrophic respiration. *Nat. Commun.* **2023**, *14*, 3452. [CrossRef] [PubMed]
82. Ping, J.; Cui, E.; Du, Y.; Wei, N.; Zhou, J.; Wang, J.; Niu, S.; Luo, Y.; Xia, J. Enhanced causal effect of ecosystem photosynthesis on respiration during heatwaves. *Sci. Adv.* **2023**, *9*, eadi6395. [CrossRef] [PubMed]
83. Odum, E.P. The strategy of ecosystem development. *Science* **1969**, *164*, 262–270. [CrossRef] [PubMed]

84. McGrath, M.J.; Schulte-Frohlinde, A.; Luyssaert, S. New ways for (in)validating the forest carbon neutrality hypothesis. *Glob. Chang. Biol.* **2024**, *30*, e16982. [CrossRef] [PubMed]
85. Gundersen, P.; Thybring, E.E.; Nord-Larsen, T.; Vesterdal, L.; Nadelhoffer, K.J.; Johannsen, V.K. *Old-Growth Forest Carbon Sinks Overestimated*; Nature Publishing Group: New York, NY, USA, 2021; Volume 591, pp. E21–E23.
86. Kirschbaum, M.U.F. Direct and indirect climate change effects on photosynthesis and transpiration. *Plant Biol.* **2004**, *6*, 242–253. [CrossRef] [PubMed]
87. Garrity, S.R.; Bohrer, G.; Maurer, K.D.; Mueller, K.L.; Vogel, C.S.; Curtis, P.S. A comparison of multiple phenology data sources for estimating seasonal transitions in deciduous forest carbon exchange. *Agric. For. Meteorol.* **2011**, *151*, 1741–1752. [CrossRef]
88. Weir, J.C.; Phillimore, A.B. Buffering and phenological mismatch: A change of perspective. *Glob. Chang. Biol.* **2024**, *30*, e17294. [CrossRef]
89. Bowling, D.R.; Schädel, C.; Smith, K.R.; Richardson, A.D.; Bahn, M.; Arain, M.A.; Varlagin, A.; Ouimette, A.P.; Frank, J.M.; Barr, A.G.; et al. Phenology of Photosynthesis in Winter-Dormant Temperate and Boreal Forests: Long-Term Observations from Flux Towers and Quantitative Evaluation of Phenology Models. *J. Geophys. Res. Biogeosci.* **2024**, *129*, e2023JG007839. [CrossRef]
90. Beamesderfer, E.R.; Arain, M.A.; Khomik, M.; Brodeur, J.J. The Impact of Seasonal and Annual Climate Variations on the Carbon Uptake Capacity of a Deciduous Forest Within the Great Lakes Region of Canada. *J. Geophys. Res. Biogeosci.* **2020**, *125*, e2019JG005389. [CrossRef] [PubMed]
91. Luo, X.; Zhao, R.; Chu, H.; Fatichi, S.; Keenan, T.; Lu, X.; Nguyen, N.; Prentice, I.; Yu, L. Deciduous forests use carbon more eciently than evergreen forests. *Preprint* **2024**. [CrossRef]
92. Terrer, C.; Vicca, S.; Hungate, B.A.; Phillips, R.P.; Prentice, I.C. Mycorrhizal association as a primary control of the CO_2 fertilization effect. *Science* **2016**, *353*, 72–74. [CrossRef]

forests MDPI

Article

Stand Age and Climate Change Effects on Carbon Increments and Stock Dynamics

Elia Vangi [1,2,*], Daniela Dalmonech [1,3], Mauro Morichetti [1], Elisa Grieco [1], Francesca Giannetti [2], Giovanni D'Amico [2], Mahdi (Andre) Nakhavali [4], Gherardo Chirici [2,3,5] and Alessio Collalti [1,3]

1 Forest Modelling Laboratory, Institute for Agriculture and Forestry Systems in the Mediterranean, National Research Council of Italy (CNR-ISAFOM), Via Madonna Alta 128, 06128 Perugia, Italy; daniela.dalmonech@cnr.it (D.D.); mauro.morichetti@isafom.cnr.it (M.M.); elisa.grieco@cnr.it (E.G.); alessio.collalti@cnr.it (A.C.)
2 geoLAB-Laboratory of Forest Geomatics, Department of Agriculture, Food, Environment and Forestry, Università degli Studi di Firenze, Via San Bonaventura 13, 50145 Firenze, Italy; francesca.giannetti@unifi.it (F.G.); giovanni.damico@unifi.it (G.D.); gherardo.chirici@unifi.it (G.C.)
3 National Biodiversity Future Centre (NBFC), Piazza Marina 61, 90133 Palermo, Italy
4 International Institute for Applied Systems Analysis (IIASA), 2361 Laxenburg, Austria; nakhavali@iiasa.ac.at
5 Fondazione per il Futuro delle Città, 50127 Firenze, Italy
* Correspondence: elia.vangi@isafom.cnr.it

Abstract: Carbon assimilation and wood production are influenced by environmental conditions and endogenous factors, such as species auto-ecology, age, and hierarchical position within the forest structure. Disentangling the intricate relationships between those factors is more pressing than ever due to climate change's pressure. We employed the 3D-CMCC-FEM model to simulate undisturbed forests of different ages under four climate change (plus one no climate change) Representative Concentration Pathways (RCP) scenarios from five Earth system models. In this context, carbon stocks and increment were simulated via total carbon woody stocks and mean annual increment, which depends mainly on climate trends. We find greater differences among different age cohorts under the same scenario than among different climate scenarios under the same age class. Increasing temperature and changes in precipitation patterns led to a decline in above-ground biomass in spruce stands, especially in the older age classes. On the contrary, the results show that beech forests will maintain and even increase C-storage rates under most RCP scenarios. Scots pine forests show an intermediate behavior with a stable stock capacity over time and in different scenarios but with decreasing mean volume annual increment. These results confirm current observations worldwide that indicate a stronger climate-related decline in conifers forests than in broadleaves.

Keywords: carbon cycle; climate change; forest age; forest management; carbon stocks

Citation: Vangi, E.; Dalmonech, D.; Morichetti, M.; Grieco, E.; Giannetti, F.; D'Amico, G.; Nakhavali, M.; Chirici, G.; Collalti, A. Stand Age and Climate Change Effects on Carbon Increments and Stock Dynamics. *Forests* **2024**, *15*, 1120. https://doi.org/10.3390/f15071120

Academic Editor: Tianxiang Yue

Received: 24 May 2024
Revised: 24 June 2024
Accepted: 25 June 2024
Published: 27 June 2024

1. Introduction

Assessing the quantity of CO_2 equivalent stored in forest ecosystems is one of the main goals for implementing the new European Forest Strategy for 2030, a key component of the European Green Deal, to achieve greenhouse gas emission neutrality by 2050. Within this framework, European forest strategies have been geared towards forest-based mitigation plans [1,2], which makes it essential to estimate the carbon sequestration capacity and potential under future climate conditions.

In the near future, Europe and Mediterranean areas will emerge as focal points ('hot spots') of climate change, characterized by heightened temperatures and environmental impacts [3,4]. Carbon assimilation and wood production are influenced by environmental conditions (e.g., precipitation, temperature, atmospheric CO_2, etc.) and endogenous factors, such as species auto-ecology, age, and hierarchical position within the forest structure. In the past decades, forest ecosystems proved to be crucial net carbon sinks [5,6], likely due to the positive fertilization effects of rising atmospheric CO_2 and temperature [7]. However, whether this effect will remain positive or be compensated by other limiting factors is still a matter of debate [8–10]. Some studies suggest that the fertilization effect on carbon storage and biomass production fades with forest aging in temperate forests [11,12] since these positive effects cannot continue indefinitely, complicating the picture of the forest response to climate changes even further. This is already the case in Europe, where forest aging and increased disturbances are causing the saturation and decline of the forest carbon sink [9]. Unfortunately, there is not yet a clear strategy to increase the mitigation potentials of forests, and the factors involved are manifold and entangled together [11,13,14].

The need to disentangle the intricate relationships between those factors is even more pressing under climate change. Our current understanding of how future climate will interact with forests of different age classes is particularly limited, especially since only a few studies have explored the relationship between age and the ecosystem's carbon balance under changing climate conditions [15].

The climate sensitivity of age cohorts is driven, among all, by different access to environmental resources, such as root depth and, therefore, access to water, as well as height, which affects leaf-level water potential and, thus, stomatal conductance [16]. Rooting depth and height jointly affect the tree's sensitivity to water scarcity, a key environmental driver of change. Future changes in environmental conditions are expected to impact the age spectrum differently [17–19].

Since forest age is determined by management practices and 75% of European forests are even aged [20,21], it is crucial to grasp and pin down the role of age in the sensitivity of forest carbon stocks to climate change to guide and inform adaptative forest management.

Process-based forest models enable the exploration of climate change impacts on various age cohorts within the same area, a task difficult to achieve through direct field measurements, which would require decades or more. In this regard, this study examines the ability of different forest age classes under the same future climate conditions to sustain high productivity and carbon stock capacity. To achieve this goal, we employed the 'Three Dimensional-Coupled Model Carbon Cycle-Forest Ecosystem Module' (3D-CMCC-FEM) [22,23], simulating undisturbed forests of different cohorts under four climate change scenarios (and including one 'no climate change' scenario), from the moderate one (RCP 2.6) up to the most severe one (RCP 8.5) coming from five Earth system models. In this context, carbon stocks and increment were simulated via total carbon woody stocks (TCWS, i.e., the standing woody biomass in MgC ha^{-1}) and the mean annual increment (MAI, in m^3 $ha^{-1}year^{-1}$), which depend mainly on age and long-term processes, such as climate trends.

The primary aim of this research is (i) to explore the direct effects of climate change on the overall carbon storage capacity across various stands, species, and age classes situated in diverse regions of Europe, and (ii) to elucidate the potential influence of forest age on stand dynamics in adapting to forthcoming climate shifts.

2. Materials and Methods

2.1. Study Sites

The study was conducted in three even-aged, previously managed European forest stands (Figure 1): (i) the Boreal Scots pine (*Pinus sylvestris* L.) forest of Hyytiälä, Finland (FI-Hyy); (ii) the wet temperate continental Norway spruce (*Picea abies* (L.) H. Karst) forest of Bílý Krìz in the Czech Republic (CZ-BK1); and (iii) the temperate oceanic European beech (*Fagus sylvatica* L.) forest of Sorø, Denmark (DK-Sor) where the 3D-CMCC-FEM (in different versions) has been already validated in the past [14,24,25]. An overview of the main site's characteristics is presented in Tables 1 and 2.

For each site, daily bias-adjusted downscaled climate data from five Earth system models (i.e., HadGEM2-ES, IPSL-CM5A-LR, MIROC-ESM-CHEM, GFDL-ESM2M, and NorESM1-M) driven by four representative concentration pathways, namely RCP 2.6, 4.5, 6.0, and 8.5 were available [26,27] (Figure S1). For more detailed information on the study site characteristics and climate data, see [14,24,25,28]. The chosen sites have been selected due to their long monitoring history and the availability of a wide range of data sources for both carbon fluxes and biometric data for model evaluation, as well as bias-corrected climate scenarios for simulations under climate change scenarios from the ISIMIP-PROFOUND initiatives (https://www.isimip.org/, accessed on 1 January 2024) [25,28]. In addition, these stands (i) represent the most common European tree species; (ii) have a current state that is the result of the legacy of past forest management; (iii) are mainly mono-specific and therefore represent interesting «living labs» to study the effects of climate change on single-species and their productivity, reducing confounding effects which otherwise make models struggle to predict forest growth and carbon dynamics (e.g., [29,30]); and (iv) they have already been investigated in the context of climate-smart-forestry silvicultural scenarios [14].

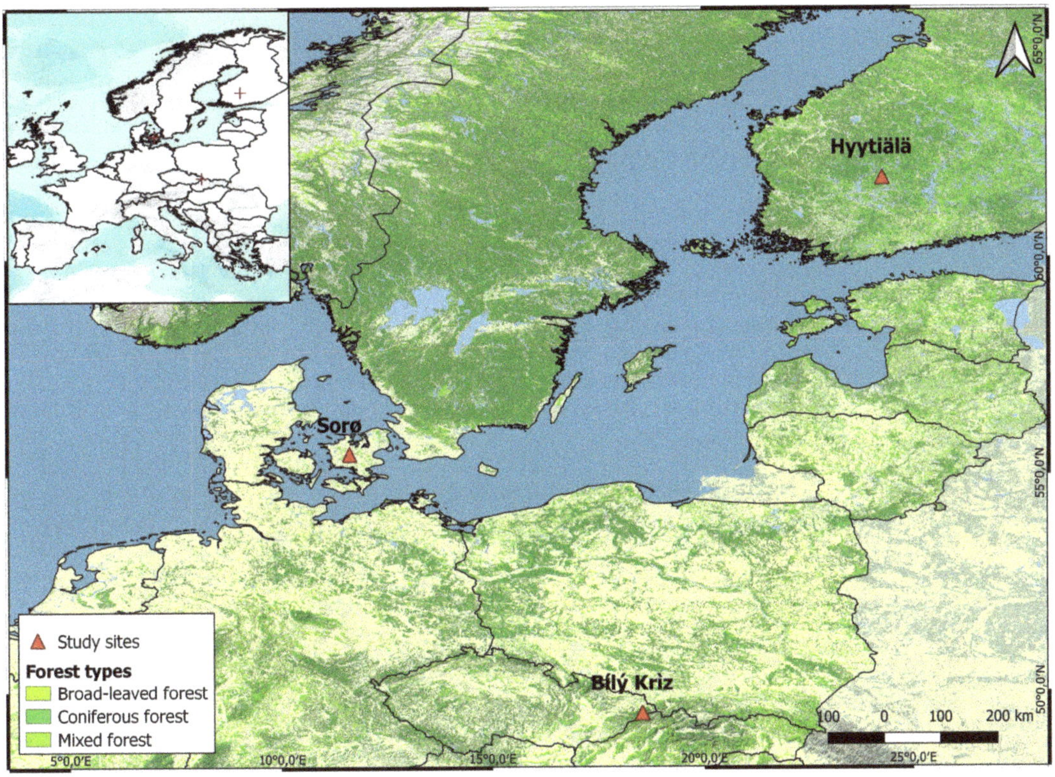

Figure 1. Test site locations in Europe. The red "+" in the overwiew panel represent the sitre locations.

Table 1. Overview of the main site characteristics provided for each forest site. Years of obs. refers to the first and last year of measurement; the temporal resolution of measurement is annual. For the stand values (DBH, height, BA, age, and tree density), the range corresponds to the first and last field measurement according to the years of obs. Column. DBH = diameter at breast height; BA = basal area.

Name	Species	Lat	Long	Aspect	Elevation	Slope	Years of Obs.	DBH (cm)	Height (m)	BA (m^2 ha^{-1})	Age	Tree Density (ha^{-1})
Bílý Krìz	Picea abies	49.3	18.32	180	875	12.5	1997–2015	8.16–20.47	6.26–15.26	10.33–36.96	16–34	2408–1252
Hyytiälä	Pinus sylvestris	61.85	24.29	180	185	2	1995–2011	15.89–20.58	12.61–18.62	12.64–18.33	34–50	870–684
Sorø	Fagus sylvatica	55.49	11.64	-	40	0	1994–2017	28.99–48.25	24.23–31.15	18.50–29.76	62–87	407–199

Table 2. Yearly averages of the daily maximum temperature (Tmax), daily minimum temperature (Tmin), daily mean temperature (Tmean), annual precipitation sum (P), daily mean relative humidity (RH), daily mean air pressure (AP), and annual sum of global radiation (R, direct + diffuse shortwave radiation) for each of the sites. The column "Years" indicates the data's acquisition year and the period the average values refer to.

Site	Source	Years	Tmax (°C)	Tmean (°C)	Tmin (°C)	P (mm)	RH (%)	R (J cm^{-2})
Bílý Krìz	Local	2000–2008	11.5	7.36	3.8	1434.56	81.99	378 774.86
Hyytiälä	Local	1996–2014	7.4	4.36	1.13	604.01	77.95	309 628.86
Sorø	Local	1996–2012	10.66	8.26	5.91	760.52	82.95	360 687.83

2.2. The Model

The 'Three Dimensional-Coupled Model Carbon Cycle-Forest Ecosystem Module' (3D-CMCC-FEM v 5.6 [12,14,22–24,31] is a biogeochemical, biophysical, process-based, stand-level forest model. The model is built to simulate carbon, nitrogen, and water cycles in forest ecosystems, even including forest dynamics, under scenarios of climate change and disturbances (e.g., forest management) and parameterized at the species level. Photosynthesis is modeled through the biogeochemical model of Farquhar von Caemmerer and Berry [32], implemented for sun and shaded leaves [33] and parametrized as in Bernacchi et al. [34,35]. Temperature acclimation of leaf photosynthesis to increasing temperature is accounted for following Kattge and Knorr [36]. Autotrophic respiration (RA) is modeled mechanistically by distinguishing the cost of maintaining already existing tissues (RM) and the cost of synthesizing new ones (RG). Maintenance respiration is controlled by the amount of nitrogen (stoichiometrically fixed fraction of live tissues) and temperature. Temperature effects on enzyme kinetics are modeled through a standard Arrhenius relationship but acclimated for temperature as described in Collalti et al. [24]. The net primary productivity (NPP) is the gross primary productivity (GPP) less RA. Not all the annual NPP goes for biomass production since the model considers the non-structural carbon (NSC) pool, an additional seventh C-pool that includes starch and sugars (undistinguished) used to buffer periods of negative carbon balance (when respiration exceeds assimilation, i.e., RA > GPP). Ultimately, the more trees respire, the more NSC is used to sustain metabolism and NSC pool replenishment, and the less NPP and BP there are (and less carbon is stocked). In the extreme case, when and if all NSCs are depleted because of metabolism without being replenished through current photosynthates, the model predicts stand mortality based on the carbon starvation hypothesis [37,38].

The phenological and allocation schemes are all described extensively in Collalti et al. [22,23,39] and Merganičová et al. [39]. The 3D-CMCC-FEM accounts for the 'age-effect' in several ways. Ecological theories of the '60s describe [40,41], and past and growing pieces of evidence suggest, that stabilization and a further slight decline follow an initial step-wise increase in forest productivity. The causes of such a decline are debated and include a decline in the GPP because of hydraulic limitation [16,42] as well as an increase in RA because of increased respiring biomass [18,19,43]. The 3D-CMCC-FEM accounts for both by including an age modifier [44], which reduces maximum stomatal conductance (and then also GPP) in the Jarvis model and increases RA because of biomass accumulation during forest development.

2.3. Virtual Stands, Model Runs, and Results Evaluation

The 3D-CMCC-FEM was first evaluated under observed climate and field data for GPP and NPP_{woody} (i.e., the NPP for woody compound; gC m^{-2} year^{-1}) and the diameter at breast height (DBH) (see 'Model validation' paragraph in Supplementary Materials; [12,14]). The model was forced with the modeled climate under different emission scenarios, corresponding to the RCP atmospheric CO_2 concentration values for the period 1997 to 2100, ranging from 421.4 μmol mol^{-1} in the 'best-case scenario' (RCP2.6) to 926.6 μmol mol^{-1}

in the 'worst-case scenario' (RCP 8.5) coming from the ISIMIP-PROFOUND initiative. For comparison purposes, we forced the forest model with a detrended and repeated meteorology and atmospheric CO_2 concentration from 1996 to 2006. The current climate (i.e., no climate change 'NoCC') is considered the baseline to compare against climate change scenarios. At the start of the simulations, we created a composite forest matrix (CFM, composed of both measured stand data and "virtual" stand data), following the approach described in Dalmonech et al. [14], to simulate the potential effect of climate stressors on stands of different ages. The 3D-CMCC-FEM has been run at each site to cover the rotation period of each species (from 1997 to 2099) amid the current climate scenario (fixed atmospheric CO_2 concentration at the year 2000 of 368.8 μmol mol^{-1}) consisting of detrended and repeated cycles of the present-day observed meteorology from 1996 to 2006 and the Business-as-Usual (BAU) management practices observed at each site (see [28] for the description of BAU applied at each site). Data required to re-initialize the model at every tenth of the rotation length were retrieved from each simulation. Hence, 10 additional stands were chosen for each age in the composite matrix and added to the CFM. This collection of virtual forest stands was used to set different starting stand ages at the present day (age_{t0}) due, ideally, to the past silvicultural practice and climate. Under this framework, a landscape of eleven different stands (in age and their relative C-pools and forest structure) for each site is created. These new stands were used, each running from 2006 to 2099, to assess the impact of climate forcing, as the model has already been shown to be sensitive to forest stand development and the relative standing biomass.

The 3D-CMCC-FEM was initialized with the structural attributes of the newly created stands from 1997, which was the starting year of all simulations and for all stands. Modeled climate change simulations under different RCP-emissions scenarios started to differentiate in 2006 (up to 2100). The simulation runs from the different stand initial conditions, corresponding to different age_{t0} classes, were carried out without forest management, as we are interested in the direct climate impact on undisturbed forest stand response, avoiding the confounding effects of forest management on the responses (for forest management effects, see [14]). A total of 825 different simulations were performed, as they combined 5 ESMs × 5 RCPs (4 RCPs + 1 current climate scenario) × 11 age_{t0} classes × 3 sites. Results are reported for MAI (mean annual increment; m^3 ha^{-1} $year^{-1}$) and TCWS (total carbon woody stocks; MgC ha^{-1}), respectively, as they are considered some of the most representative and fundamental variables in the carbon cycle and forestry. Following the methodology reported [14] (see Table S1 in Supplementary Materials), we evaluated the model forced with the modeled climate. We compared GPP and NPP_{woody} against eddy covariance estimates and ancillary data for the years 1997–2005 for DK-Sor and FI-Hyy and 2000–2005 for CZ-BK1. We also compared the diameter at breast height (DBH) in all sites with field measures (see Supplementary Materials).

3. Results

Effect of Age Classes and Climate Change on Total Carbon Woody Stock and Increments

Norway spruce at CZ-BK1 shows mean TCWS values ranging between ~70 and ~140 MgC ha^{-1} under the NoCC scenario over the century, and from ~70 to ~130 MgC ha^{-1} with a decreasing pattern across all RCPs (Figure 2). In the Norway spruce stands under some ESMs climate forcing (HadGEM2-ES and GFDL-ESM 2M mostly) and under all climate change scenarios, the 3D-CMCC-FEM simulates mortality events for carbon starvation, which increase across stands under gradually warmer climate scenarios and from the oldest stands to the progressively youngest ones.

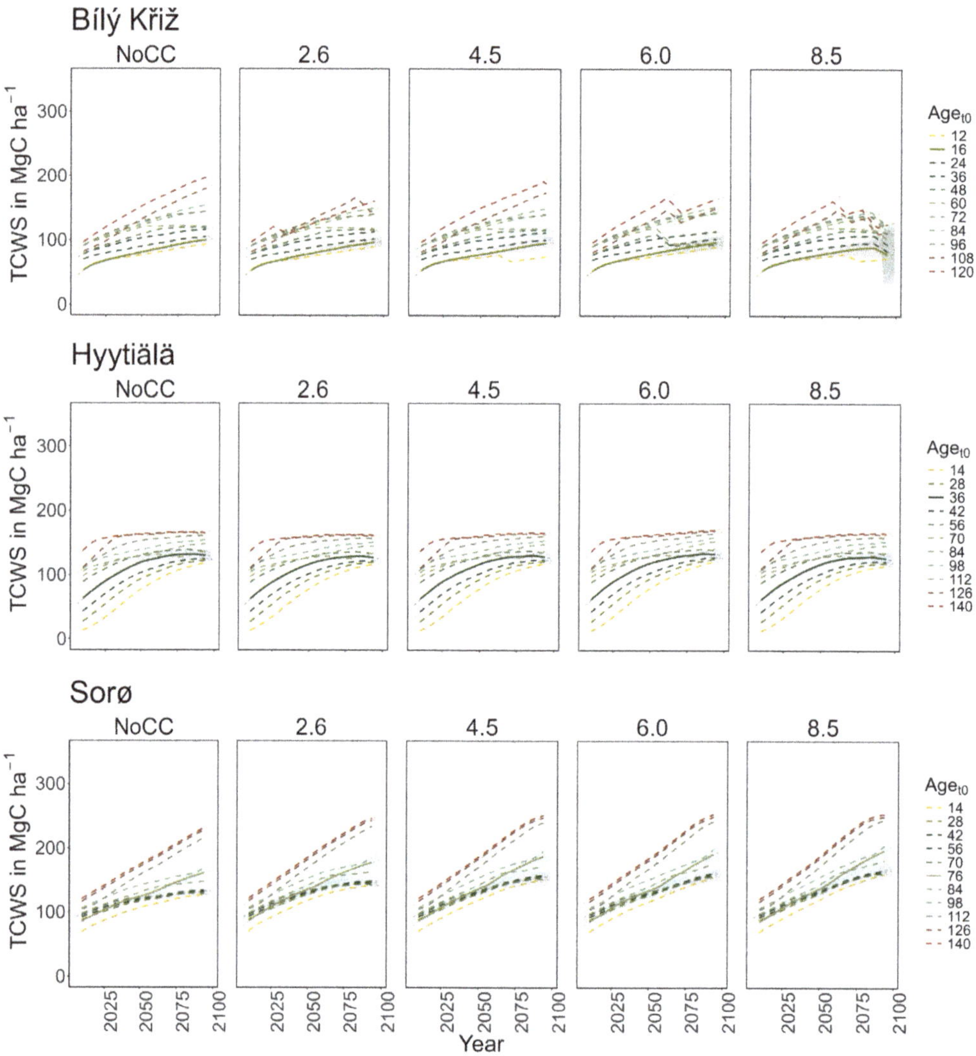

Figure 2. Modeled total carbon woody stock (TCWS) (MgC ha^{-1}) for age classes at the three sites in all scenarios along the simulation period (2006–2099). Lines represent the moving average of 10 years. The solid line corresponds to the real stand, while the dotted lines correspond to the virtual ones. The shaded area represents two standard deviations from the mean predictions with the results from the five ESMs' climate change scenarios.

Under RCP 8.5, all classes show signs of decay at the end of the century. In the youngest age$_{t0}$ classes, a sharp decrease in MAI was observed (from 8 to 4 m^3 ha^{-1} year^{-1}), while in the older ones, it holds steady to ~3 m^3 ha^{-1} year^{-1} with a peak around 2075 (Figure 3). At FI-Hyy, younger age$_{t0}$ classes (14- to 42-year-old) showed the fastest increase in TCWS (reaching 120–130 MgC ha^{-1} at the end of the century under all scenarios), also reflected in the pattern of MAI. Older age$_{t0}$ classes showed a more stable trend throughout the simulation (Figure 2), culminating at ~150 MgC ha^{-1}, with MAI steadily declining from 2.5 to 2 m^3 ha^{-1} year^{-1}. In all scenarios, the Scots pine peaked in the 126 and 56 age$_{t0}$ in TCWS and MAI, respectively. Minor differences were found in mean TCWS between the NoCC and other RCP scenarios, ranging from −1.6% (140-year-old class under RCP

2.6) to +2.8% (14-year-old class under RCP 6.0). At DK-Sor, the results for TCWS show different patterns to other sites, with the highest values ranging between ~240 MgC ha^{-1} (under NoCC) and ~255 MgC ha^{-1} (under RCP 8.5) at the end of the century, with the least TCWS under NoCC. The younger classes showed a shallow increase in TCWS during the simulation period, stabilizing at the end of the century, while the older ones kept growing (Figure 4). DK-Sor was the only site where the tightening of the climate conditions caused a positive effect on the MAI, particularly in the younger classes, reversing the trend from negative to positive at the end of the century.

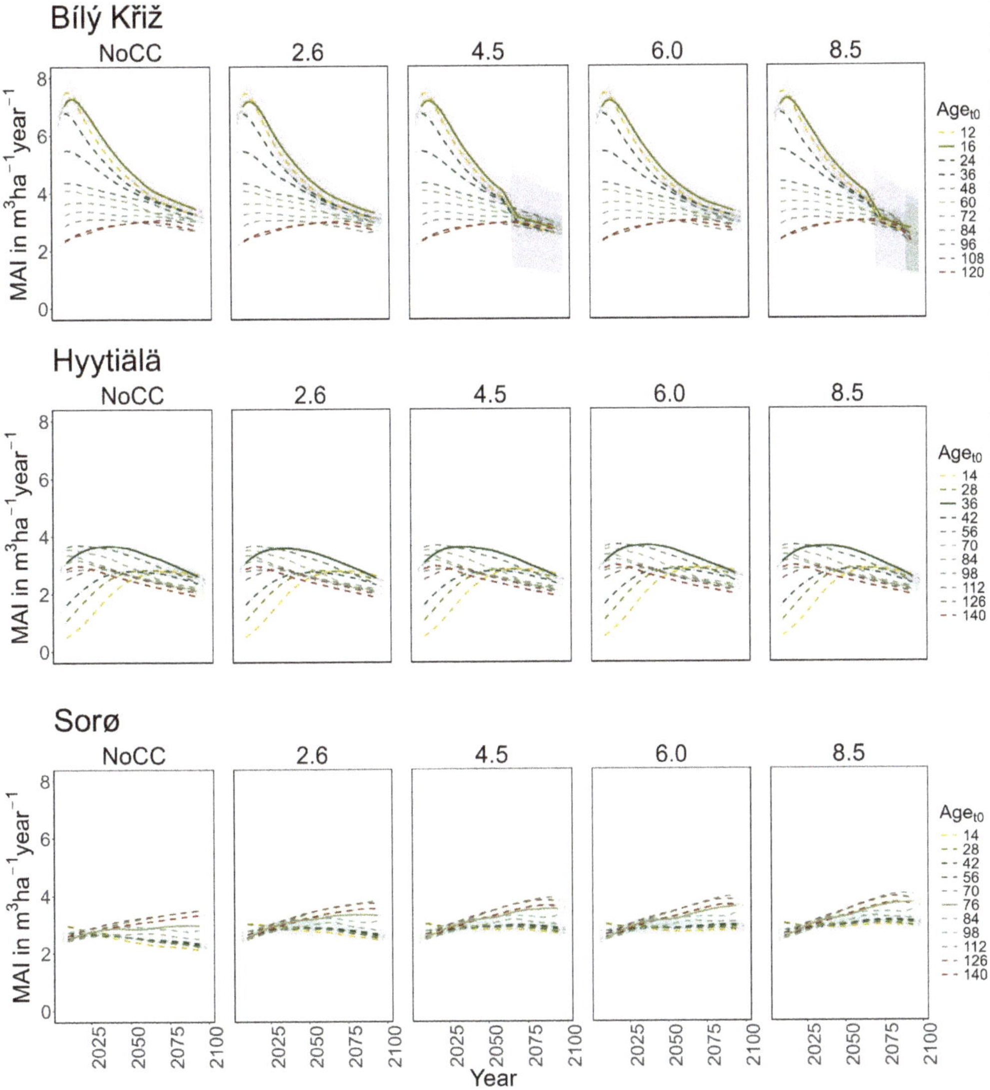

Figure 3. Modeled mean annual increment (MAI) (m^3 ha^{-1} year^{-1}) for age classes at the three sites in all scenarios along the simulation period (2006–2099). Lines represent the moving average of 10 years. The solid line corresponds to the real stand, while the dotted lines correspond to the virtual ones. The shaded area represents two standard deviations from the mean predictions with the results from the five ESMs climate change scenarios.

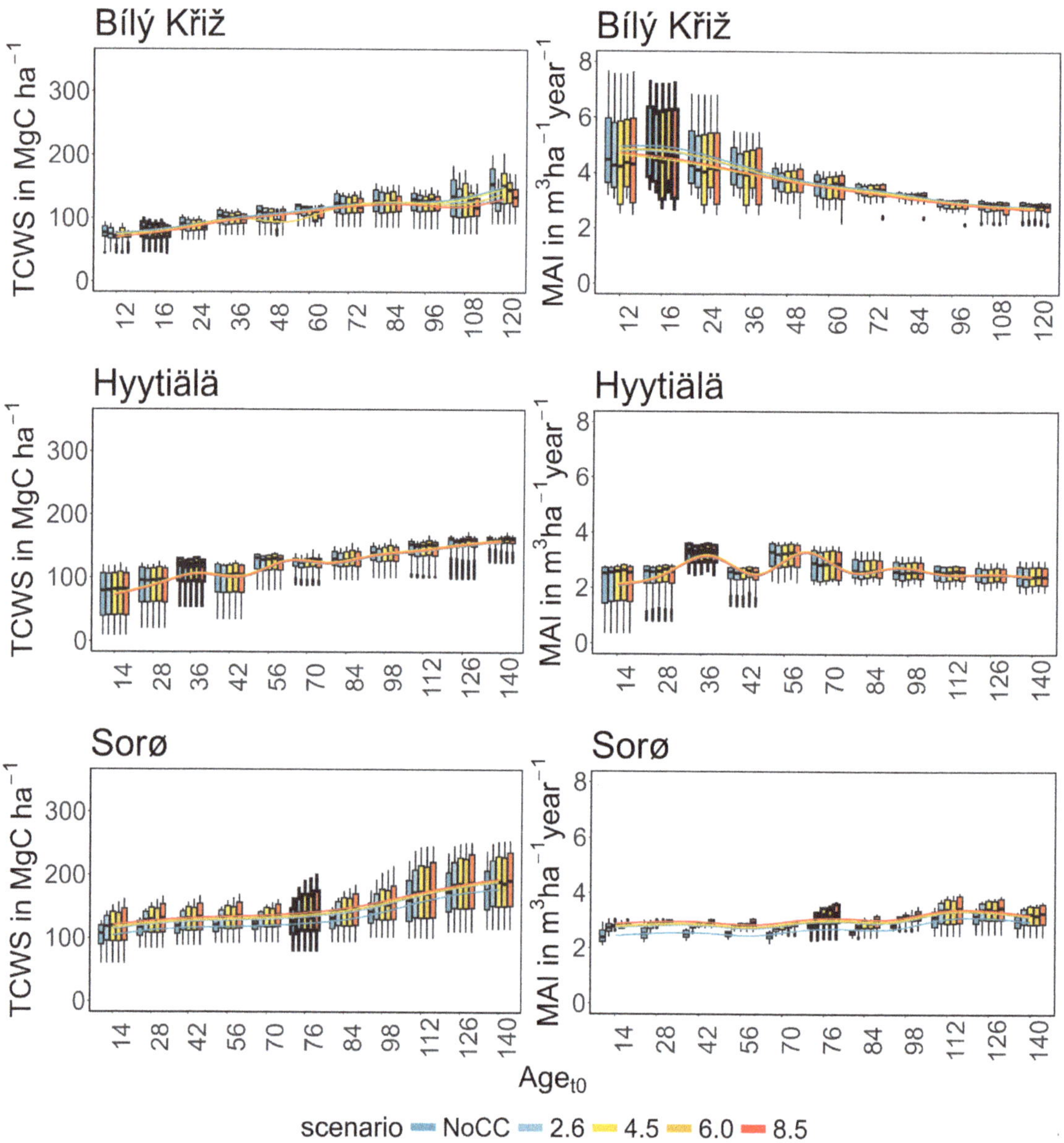

Figure 4. Boxplot of modeled total carbon woody stock (TCWS) (left, MgC ha^{-1}) and mean annual increment (MAI) (right, m^3 ha^{-1} year^{-1}) for age classes at the three sites in the four RCPs scenarios compared to the NoCC (no climate change) scenario. Boxplots with thick borders correspond to the real stand. Lines are fitted throughout the median of the values of the variables using a generalized additive model.

In summary, a positive growth trend of TCWS over time was found in all sites, with the oldest age$_{t0}$ classes accounting for the most carbon accumulation. Both conifer stands show a plateau with a reduction in growth at the end of the simulation, which is more pronounced and more severe in the warmest climate scenario. Conversely, the beech stands show a positive growth pattern in all scenarios. Similar results were obtained for MAI, where the conifers showed a decreasing trend over the simulation period despite different magnitudes and patterns among age_{t0} classes. The beech stands exhibited smaller variations among age_{t0} than among scenarios concerning other sites. In Table 3, we report the mean value of TCW and MAI over the simulation period for each site and climate scenario.

Table 3. Mean values of total carbon woody stock (TCWS) and mean annual increment (MAI) over the simulation period (2006–2099) for each scenario and age class. CZ-BK1 = Bílý Krìz; FI-Hyy = Hyytiälä; DK-Sor = Sorø.

		Scenario					Scenario				
		NoCC	RCP 2.6	RCP 4.5	RCP 6.0	RCP 8.5	NoCC	RCP 2.6	RCP 4.5	RCP 6.0	RCP 8.5
	Age	TCWS (Mg Cha^{-1})					MAI (m^3 ha^{-1})				
CZ-BK1	12	72.22	69.92	65.04	70.42	65.93	4.89	4.75	4.57	4.80	4.63
	16	75.71	73.60	73.20	74.10	71.88	5.03	4.86	4.67	4.91	4.72
	24	86.53	84.47	84.06	85.02	82.70	4.73	4.60	4.41	4.63	4.45
	36	97.46	95.37	95.00	96.13	93.64	4.30	4.19	4.01	4.22	4.07
	48	101.76	99.69	99.33	91.00	99.58	3.81	3.71	3.71	3.74	3.68
	60	110.19	108.17	107.88	99.34	107.00	3.58	3.51	3.50	3.53	3.47
	72	118.19	116.00	115.64	117.06	114.55	3.42	3.35	3.34	3.37	3.31
	84	121.56	119.38	119.04	120.58	117.50	3.18	3.12	3.11	3.14	3.08
	96	120.19	118.12	117.98	119.70	116.90	2.88	2.83	2.83	2.85	2.80
	108	126.78	121.26	124.28	117.57	112.33	2.74	2.70	2.70	2.71	2.66
	120	145.50	124.43	142.41	135.02	127.84	2.72	2.69	2.68	2.69	2.65
FI-Hyy	14	66.42	66.62	67.39	68.49	66.98	1.91	1.93	1.95	1.99	1.94
	28	79.83	79.00	79.73	81.05	79.37	2.21	2.19	2.21	2.25	2.20
	36	104.34	102.42	103.17	104.75	102.86	3.20	3.14	3.16	3.21	3.15
	42	89.82	88.55	89.35	90.79	89.08	2.35	2.31	2.33	2.37	2.33
	56	118.76	116.62	117.57	119.20	117.36	3.20	3.14	3.17	3.21	3.17
	70	119.29	117.43	118.61	120.25	118.65	2.91	2.86	2.89	2.92	2.89
	84	125.92	123.95	125.19	126.85	125.27	2.76	2.72	2.74	2.77	2.74
	98	133.33	131.06	132.29	134.12	132.33	2.66	2.62	2.64	2.67	2.64
	112	141.29	138.87	140.00	141.92	140.19	2.57	2.53	2.55	2.57	2.55
	126	148.26	145.43	146.69	148.63	146.91	2.49	2.45	2.47	2.49	2.47
	140	155.77	153.11	154.50	156.52	154.79	2.47	2.43	2.45	2.47	2.45
DK-Sor	14	98.53	107.89	110.48	109.94	113.32	2.46	2.72	2.79	2.77	2.86
	28	110.94	120.06	122.68	122.18	125.53	2.60	2.86	2.92	2.90	2.99
	42	113.56	122.68	125.31	124.82	128.16	2.54	2.78	2.85	2.83	2.92
	56	115.97	125.07	127.73	127.23	130.58	2.52	2.76	2.82	2.80	2.89
	70	116.15	125.14	127.79	127.30	130.62	2.47	2.69	2.75	2.73	2.81
	76	120.65	130.45	133.16	132.55	136.05	2.76	3.00	3.06	3.04	3.13
	84	124.68	133.99	136.65	136.14	139.53	2.56	2.78	2.84	2.82	2.90
	98	135.87	145.41	148.06	147.49	150.93	2.82	3.02	3.07	3.06	3.13
	112	154.13	164.32	166.58	165.71	168.77	3.00	3.21	3.25	3.23	3.30
	126	166.84	176.33	178.38	177.63	180.32	3.06	3.24	3.28	3.26	3.31
	140	170.49	180.08	181.84	181.26	184.05	2.91	3.08	3.11	3.09	3.14

4. Discussion

Age-Dependent Impacts of Climate Change on Forests' Increment and C-Stocks

The successional stage, represented by forest age, was the main driver controlling C-storage capacity and biomass accumulation, as already known from previous studies [45–47], with differences greater among different age cohorts under the same scenario than among different climate scenarios under the same age class [12,14]. The evidence that the carbon budget is mainly controlled by stand age suggests that the effects of climate change on forest cohorts are generally less significant than the effect of age, mainly in terms of the amount of standing biomass. In this sense, age represents multiple and interacting processes, such as tree size [48,49], forest structural traits (canopy closure and LAI), reduction in stomatal conductance [16], and adaptation to specific environmental conditions which, in turn, make it possible to increases the above-ground biomass (AGB) [50]. The model could reproduce the expected behavior of biomass (and thus carbon) accumulation, simulating rapid growth at a young age and saturation for the oldest age class, but not necessarily at the end of the simulation period. Approaching the physiological optima for the species may benefit the biomass synthesis through an augmented photosynthate supply but may eventually increase the respiratory costs of tissue growth and maintenance despite a strong acclimation capacity [18]. High respiratory costs in warm climates with low precipitation regimes, in the older age classes, lead to C-starvation and mortality phenomena, as modelled for the Norway spruce at the CZ-BK1 site. This indicates that the environment has reached its carrying capacity and that competition for limited resources, such as light and water, is excessively high to sustain more biomass in the oldest age classes.

We found different C-accumulation patterns under climate change between coniferous stands and broadleaves. As expected, increasing temperature and changes in precipitation patterns led to a decline in above-ground biomass in spruce stands, especially in the older age classes. On the contrary, the results show that beech forests at DK-Sor will maintain and even increase C-storage rates under most RCP scenarios. Scots pine forests show an intermediate behavior with a stable stock capacity over time and in different scenarios but with decreasing MAI. These results confirm current observations worldwide that indicate a stronger climate-related decline in conifers forests than in broadleaves [51–53]. This contrasting response is explained by the different characteristics of the two *phyla*, in particular, it is due to the temperature adaptation, with generally lower optimum temperature in conifer in addition to its lower sensitivity to the length of the growing season. Similarly, conifers also show lower efficiency in water management because of the shallower root system, which increases the sensitivity to soil aridity and its vulnerability to drought events [54]. Recent studies confirm that growth decline is more pronounced in conifers than in broadleaf, especially beech forests, in the most northern species distribution [55]. Our results confirm the same growth patterns found by recent studies [47,53,56], where broadleaves outperform conifers in productivity, and climate warming will probably exacerbate these opposite growth patterns.

However, despite some studies suggesting that age modulates different adaptation strategies to some extent, it remains unclear whether younger trees may be more affected by climate change than older ones. Bennett et al. [57], in a global analysis, found that droughts consistently had more severe impacts on larger (older) trees, while Wang et al. [11] observed a more substantial and sharper decline in basal area increment in young Korean pine in China. Hogg et al. [58] found that the percentage decrease in biomass growth was not significantly different for young, productive stands compared to older, less productive ones. Our study suggests that warmer and drier conditions and extended growing seasons will affect younger stands more than older ones, but with different trends among species. In particular, MAI will be positively affected in younger beech forests, while it will remain stable in older stands. On the contrary, climate change will strongly impact the growth rate of young conifers stands more than older ones. Older forests tend to be more stable and resilient than younger ones due to their rugged and stable interaction with climate triggers and better responsiveness to environmental changes. The year-to-year climate variability is buffered by larger carbon pools in sapwood and reservoirs in older trees, leading to higher long-term stability than younger trees [12]. In this sense, ages represent the "memory" of the forest to past climate and disturbance regimes, which align the species-specific traits to the environmental conditions in which they grow, creating the niches in which AGB accumulates [52,59].

Despite numerous efforts to decipher forests' response to climate change, the intricate methods employed by tree species to withstand extreme climates still need to be fully unveiled. Further research exploiting ecophysiological models explicitly accounting for age, tree-ring experiments, and remote sensing will be critical to understanding forest ecosystems' adaptation strategies to climate change, particularly in the face of rapid warming and extreme disturbances. A better understanding of the interaction between forests and climate can inform better forest management strategies, ultimately dampening the impacts of climate change on forest ecosystems.

5. Limitations

The presented modeling framework has some limitations that should be considered. Firstly, natural disturbances as consequences of climate change, such as windstorms, forest fires, and insect outbreaks, were not simulated. These disturbances cause changes in carbon stocks, nutrients, and soil conditions and contribute to the global release of CO_2 in the atmosphere, ultimately leading to increasing temperature and radiation. In contrast, climate extreme events are considered to be already included in the climate scenarios used to force the model and, thus, already accounted for in the model outputs. Additionally, other indirect alterations due to climate change of key drivers, such as nitrogen deposition, phosphorus, or ozone, which can somewhat amplify or reduce our results, were not assessed. Nonetheless, some studies (e.g., [60]) lend credence to the notion that this phenomenon may not be applicable across the board. They highlight the significant responsiveness of various tree species to CO_2 fertilization across a wide range of nutrient availability. Finally, no allowance was made for the possibility of species migration to and from the study areas. However, these dynamics may require longer timescales than those simulated in this study.

6. Conclusions

Forest age is confirmed to be a significant factor in determining the carbon storage capacity and biomass accumulation in forest ecosystems, especially in the context of future climate uncertainty. The effects of species, site location, stand-level characteristics, and development stage vary significantly and are contingent on specific factors. We observed that differences in biomass accumulation were more pronounced among different age cohorts than among different climate scenarios within the same age class, with contrasting carbon accumulation patterns under climate change between coniferous and broadleaf forests. Furthermore, our findings shed light on the differential impacts of climate change on younger versus older forest stands. Warmer and drier conditions are projected to affect younger stands more severely, particularly in coniferous forests. However, older forests will likely exhibit greater stability and resilience due to their accumulated carbon pools and enhanced adaptability to environmental changes. While our study provides valuable insights, it also underscores the need for further research to unravel the complex mechanisms by which forests adapt to climate change. This deeper understanding can inform more effective forest management strategies, helping to mitigate the impacts of climate change on forest ecosystems in the future. The varying responses of different tree species highlight the need for tailored management approaches and conservation efforts to enhance the resilience of our forests.

Supplementary Materials: The following supporting information can be downloaded at: https://www.mdpi.com/article/10.3390/f15071120/s1, Figure S1: Evaluation of monthly seasonal GPP (gC m−2 month−1) fluxes (left column) and annual (gC m−2 year−1) fluxes (central column) for the sites of Sorø, Bily Kriz, and Hyytiala (rows). Quality-checked and -filtered GPP values evaluated at the sites by the eddy covariance technique are reported as black dots. The shaded area for seasonal values reports the maximum and minimum monthly values recorded in the time series. The shaded area for annual data represents the relative uncertainty bounds. In the third column, a comparison of the predicted annual DBH increment (cm y−1) with site observations at the three sites is reported. Measured data are shown as black dots. Simulated data are reported as continuous lines. Table S1: Performance statistics (coefficient of determination R2, relative root mean square error RMSE (gC m−2 day−1) and Fractional Mean Bias, FMB) computed from monthly seasonal values and annual series of model gross primary productivity, GPP, against eddy covariance estimated and diametric annual increment data, DBH increment, against measured data. Results are reported for simulations forced with local and modeled climate (i.e., ESM) (ESM1, 2, 3, 4, 5 refer to HadGEM2-ES, IPSL-CM5A-LR, MIROC-ESM-CHEM, GFDL-ESM 2M, and NorESM1-M, respectively).

Author Contributions: E.V.: Data curation, Formal analysis, Investigation, Writing—original draft, Writing—review & editing; D.D.: Data curation, Formal analysis, Investigation, Writing—review & editing; M.M.: Writing—review & editing; E.G.: Writing—review & editing; F.G.: Writing—review & editing; G.D.: Writing—review & editing; M.N.: Writing—review & editing; G.C.: Writing—review & editing; A.C.: Formal analysis, Investigation, Writing—original draft, Writing—review & editing, Conceptualization. All authors have read and agreed to the published version of the manuscript.

Funding: E.V. and A.C. acknowledge funding by the project "FORESTNAVIGATOR" Horizon Europe research and innovation program under grant agreement No. 101056875. M.M., E.G., F.G. and A.C. acknowledge funding by the project "OptForEU" Horizon Europe research and innovation program under grant agreement No. 101060554. D.D. and A.C. also acknowledge the project funded under the National Recovery and Resilience Plan (NRRP), Mission 4 Component 2 Investment 1.4-Call for tender No. 3138 of 16 December 2021, rectified by Decree n.3175 of 18 December 2021 of Italian Ministry of University and Research funded by the European Union–NextGenerationEU under award Number: Project code CN_00000033, Concession Decree No. 1034 of 17 June 2022 adopted by the Italian Ministry of University and Research, CUP B83C22002930006, Project title "National Biodiversity Future Centre-NBFC". E.V. and A.C. also acknowledge funding from the MIUR Project (PRIN 2020) "Unraveling interactions between WATER and carbon cycles during drought and their impact on water resources and forest and grassland ecosySTEMs in the Mediterranean climate (WATERSTEM)" (Project number: 20202WF53Z), "WAFER" at CNR (Consiglio Nazionale delle Ricerche), and by PRIN 2020 (cod 2020E52THS)-Research Projects of National Relevance funded by the Italian Ministry of University and Research entitled: "Multi-scale observations to predict Forest response to pollution and climate change" (MULTIFOR, project number 2020E52THS).

Data Availability Statement: The 3D-CMCC-FEM v.5.6 model code is publicly available and can be found on the GitHub platform at: https://github.com/Forest-Modelling-Lab/3D-CMCC-FEM (accessed on 24 June 2024). The raw data supporting the conclusions of this article will be made available by the authors on request.

Acknowledgments: We thank the ISIMIP project and the COST Action FP1304 "PROFOUND" (Towards Robust Projections of European Forests under Climate Change), supported by COST (European Cooperation in Science and Technology) for providing us the climate historical scenarios and site data used in this work. This work used eddy covariance data acquired and shared by the "FLUXNET" community, including these networks: AmeriFlux, AfriFlux, AsiaFlux, CarboAfrica, CarboEurope-IP, CarboItaly, CarboMont, ChinaFlux, Fluxnet-Canada, GreenGrass, ICOS, KoFlux, LBA, NECC, OzFlux-TERN, TCOS-Siberia, and USCCC. The ERA-Interim reanalysis data are provided by ECMWF and processed by LSCE. The FLUXNET eddy covariance data processing and harmonization was carried out by the European Fluxes Database Cluster, AmeriFlux Management Project, and Fluxdata Project of FLUXNET, with the support of CDIAC and ICOS Ecosystem Thematic Center, and the OzFlux, ChinaFlux, and AsiaFlux offices. We acknowledge the World Climate Research Programme's Working Group on Coupled Modelling, which is responsible for CMIP, and we thank the respective climate modeling groups for producing and making available their model output. The U.S. Department of Energy's Program for Climate Model Diagnosis and Intercomparison at Lawrence Livermore National Laboratory provides coordinating support for CMIP and led the development of software infrastructure in partnership with the Global Organization for Earth System Science Portals.

Conflicts of Interest: The authors declare no conflicts of interest.

References

1. Favero, A.; Mendelsohn, R.; Sohngen, B.; Stocker, B. Assessing the long-term interactions of climate change and timber markets on forest land and carbon storage. *Environ. Res. Lett.* **2021**, *16*, 014051. [CrossRef]
2. Vangi, E.; D'Amico, G.; Francini, S.; Giannetti, F.; Lasserre, B.; Marchetti, M.; McRoberts, R.E.; Chirici, G. The effect of forest mask quality in the wall-to-wall estimation of growing stock volume. *Remote Sens.* **2021**, *13*, 1038. [CrossRef]
3. Noce, S.; Collalti, A.; Valentini, R.; Santini, M. Hot spot maps of forest presence in the Mediterranean basin. *iForest Biogeosci. For.* **2016**, *9*, 766. [CrossRef]
4. Lionello, P.; Scarascia, L. The relation between climate change in the Mediterranean region and global warming. *Reg. Environ. Chang.* **2018**, *18*, 1481–1493. [CrossRef]

5. Pretzsch, H.; Biber, P.; Schütze, G.; Uhl, E.; Rötzer, T. Forest stand growth dynamics in Central Europe have accelerated since 1870. *Nat. Commun.* **2014**, *5*, 4967. [CrossRef] [PubMed]
6. He, Y.; Liu, Y.; Lei, L.; Terrer, C.; Huntingford, C.; Peñuelas, J.; Xu, H.; Piao, S. CO_2 fertilization contributed more than half of the observed forest biomass increase in northern extra-tropical land. *Glob. Chang. Biol.* **2023**, *29*, 4313–4326. [CrossRef]
7. Roebroek, C.T.J.; Duveiller, G.; Seneviratne, S.I.; Davin, E.L.; Cescatti, A. Releasing global forests from human management: How much more carbon could be stored? *Science* **2023**, *380*, 749–753. [CrossRef] [PubMed]
8. Duffy, K.A.; Schwalm, C.R.; Arcus, V.L.; Koch, G.W.; Liang, L.L.; Schipper, L.A. How close are we to the temperature tipping point of the terrestrial biosphere? *Sci. Adv.* **2021**, *7*, eaay1052. [CrossRef]
9. Nabuurs, G.J.; Verkerk, P.J.; Schelhaas, M.; González-Olabarria, J.R.; Trasobares, A.; Cienciala, E. *Climate-Smart Forestry: Mitigation Impacts in Three European Regions*; From Science to Policy 6; European Forest Institute: Joensuu, Finland, 2018.
10. Wang, X.; Pederson, N.; Chen, Z.; Lawton, K.; Zhu, C.; Han, S. Recent rising temperatures drive younger and southern Korean pine growth decline. *Sci. Total Environ.* **2019**, *649*, 1105–1116. [CrossRef]
11. Gregor, K.; Krause, A.; Reyer, C.P.O.; Knoke, T.; Meyer, B.F.; Suvanto, S.; Rammig, A. Quantifying the impact of key factors on the carbon mitigation potential of managed temperate forests. *Carbon Balance Manag.* **2024**, *19*, 10. [CrossRef]
12. Vangi, E.; Dalmonech, D.; Cioccolo, E.; Marano, G.; Bianchini, L.; Puchi, P.F.; Grieco, E.; Cescatti, A.; Colantoni, A.; Chirici, G.; et al. Stand age diversity and climate change affect forests' resilience and stability, although unevenly. *bioRxiv* **2023**. [CrossRef]
13. Erb, K.; Haberl, H.; Le Noë, J.; Tappeiner, U.; Tasser, E.; Gingrich, S. Changes in perspective needed to forge 'no-regret' forest-based climate change mitigation strategies. *GCB Bioenergy* **2022**, *14*, 246–257. [CrossRef] [PubMed]
14. Dalmonech, D.; Marano, G.; Amthor, J.; Cescatti, A.; Lindner, M.; Trotta, C.; Collalti, A. Feasibility of enhancing carbon sequestration and stock capacity in temperate and boreal European forests via changes to forest management. *Agric. For. Meteorol.* **2022**, *327*, 109203. [CrossRef]
15. Anderson-Teixeira, K.J.; Miller, A.D.; Mohan, J.E.; Hudiburg, T.W.; Duval, B.D.; Delucia, E.H. Altered dynamics of forest recovery under a changing climate. *Glob. Chang. Biol.* **2013**, *19*, 2001–2021. [CrossRef] [PubMed]
16. Ryan, M.G.; Binkley, D.; Fownes, J.H. Age-related decline in forest productivity: Pattern and process. *Adv. Ecol. Res.* **1997**, *27*, 213–262. [CrossRef]
17. Goulden, M.L.; McMillan, A.M.; Winston, G.C.; Rocha, A.V.; Manies, K.L.; Harden, J.W.; Bond-Lamberty, B.P. Patterns of NPP, GPP, respiration, and NEP during boreal forest succession. *Glob. Chang. Biol.* **2011**, *17*, 855–871. [CrossRef]
18. Collalti, A.; Tjoelker, M.G.; Hoch, G.; Mäkelä, A.; Guidolotti, G.; Heskel, M.; Petit, G.; Ryan, M.G.; Battipaglia, G.; Prentice, I.C. Plant respiration: Controlled by photosynthesis or biomass? *Glob. Chang. Biol.* **2020**, *26*, 1739–1753. [CrossRef] [PubMed]
19. Collalti, A.; Ibrom, A.; Stockmarr, A.; Cescatti, A.; Alkama, R.; Fernandez-Martínez, M.; Matteucci, G.; Sitch, S.; Friedlingstein, P.; Ciais, P.; et al. Forest production efficiency increases with growth temperature. *Nat. Commun.* **2020**, *11*, 5322. [CrossRef] [PubMed]
20. FOREST EUROPE. *State of Europe's Forests 2018*; 2018.
21. FOREST EUROPE. *State of Europe's Forests 2020*; 2020; Available online: https://foresteurope.org/wp-content/uploads/2016/08/SoEF_2020.pdf (accessed on 23 May 2024).
22. Collalti, A.; Marconi, S.; Ibrom, A.; Trotta, C.; Anav, A.; D'Andrea, E.; Matteucci, G.; Montagnani, L.; Gielen, B.; Mammarella, I.; et al. Validation of 3D-CMCC Forest Ecosystem Model (v.5.1) against eddy covariance data for ten European forest sites. *Geosci. Model Dev.* **2016**, *9*, 479–504. [CrossRef]
23. Collalti, A.; Dalmonech, D.; Vangi, E.; Marano, G.; Puchi, P.F.; Morichetti, M.; Saponaro, V.; Orrico, M.R.; Grieco, E. *Monitoring and Predicting Forest Growth and Dynamics*; CNR Edizioni: Rome, Italy, 2024; ISBN 978-88-8080-655-4. [CrossRef]
24. Collalti, A.; Trotta, C.; Keenan, T.F.; Ibrom, A.; Bond-Lamberty, B.; Grote, R.; Vicca, S.; Reyer, C.P.O.; Migliavacca, M.; Veroustraete, F.; et al. Thinning can reduce losses in carbon use efficiency and carbon stocks in managed forests under warmer climate. *J. Adv. Model. Earth Syst.* **2018**, *10*, 2427–2452. [CrossRef]
25. Mahnken, M.; Cailleret, M.; Collalti, A.; Trotta, C.; Biondo, C.; D'Andrea, E.; Dalmonech, D.; Marano, G.; Mäkelä, A.; Minunno, F.; et al. Accuracy, realism and general applicability of European forest models. *Glob. Chang. Biol.* **2022**, *28*, 6921–6943. [CrossRef] [PubMed]
26. Moss, R.; Edmonds, J.; Hibbard, K. The next generation of scenarios for climate change research and assessment. *Nature* **2010**, *463*, 747–756. [CrossRef] [PubMed]
27. Van Vuuren, D.P.; Edmonds, J.; Kainuma, M.; Riahi, K.; Thomson, A.; Hibbard, K.; Hurtt, G.C.; Kram, T.; Krey, V.; Lamarque, J.-F.; et al. The representative concentration pathways: An overview. *Clim. Chang.* **2011**, *109*, 5–31. [CrossRef]
28. Reyer, C.P.O.; Silveyra Gonzalez, R.; Dolos, K.; Hartig, F.; Hauf, Y.; Noack, M.; Lasch-Born, P.; Rötzer, T.; Pretzsch, H.; Meesenburg, H.; et al. The PROFOUND Database for evaluating vegetation models and simulating climate impacts on European forests. *Earth Syst. Sci. Data* **2020**, *12*, 1295–1320. [CrossRef]
29. Vacchiano, G.; Magnani, F.; Collalti, A. Modeling Italian forests: State of the art and future challenges. *iForest* **2012**, *5*, 113–120. [CrossRef]
30. Collalti, A.; Perugini, L.; Santini, M.; Chiti, T.; Nolè, A.; Matteucci, G.; Valentini, R. A process-based model to simulate growth in forests with complex structure: Evaluation and use of 3D-CMCC Forest Ecosystem Model in a deciduous forest in Central Italy. *Ecol. Model.* **2014**, *272*, 362–378. [CrossRef]
31. Farquhar, G.; von Caemmerer, S.; Berry, J. A biogeochemical model of photosynthetic CO_2 assimilation in leaves of C3 species. *Planta* **1980**, *149*, 78–90. [CrossRef] [PubMed]

32. de Pury, D.G.G.; Farquhar, G.D. Simple scaling of photosynthesis from leaves to canopies without the errors of big-leaf models. *Plant Cell Environ.* **1997**, *20*, 537–557. [CrossRef]
33. Bernacchi, C.J.; Singsaas, E.L.; Pimentel, C.A.R.L.O.S.; Portis Jr, A.R.; Long, S.P. Improved temperature response functions for models of Rubisco-limited photosynthesis. *Plant Cell Environ.* **2001**, *24*, 253–259. [CrossRef]
34. Bernacchi, C.J.; Calfapietra, C.A.R.L.; Davey, P.A.; Wittig, V.E.; Scarascia-Mugnozza, G.E.; Raines, C.A.; Long, S.P. Photosynthesis and stomatal conductance responses of poplars to free-air CO_2 enrichment (PopFACE) during the first growth cycle and immediately following coppice. *New Phytol.* **2003**, *159*, 609–621. [CrossRef]
35. Kattge, J.; Knorr, W. Temperature acclimation in a biochemical model of photosynthesis: A reanalysis of data from 36 species. *Plant Cell Environ.* **2007**, *30*, 1176–1190. [CrossRef] [PubMed]
36. McDowell, N. Mechanism linking drought, hydraulics, carbon metabolism, and vegetation mortality. *Plant Physiol.* **2011**, *155*, 1051–1059. [CrossRef]
37. Rowland, L.; da Costa, A.C.L.; Galbraith, D.R.; Oliveira, R.S.; Binks, O.J.; Oliveira, A.A.R.; Pullen, A.M.; Doughty, C.E.; Metcalfe, D.B.; Vasconcelos, S.S.; et al. Death from drought in tropical forests is triggered by hydraulics not carbon starvation. *Nature* **2015**, *528*, 119–122. [CrossRef] [PubMed]
38. Collalti, A.; Biondo, C.; Buttafuoco, G.; Maesano, M.; Caloiero, T.; Lucà, F.; Pellicone, G.; Ricca, N.; Salvati, R.; Veltri, A.; et al. Simulation, calibration and validation protocols for the model 3D-CMCC-CNR-FEM: A case study in the Bonis' watershed (Calabria), Italy. *For. J. Silvic. For. Ecol.* **2017**, *14*, 247–256. [CrossRef]
39. Merganičová, K.; Merganič, J.; Lehtonen, A.; Vacchiano, G.; Sever, M.Z.O.; Augustynczik, A.L.D.; Grote, R.; Kyselová, I.; Mäkelä, A.; Yousefpour, R.; et al. Forest carbon allocation modelling under climate change. *Tree Physiol.* **2019**, *39*, 1937–1960. [CrossRef] [PubMed]
40. Kira, T.; Shidei, T. Primary production and turnover of organic matter in different forest ecosystems of the western Pacific. *Jpn. J. Ecol.* **1967**, *17*, 70–87.
41. Odum, E.P. The Strategy of Ecosystem Development: An understanding of ecological succession provides a basis for resolving man's conflict with nature. *Science* **1969**, *164*, 262–270. [CrossRef]
42. Tang, J.; Luyssaert, S.; Richardson, A.D.; Kutsch, W.; Janssens, I.A. Steeper declines in forest photosynthesis than respiration explain age-driven decreases in forest growth. *Proc. Natl. Acad. Sci. USA* **2014**, *111*, 8856–8860. [CrossRef] [PubMed]
43. Mäkelä, A.; Valentine, H. The ratio of NPP to GPP: Evidence of change over the course of stand development. *Tree Physiol.* **2000**, *21*, 1015–1030. [CrossRef]
44. Landsberg, J.J.; Waring, R.H. A generalised model of forest productivity using simplified concepts of radiation-use efficiency, carbon balance and partitioning. *For. Ecol. Manag.* **1997**, *95*, 209–228. [CrossRef]
45. Anderson, K.J.; Allen, A.P.; Gillooly, J.F.; Brown, J.H. Temperature-dependence of biomass accumulation rates during secondary succession. *Ecol. Lett.* **2006**, *9*, 673–682. [CrossRef]
46. Cook-Patton, S.C.; Leavitt, S.M.; Gibbs, D.; Harris, N.L.; Lister, K.; Anderson-Teixeira, K.J.; Briggs, R.D.; Chazdon, R.L.; Crowther, T.W.; Ellis, P.W.; et al. Mapping carbon accumulation potential from global natural forest regrowth. *Nature* **2020**, *585*, 545–550. [CrossRef]
47. Anderson-Teixeira, K.J.; Herrmann, V.; Morgan, R.B.; Bond-Lamberty, B.; Cook-Patton, S.C.; E Ferson, A.; Muller-Landau, H.C.; Wang, M.M. Carbon cycling in mature and regrowth forests globally. *Environ. Res. Lett.* **2021**, *16*, 053009. [CrossRef]
48. Ouyang, S.; Xiang, W.; Wang, X.; Xiao, W.; Chen, L.; Li, S.; Sun, H.; Deng, X.; Forrester, D.I.; Zeng, L.; et al. Effects of stand age, richness and density on productivity in subtropical forests in China. *J. Ecol.* **2019**, *107*, 2266–2277. [CrossRef]
49. Ullah, F.; Gilani, H.; Sanaei, A.; Hussain, K.; Ali, A. Stand structure determines aboveground biomass across temperate forest types and species mixture along a local-scale elevational gradient. *For. Ecol. Manag.* **2021**, *486*, 118984. [CrossRef]
50. Lee, Y.-J.; Park, G.-E.; Lee, H.-I.; Lee, C.-B. Stand age-driven tree size variation and stand type regulate aboveground biomass in alpine-subalpine forests, South Korea. *Sci. Total. Environ.* **2024**, *915*, 170063. [CrossRef] [PubMed]
51. Hlásny, T.; Barka, I.; Kulla, L.; Bucha, T.; Sedmák, R.; Trombik, J. Sustainability of forest management in a Central European mountain forest: The role of climate change. *Reg. Environ. Chang.* **2015**, *17*, 65–77. [CrossRef]
52. Dymond, C.C.; Beukema, S.; Nitschke, C.R.; Coates, K.D.; Scheller, R.M. Carbon sequestration in managed temperate coniferous forests under climate change. *Biogeosciences* **2016**, *13*, 1933–1947. [CrossRef]
53. Pretzsch, H.; del Río, M.; Arcangeli, C.; Bielak, K.; Dudzinska, M.; Forrester, D.I.; Klädtke, J.; Kohnle, U.; Ledermann, T.; Matthews, R.; et al. Forest growth in Europe shows diverging large regional trends. *Sci. Rep.* **2023**, *13*, 15373. [CrossRef] [PubMed]
54. Krejza, J.; Cienciala, E.; Světlík, J.; Bellan, M.; Noyer, E.; Horáček, P.; Štěpánek, P.; Marek, M.V. Evidence of climate-induced stress of Norway spruce along elevation gradient preceding the current dieback in Central Europe. *Trees* **2021**, *35*, 103–119. [CrossRef]
55. del Castillo, E.M.; Zang, C.S.; Buras, A.; Hacket-Pain, A.; Esper, J.; Serrano-Notivoli, R.; Hartl, C.; Weigel, R.; Klesse, S.; de Dios, V.R.; et al. Climate-change-driven growth decline of European beech forests. *Commun. Biol.* **2022**, *5*, 163. [CrossRef] [PubMed]
56. Gong, X.; Yuan, D.; Zhu, L.; Li, Z.; Wang, X. Long-term changes in radial growth of seven tree species in the mixed broadleaf-Korean pine forest in Northeast China: Are deciduous trees favored by climate change? *J. For. Res.* **2024**, *35*, 70. [CrossRef]
57. Bennett, A.C.; McDowell, N.G.; Allen, C.D.; Anderson-Teixeira, K.J. Larger trees suffer most during drought in forests worldwide. *Nat. Plants* **2015**, *1*, 15139. [CrossRef] [PubMed]

58. Hogg, E.H.; Michaelian, M.; Hook, T.I.; Undershultz, M.E. Recent climatic drying leads to age-independent growth reductions of white spruce stands in western Canada. *Glob. Chang. Biol.* **2017**, *23*, 5297–5308. [CrossRef]
59. Reyer, C.P.O. Forest Productivity Under Environmental Change—A Review of Stand-Scale Modeling Studies. *Curr. For. Rep.* **2015**, *1*, 53–68. [CrossRef]
60. Terrer, C.; Vicca, S.; Hungate, B.; Phillips, R.P.; Prentice, I.C. Mycorrhizal association as a primary control of the CO_2 fertilization effect. *Science* **2016**, *353*, 72–74. [CrossRef]

forests

MDPI

Article

Soil Temperature, Organic-Carbon Storage, and Water-Holding Ability Should Be Accounted for the Empirical Soil Respiration Model Selection in Two Forest Ecosystems

Sergey Kivalov *, Valentin Lopes de Gerenyu, Dmitry Khoroshaev, Tatiana Myakshina, Dmitry Sapronov, Kristina Ivashchenko and Irina Kurganova

Institute of Physicochemical and Biological Problems in Soil Science, Russian Academy of Sciences, Institutskaya 2, Pushchino 142290, Russia
* Correspondence: kivalov@pbcras.ru

Abstract: Soil respiration (SR) is a main component of the carbon cycle in terrestrial ecosystems, and being strongly affected by changes in the environment, it is a good indicator of the ecosystem's ability to cope with climate change. This research aims to find better empirical SR models using 25-year-long SR monitoring in two forest ecosystems formed on sandy Entic Podzol and loamy Haplic Luvisol. The following parameters were considered in the examined models: the mean monthly soil or air temperatures (T_{soil} or T_{air}), the amount of precipitation during the current (P) and the previous (PP) months, and the storage of soil organic carbon (*SOC*). The weighted non-linear regression was used for model parameter estimations for the normal, wet, and dry years. To improve the model resolutions by magnitude, we controlled the slope and intercept of the linear model comparison between the measured and modeled data through the change in R_0—SR at zero soil temperature. The mean bias error (MBE), root-mean-square error (RMSE), and determination coefficient (R^2) were used for the estimation of the goodness of model performances. For the sandy Entic Podzol, it is more appropriate to use the models dependent on *SOC* (TPPC). While for the loamy Haplic Luvisol, the Raich–Hashimoto model (TPPrh) with the quadratic T_{soil} or T_{air} dependency shows the better results. An application of T_{soil} for the model parameterization gives better results than T_{air}: the TPPC model was able to adequately describe the cold-period SR ($T_{soil} \leq 2$ °C); the TPPrh model was able to avoid overestimations of the warm-period SR ($T_{soil} > 2$ °C). The TPPC model parameterized with T_{soil} can be used for the quality control of the cold-period SR measurements. Therefore, we showed the importance of accounting for *SOC* and the water-holding ability when the optimal SR model is chosen for the analysis.

Keywords: CO_2 emission; forest soils; hydrothermal regime; carbon content; long-term observations; humidity/aridity level; climate change; statistical modeling

Citation: Kivalov, S.; Lopes de Gerenyu, V.; Khoroshaev, D.; Myakshina, T.; Sapronov, D.; Ivashchenko, K.; Kurganova, I. Soil Temperature, Organic-Carbon Storage, and Water-Holding Ability Should Be Accounted for the Empirical Soil Respiration Model Selection in Two Forest Ecosystems. *Forests* **2023**, *14*, 1568. https://doi.org/10.3390/f14081568

Academic Editors: Daniela Dalmonech, Alessio Collalti and Gina Marano

Received: 26 June 2023
Revised: 21 July 2023
Accepted: 27 July 2023
Published: 31 July 2023

1. Introduction

Soil respiration (SR) is the main pathway through which carbon exits terrestrial ecosystems [1–3], changing soil organic carbon (*SOC*) storage and its allocation in soil [4]. It is the integral part of the ecosystem carbon balance (net ecosystem production, $NEP = GPP - Re$), defined as the difference between the gross primary production or photosynthesis (*GPP*) and the total respiration (*Re*), which is the direct indicator of the ecosystem's well being [5,6], and a useful indicator of plant metabolism [1]. That is why SR—a main component of *Re*—should be monitored for estimation of the ecosystem ability to withstand environmental stresses due to adverse changes in the environment or due to climate change [2,3], and for reporting the annual greenhouse-gas inventory [7].

For the separation of the ecosystem-related and soil-related parameters in *NEP*, the total respiration (*Re*) should be split to ecosystem respiration related to its growth (autotrophic respiration, *Ra*) and the soil respiration related to the microbial activity for the

SOC decomposition (heterotrophic respiration, *Rh*) [8,9]. This respiration separation is often modeled from the eddy-covariance data [1,10–14] that could lead to inconsistent results [15,16], due to the high variability of the *Ra* contribution to *Re* (10%–90%), depending on the seasonality and vegetation types [17].

The direct methods for SR measurements employ operation of soil chambers of the closed type [18,19]. As a rule, these methods require regular travelling to the site to conduct measurements to assure the adequate representation of SR estimations during the year. This represents the significant labor intensity of the soil-chamber measurement procedure [20]. The quality of these measurements can be altered by the size and the installation of the chambers as well as by the soil heterogeneity [4], introducing high measurement variability, and the presence of snow or vegetation covering the soil surface and interrupting gas exchange [21,22]. Moreover, such a monitoring on a country level looks unrealistic because of a prohibitively high cost [4].

The viable alternative to the measurement methods is to focus on the modeling of SR including both simple empirical models [14,23–26] and more sophisticated process-based dynamical models [4,27–29].

A number of research studies focusing on both SR measurements and applications of empirical SR models highlight a strong dependency of SR on the following parameters: (i) temperature [24,30–33] due to change in microbial activity; (ii) soil moisture [24,34–37] due to change in soil porosity and accessibility of atmospheric oxygen; (iii) precipitation [20,24,38–42] as a simpler way for the soil-moisture estimation; (iv) change of water level in soil [26,43,44] blocking below-water-level SOC oxidation; and (v) allocation of above-ground biomass [23,45]. Several studies link changes in the amount of *SOC* stored in ecosystem soils to differences in SR [1,8,46–49].

The empirical models usually use one of two temperature sources for parameterization: the soil temperature (T_{soil}) [23,26,31–33] or the air temperature (T_{air}) [20,24,25,42]. Raich and Potter (1995) [24] note that the soil-temperature application is more consistent from the Q_{10} temperature-coefficient behavior's point of view. They justified it by the direct response of the microbial activity on changes in T_{soil}. On the other hand, Suhoveeva and Karelin (2022) [42] showed that the Raich–Hashimoto model [25] with the quadratic temperature dependency gives good results when it is parameterized by T_{air}.

It should be noted that, as a rule, the choice of T_{soil} or T_{air} stays on investigators' judgment, and until now, no analysis on the preferred temperature sources for the model parameterizations for different soil conditions have been conducted. The second remark on the empirical SR modeling is often the presence of an insufficient magnitude resolution of the modeled data in comparison to the measurements [20,31,42]—leaving the extreme (summer or winter) measurements without adequate coverage by modeling, due to the lack of their representativity in time series in comparison to the intermediately measured values.

Following Raich and Potter's (1995) [24] notes from above and paying attention to the high variability of the winter-time measurements, we hypothesize that using T_{soil} in cold periods, generally not limited with water availability, could significantly improve the SR modeling results in comparison to using T_{air}. On the other hand, following Maier et al. (2010) [36], we hypothesize that in dryer warm periods, lacking a persistent amount of moisture, soil structure and porosity affect the atmospheric oxygen availability in soil, which together with the different amounts of *SOC* can affect the magnitude of SR.

The current research aims to address these hypotheses by identifying the better versions of the empirical models parameterized by the monthly averages of (i) soil or air temperature, (ii) the amount of precipitation, and (iii) the amount of *SOC* in application to sandy Entic Podzol and loamy Haplic Luvisol. To do this, the weighted non-linear regression was used to estimate the model coefficients for the normal, wet, and dry years separately to ensure an adequate coverage of different climatic periods. By controlling the slope and intercept of the linear-model comparison between the measured and modeled values, the selected models are being re-adjusted to adequately represent the measurement range during the year.

2. Materials and Methods

2.1. The Sites, Measurements, and Soil Properties

The research is focused on two forest ecosystems situated on the opposite banks of Oka River near Pushchino town, Moscow region. The first site (54°50′ N, 37°35′ E) with the sandy Entic Podzol (Arenic) [50] is located in the zone of coniferous-deciduous forests in Prioksko-Terrasny Nature Biosphere Reserve on the left (northern) bank of Oka River (top, Figure 1). The landscapes are plain sandy terraces formed as the result of modern and ancient erosion processes located above Oka-River flood plain—the low plains with the gentle southern slopes toward the River. The second site (54°20′ N, 37°37′ E) with the loamy Haplic Luvisol (Siltic) [50] is located in the zone of deciduous forests on the right (southern) bank of Oka River (bottom, Figure 1). The landscape is hilly with about 150-m of elevation above the River. Oka River serves as the boundary between the forest zones. The cross distance between the sites is about 8.6 km in the north–south direction.

Figure 1. Study sites, soil profiles (Entic Podzol and Haplic Luvisol) and site locations on the opposite banks of Oka River; Prioksko-Terrasny Reserve is north from the River ("Prioksko-Terrasny Reserve", 54°52′ N, 37°35′ E, Google Earth. November 2021. 25 May 2023); top left—Entic Podzol and bottom left—Haplic Luvisol soil profiles.

Both forest sites are located in the same moderately-continental climate zone with warm summers and moderately cold winters. Long-term meteorological observations (the Complex Background Monitoring Station, settlement of Danki, Serpukhov district, Moscow region; 54°50′ N, 37°35′ E) for the last 10 years (https://pt-zapovednik.org, accessed on 1 June 2023) report the following: the average annual air temperature is 4.8 °C, the average summer temperature is +17.6 °C (max 38–39 °C), and the average winter temperature is −8.3 °C (min −43 °C in 1978); the average precipitation is 671 mm (max 91 mm in July); the duration of the seasonal snow cover period is 133 days with the average snow depth being 52 cm; the vegetation season lasts 186 days.

Due to the specifics of geomorphology, the soil properties of these sites are quite different (left, Figure 1; Table 1). The higher concentration of fine particles (silt and clay) in the Haplic Luvisol and consequently, smaller pores, explain its higher water-holding capacity and as the result is the lower permittivity to the atmospheric oxygen, reducing *SOC* oxidation [36], which is directly associated with its higher carbon storage. This process is well investigated on the examples of the wet-meadow and bog soil, having high *SOC* storage as well [26,37,43,51].

Table 1. Site description and soil properties of the forest ecosystems.

		Entic Podzol	Haplic Luvisol
Forest	Forest zone	coniferous-deciduous	Deciduous
	Forest type	mature mixed with pine, linden, aspen, birch, and oak, the age of which reaches 90–120 years [2]	secondary deciduous with aspen, linden, and maple of an average tree age of 50–70 years [2]
Soil	Texture	sandy-loamy [3]	loamy [3]
	granulometry (sand:silt:clay)	11.6:1.0:1.3 [1]	4:4:2 [2]
	pH_{KCl}	3.67 [1]	5.56 [2]
	C/N	15.3 [1]	12.8 [2]
	SOC storage [kg C/m^2]	1.23 (0–20 cm) [4]	5.02 (0–20 cm) [4]
	Water-holding capacity [%]	40.5 [2]	57.5 [2]

[1] [20]; [2] [40]; [3] [52]; [4] [53].

For the current research, we use a 25-year-long SR-measurement time series conducted by the chambers of the closed type (SR—top, Figure 2) once a week. The standard chamber-measurement approach is described in [20]. Firstly, the repetitive with 10-min interval gas measurements using syringe-sample collection are conducted at sites at five nearby locations to account for SR heterogeneity. Secondly, these samples are analyzed for CO_2 concentrations in the laboratory using a gas chromatograph (KrystaLLyuks-4000 M, Meta-Chrom, Yoshkar-Ola, Russia). The obtained changes in CO_2 concentration are recalculated into the SR fluxes by applying the chamber volumetric correction [19,22]. Simultaneously with the SR, the soil temperature at 5-cm depth and the air temperature at 1-m height were measured at the sites (T_{soil}—brown, T_{air}—red; middle, Figure 2). The monthly averaged data for the air temperature (T_{air}) and precipitation (Prec) were collected from the Complex Background Monitoring Station, which is situated nearby to the coniferous-deciduous forest site (blue; bottom, Figure 2). All the data were quality checked and averaged on a monthly base to be fitted into the models.

The monthly averaged temperature (T) and precipitation (P) data have also been used for the separation of the years of the measurements into "wet", "dry", and "normal" (top, Figure 2) by an application of the following indexes to the data:

- Selyaninov hydrothermal coefficient—HTC $= \sum P / \sum T/10$, when $T > 10$ °C (HTC$_{6\text{–}8}$—summer period, June to August months) [53];
- Wetness Indexes—WI $= lg(\sum P / \sum T)$ (WI$_{5\text{–}8}$ and WI$_{5\text{–}9}$ for May to August and May to September periods, respectively) [53].

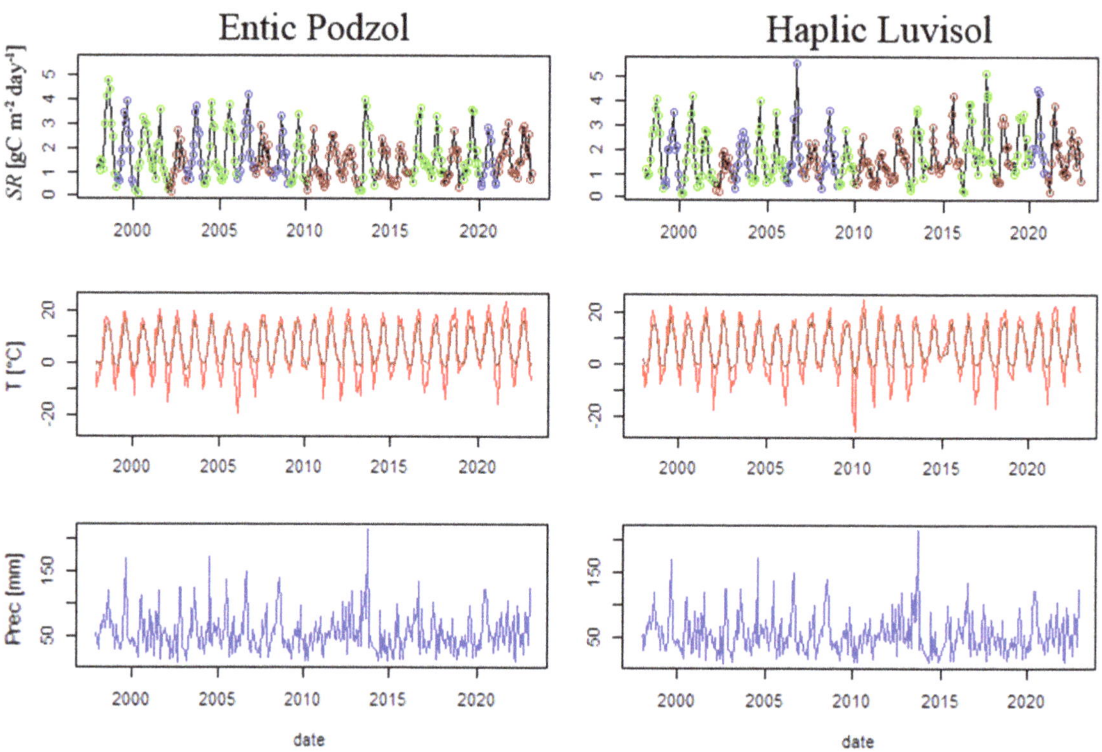

Figure 2. The 25-year-long measurements of soil respiration (SR, **top**), temperatures of air and soil (T_{air}—red and T_{soil}—brown, **middle**), and precipitation (Prec, **bottom**); on the SR graph: normal conditions—green, wet conditions—blue, dry conditions—brown.

If the values of any of these indexes differ (higher or lower) more than a standard deviation (STD) from the averages for the chosen measurement period, the year was placed into the "wet" or "dry" datasets, respectively. The following 5 years were classified as "wet"—1999, 2003, 2006, 2008, 2020; the following 9 years were "dry"—2002, 2007, 2010, 2011, 2014, 2015, 2018, 2021, 2022; and the remaining 11 years stayed as "normal"—1998, 2000, 2001, 2004, 2005, 2009, 2012, 2013, 2016, 2017, 2019.

For the initialization of the empirical models dependent on *SOC*, the estimations of the *SOC* storage in 20-cm layers of Entic Podzol and Haplic Luvisol were used (Table 1).

2.2. Empirical Soil Respiration Models

The current research of the SR estimations is focused on two groups of empirical models connecting SR with the temperature and amount of precipitation, serving as a proxy for the soil moisture [20,24,39–41,54] and by this, estimating SR affected by the climatic conditions. The first group includes the models dependent on T_{soil} [26,31–33], and the second group are the models dependent on T_{air} [20,24,25,42]. The comparison between the groups shows that both groups are generally based on the same formulations but with using different sources of temperature: T_{soil} or T_{air}.

Temperature and Temperature—Precipitation models [20,24,26,31–33]:

$$SR_T = R_0 e^{\,QT} \tag{1}$$

$$SR_{TP} = R_0 e^{\,QT}\left(\frac{P}{K+P}\right) \tag{2}$$

Temperature—Precipitation model dependent on the amounts of precipitation for the current (P) and previous (P_{m-1}) months [25,42]:

$$SR_{TPP} = R_0 e^{\,QT}\left(\frac{\alpha P + (1-\alpha)P_{m-1}}{K + \alpha P + (1-\alpha)P_{m-1}}\right) \tag{3a}$$

$$SR_{TPPrh} = R_0 e^{\,(QT - Q_2 T^2)}\left(\frac{\alpha P + (1-\alpha)P_{m-1}}{K + \alpha P + (1-\alpha)P_{m-1}}\right) \tag{3b}$$

Temperature—Precipitation—*SOC* model [48]:

$$SR_{TPC} = R_0 e^{\,QT}\left(\frac{P}{K + P}\right)\left(\frac{SOC}{\psi + SOC}\right) \tag{4}$$

As an extension from the previous models (Equations (3a) and (4)), we suggest to look at the combined Temperature—Precipitation—*SOC* model dependent on the amounts of precipitation of two months:

$$SR_{TPPC} = R_0 e^{QT}\left(\frac{\alpha P + (1-\alpha)P_{m-1}}{K + \alpha P + (1-\alpha)P_{m-1}}\right)\left(\frac{SOC}{\psi + SOC}\right) \tag{5}$$

In all models, T is the average monthly temperature of the soil surface layer [26,31–33] or the average monthly temperature of air [20,24,25,42]; P is the average monthly amount of precipitation; *SOC* is the organic-carbon storage in the top 20-cm of the soil.

The R_0 (g C/m^2day) is SR at 0 °C in normal soil-humidity conditions—it is usually estimated from the measurements as an average SR for the not-frozen top-soil level. After R_0 is identified, the non-linear regression is used to estimate other parameters of the models: Q and Q_2 are the exponential-relationship temperature coefficients; K (cm) is the half-saturation constant of the hyperbolic relationship between SR and the amount of precipitation; α is the redistribution coefficient between the amounts of precipitation for the current (P) and previous (P_{m-1}) months; and ψ (kg C/m^2) is the half-saturation constant of the hyperbolic relationship between SR and *SOC*.

Practically all the empirical models described above, T (Equation (1))–TP (Equation (2))–TPP (Equation (3a))–TPC (Equation (4))–TPPC (Equation (5)), use the linear temperature dependency in the exponential term and because of it, can be put into the same class of the temperature relationship with SR. On the other hand, the model TPPrh (Equation (3b)) uses the quadratic temperature dependency in the exponential term, which obtains good results when the model was parameterized with the air temperature [25,42], and it was taken to compare with other models.

The quality control of modeling was based on the comparison of the following statistics: the mean bias error (MBE), the root mean square error (RMSE), the slope and intercept of the linear regression (lm) between the measured and modeled data, and the lm determination coefficient (R^2).

3. Results and Discussion

3.1. Choice of the R_0

One of the key parameters of all the empirical models described above is R_0—the soil respiration at 0 °C. It significantly affects the modeling quality because it defines the low boundary of the modeling data (winter periods) and the intercept with the ordinate axis of the linear model (lm), comparing modeling and measured data. In the ideal model, the intercept of lm should lay in the origin of the coordinate system and the slope of lm should be equal to 1. However, as a rule for the real models, the slope < 1 and intercept > 0 due to an insufficient magnitude of the modeled data in comparison with the measurements—the extreme values are not represented well by the modeling [20,31,42].

In our research, R_0 was estimated from modeling by the T–TP–TPP–TPC–TPPC models in the following way: R_0 is directly interconnected with the intercept of lm and by controlling and lowing R_0, the intercept of lm is being readjusted to become positive and closer to zero and at the same time, the slope of lm increases and becomes closer to unity. With the negative intercept of lm, there will be an underestimation of the soil respiration in the winter periods.

The obtained R_0 values (colored lines, Figure 3) are two–three times smaller than the earlier obtained values by Raich and Potter (1995) [24] and Kurganova et al. (2020) [20] for the Entic Podzol. They can be directly compared with the SR measurements at $0 \leq T_{soil} < 1\ ^\circ$C in the autumn–winter period (colored dots with labels, Figure 3) when there is not any freezing of the top-soil level, and by the selection of the temperature interval, they should be located closer to the lower SR observed at the lower temperatures ($T_{soil} \approx 0$).

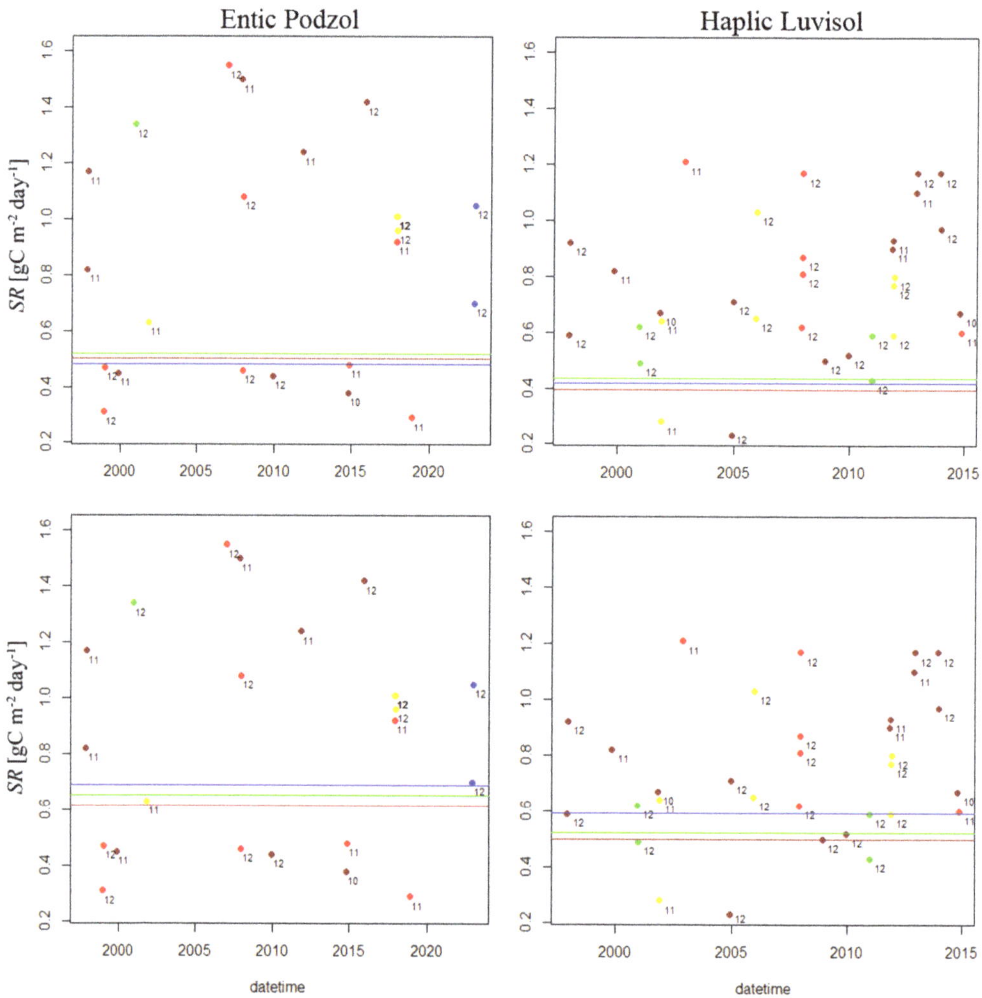

Figure 3. Parameter R_0—soil respiration in autumn–winter period, $T_{soil} \approx 0\ ^\circ$C—values (colored horizontal lines) obtained during modeling by T–TP–TPP–TPC–TPPC models (intercept → 0+) with the parameterization by the soil temperature (**top**) and the air temperature (**bottom**, monitoring station) in the normal (green line), wet (blue line), and dry (brown line) years; colored dots with labels—individual SR measurements, where the labels are the measurement-month numbers and the dot colors for the monthly-precipitation amount are red ($11 < P < 34$ mm), brown ($34 < P < 57$ mm), yellow ($57 < P < 80$ mm), green ($80 < P < 103$ mm), and blue ($103 < P < 126$ mm).

As the result of such a selection (intercept → 0+), the R_0 values obtained for the different set of years—normal (n), wet (w), and dry (d)—demonstrate a weak linear dependency from the soil moisture (colored lines, Figure 3) with the minimal R_0 values corresponding mainly with the dry period (brown lines). These R_0 maximize the slope of lm (see Tables A1–A4 in the Appendix A) as well.

It should be noted that the R_0 values obtained from the models parameterized by T_{air} are about 15% higher than the values obtained from the models parameterized by T_{soil}. This observation can be explained from the differences between T_{air} and T_{soil} for these conditions: the average soil temperature ($T_{soil} \approx 2$ °C, brown vertical line, Figure 4) is about two degrees higher than the simultaneously observed air temperature ($T_{air} = 0$ °C, blue horizontal line, Figure 4), which is associated with the more intensive SR for $T_{air} = 0$ °C than for $T_{soil} = 0$ °C.

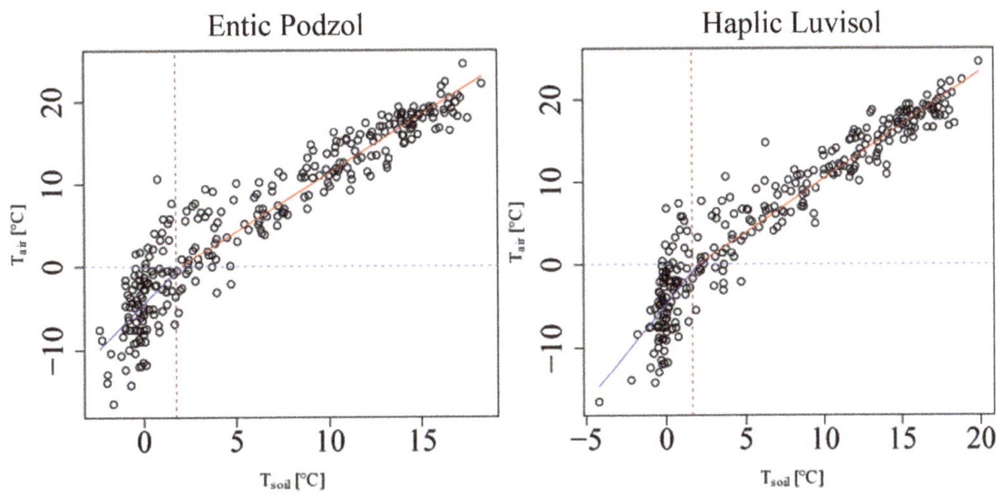

Figure 4. The interrelationship between T_{soil} measured at the sites and T_{air} obtained from the monitoring station; horizontal blue line—$T_{air} = 0$ °C; vertical brown line—average soil temperature $T_{soil} \approx 2$ °C for $T_{air} \approx 0$ °C; red line—warm-period; and blue line—cold-period linear fits.

The R_0 values obtained for different conditions (colored horizontal lines, Figure 3) are close to each other when the parameterization with T_{soil} was done. This finding signals that there are more accurate estimations with T_{soil} and agrees with Raich and Potter's (1995) [24] notes on the Q_{10} temperature coefficients that the microbial biomass responsible for SR is better reacted to the T_{soil} changes—the immediate substrate—than to the T_{air} changes.

This conclusion also agrees well with the observed (Figure 4) significant spread of the air temperatures ($T_{air} \approx -12$–$+2$ °C) for the soil temperatures close to zero Celsius ($T_{soil} = 0$ °C), seen at both sites and fewer smaller spreads of the soil temperatures ($T_{soil} \approx 0$–$+5$ °C) for $T_{air} = 0$ °C. On the other hand, the $T_{soil} \approx 2$ °C threshold serves as a clear indicator of the temperature-regime change. When $T_{soil} > 2$ °C, air and soil temperatures are in a close coupling with each other (red line, Figure 4), while for $T_{soil} \leq 2$ °C, this coupling behavior has been effectively broken due to the strong influence of liquid water keeping soil from freezing.

The individual values of the measured SR at near zero temperatures, when there is not any freezing of the top-soil level occurred (colored dots, Figure 3), are generally associated with the end-of-the-year cold periods with not very large monthly precipitation (Figure 2). Investigating the large scattering of these SR values, we found some evidence that they depend both on the monthly precipitation and soil properties together. For the Entic Podzol (sandy soil with poor water-holding ability and larger pores), the lower precipitation periods (red and brown dots, Figure 3 left) have the extremely low SR—good drainage easily dries out this soil. However, for Haplic Luvisol (loamy soil with high water-holding ability and smaller pores), the lower precipitation periods (red and brown dots, Figure 3 right) are actually associated with the higher SR—this soil is over saturated with water [36] in cold periods. These evidences point out the importance accounting for the soil properties in SR modeling.

All those mentioned above ensure us that the R_0 values we obtained from the modeling (colored horizontal lines, Figure 3) behave as expected in comparison to the measurements.

3.2. Modeling Results

The modeling with the T (Equation (1))–TP (Equation (2))–TPP (Equation (3a))–TPC (Equation (4))–TPPC (Equation (5)) models was conducted with using the non-linear regression for model fitting on the monthly-averaged measured SR, T_{soil} or T_{air}, and precipitation datasets where the winter and summer values were double weighted to obtain the representativeness of the models regardless of the time of the year. In addition to this modeling, the results of the TPPrh (Equation (3b)) modeling with the quadratic temperature dependency were used for comparisons. The temperature-related coefficients R_0 и Q were estimated from the T model with using the intercept $\rightarrow$ 0+ constraint for the lm, comparing the measurements with the modeled data (T, TP, TPP, TPC, TPPC). These coefficient values were taken as the base for further modeling with more complex TP, TPP, TPC, TPPC, and also TPPrh (R_0, Q, and Q_2) models. This way, we separate the temperature-related effects (R_0, Q, and Q_2) and, sequentially, focus on the precipitation (K) and SOC (ψ) effects, and also on the effect of the precipitation redistribution between months (α).

Figures 5 and 6 show that the smallest SR values are observed during the dry years, while the maximal 30% larger values are reached in the normal years, and the SR becomes smaller again during the wet years. These observations agree well with the obtained R_0 dependency from the soil moisture (colored lines, Figure 3) and also with the previous research results [55–57]. They can be explained by the reduced microbial activity when there is not enough water presented in soil in dry years [32,58,59] and also by the lack of available oxygen for the SOC oxidation when it is saturated with water in wet years [26,36,37].

After fixing the temperature coefficients (R_0 и Q) determined by the T–TP–TPP–TPC–TPPC modeling by the method described above (Figures 5 and 6), the comparison among the modeling results shows the following similarities and differences among the models depending on (i) the soil type: Entic Podzol or Haplic Luvisol, and (ii) the sources of the temperature used for parameterization: T_{soil} or T_{air}.

For Entic Podzol and T_{soil}:

- The best slope-lm values (slope, Figure 7) were observed with the TPPC model in a dry environment (slope $\approx$ 0.9), with the TPC model in a normal environment (slope $\approx$ 0.9), and with the TPP model in a wet environment (slope $\approx$ 0.9); the TPPC and TPPrh models show the slope > 0.85 for most of the conditions;
- The best R^2-lm values (R^2, Figure 7) were observed with the TPPC model in a dry environment ($R^2 \approx 0.7$) and with the TPPrh model in normal and wet environments ($R^2 \approx 0.75$); the TPPC and TPP models show the $R^2 > 0.7$ for all moisture conditions;
- The best MBE values of the comparison between the models and measurements (|MBE|, Figure 7) were observed with the TPC and TPPC models in a dry environment (|MBE| $\approx$ 0.15), and with the TPPrh model in normal and wet environments (|MBE| $\approx$ 0.08); the TPPC model shows |MBE| < 0.17 for all moisture conditions.
- The best RMSE values of the comparison between the models and measurements (RMSE, Figure 7) were observed with the TPC и TPPC in a dry environment (RMSE $\approx$ 0.45), and with the TPPrh model in normal and wet environments (RMSE $\approx$ 0.55); the TPPC model shows RMSE < 0.63 for all moisture conditions.

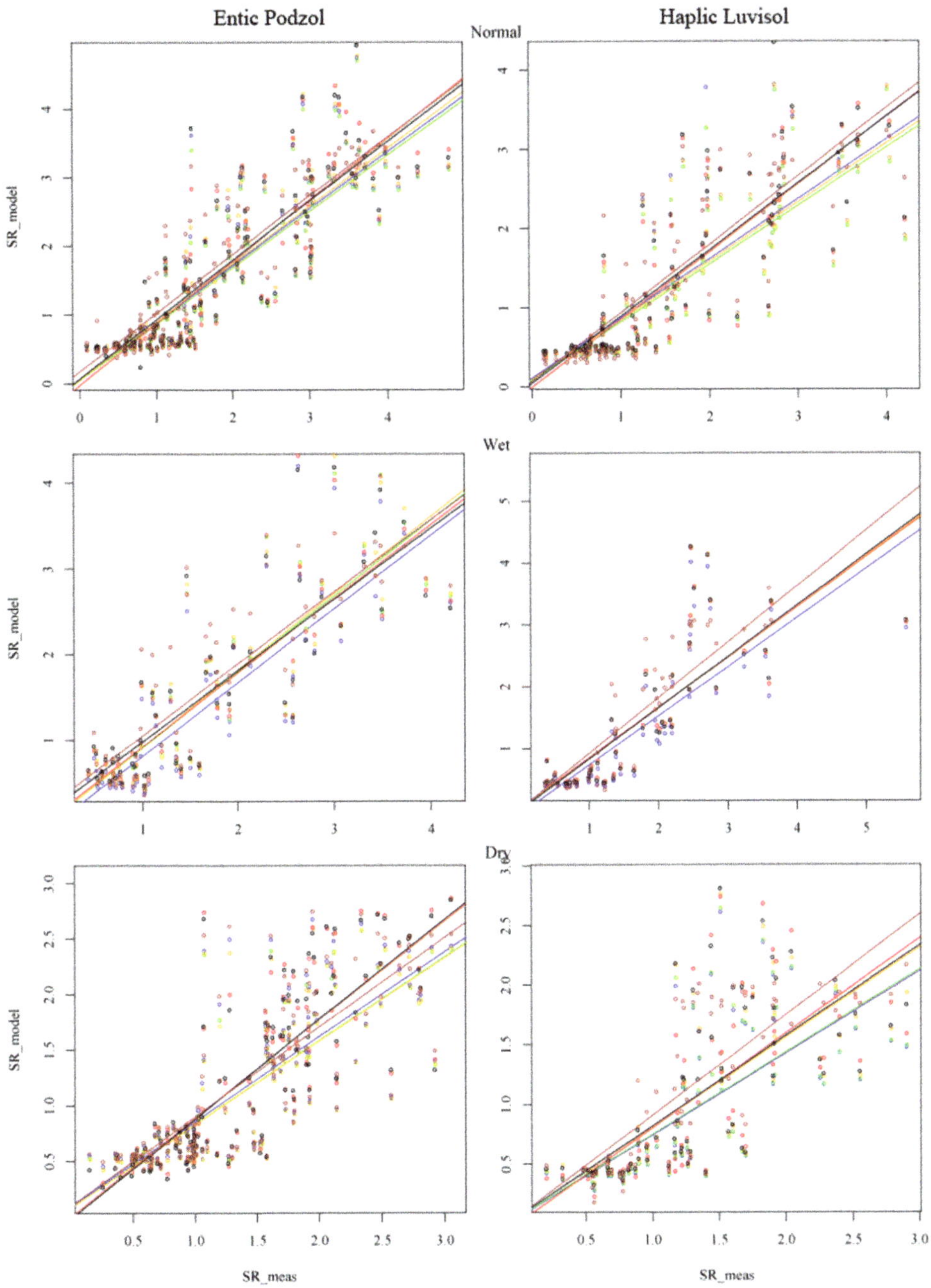

Figure 5. The comparison between the modeled (SR_model) and measured (SR_meas) SR values ($gCm^{-2}day^{-1}$) for T (blue, Equation (1)) –TP (green, Equation (2))–TPP (orange, Equation (3a))–TPC (red, Equation (4))–TPPC (black, Equation (5))–TPPrh (brown, Equation (3b)) models; lines—linear regression to determine the slope and intercept values; T_{soil} (local temperature) used for parameterizations.

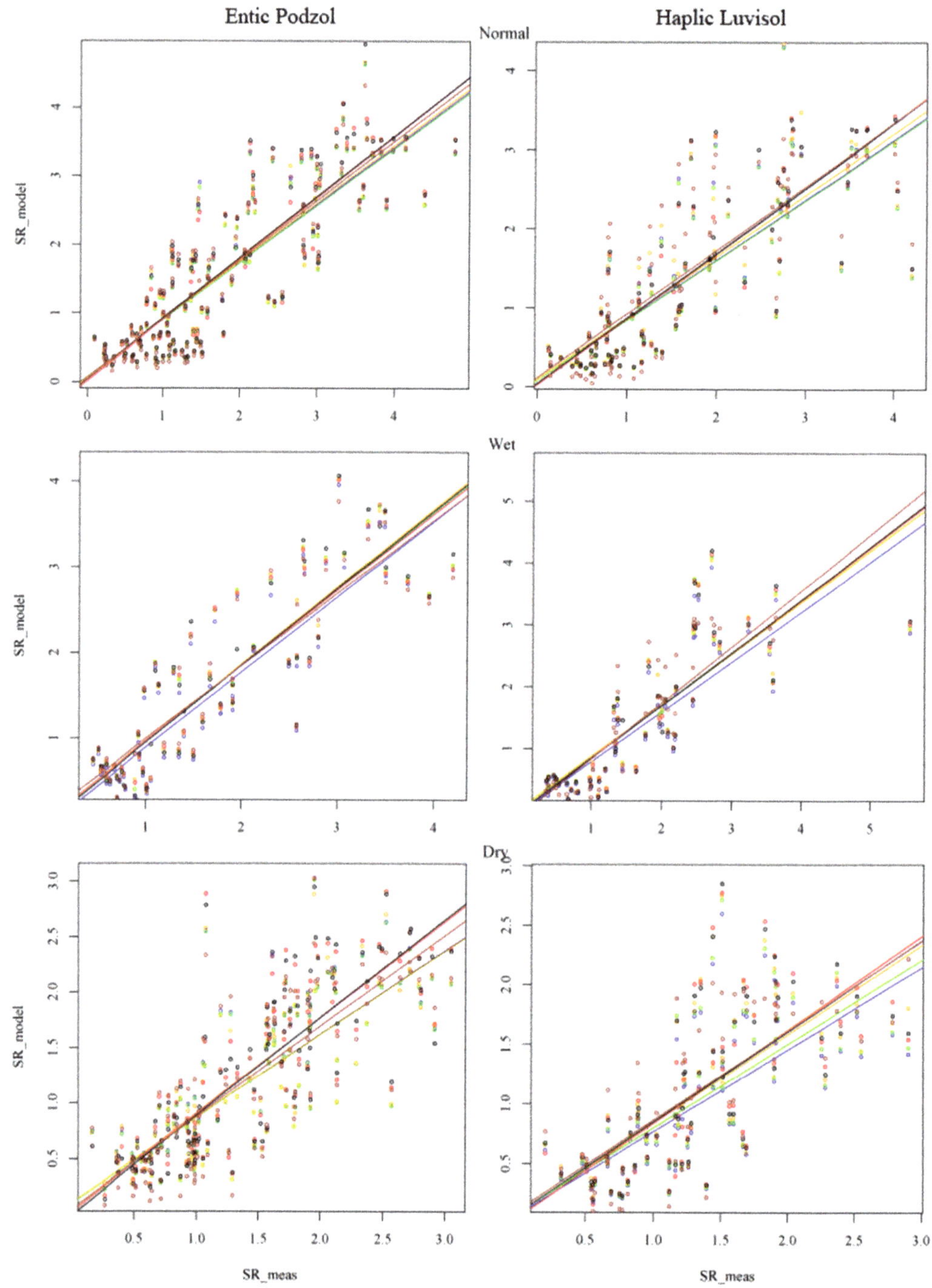

Figure 6. The comparison between the modeled (SR_model) and measured (SR_meas) SR values ($gCm^{-2}day^{-1}$) for T (blue, Equation (1)) –TP (green, Equation (2))–TPP (orange, Equation (3a))–TPC (red, Equation (4))–TPPC (black, Equation (5))–TPPrh (brown, Equation (3b)) models; lines—linear regression to determine the slope and intercept values; T_{air} (monitoring station) used for parameterization.

For Entic Podzol and T_{air}:

- The best slope lm values (slope, Figure 7) were observed with the TPPC model in dry and wet environments (slope $\approx$ 0.88–0.9), and with the TPPrh model in a normal environment (slope $\approx$ 0.88);
- The best R^2-lm values (R^2, Figure 7) were observed with the TPPC model for all moisture conditions: $R^2 \approx 0.67$ for dry, $R^2 \approx 0.77$ for wet, and $R^2 \approx 0.74$ for normal;
- The best MBE values (|MBE|, Figure 7) were observed with the TPPC model in normal and dry environments (|MBE| $\approx$ 0.15), while the TPP model gives the smallest |MBE| $\approx$ 0.11 in a wet environment;
- The best RMSE values (RMSE, Figure 7) were observed with the TPPC model for all moisture conditions: RMSE $\approx$ 0.47 for dry, RMSE $\approx$ 0.53 for wet, and RMSE $\approx$ 0.63 for normal.

For Haplic Luvisol and T_{soil}:

- The best slope-lm values (slope, Figure 7) were observed with the TPPrh for all moisture conditions (slope $\approx$ 0.85–0.9);
- The best R^2-lm values (R^2, Figure 7) were observed with the TPPrh for all moisture conditions ($R^2 \approx$ 0.65–0.75);
- The best MBE values (|MBE|, Figure 7) were observed with the TPPrh for all moisture conditions (|MBE| $\approx$ 0.15);
- The best RMSE values (RMSE, Figure 7) were observed with the TPPrh for all moisture conditions (RMSE $\approx$ 0.43–0.53).

For Haplic Luvisol and T_{air}:

- The best slope-lm values (slope, Figure 7) were observed with the TPPrh model in dry and wet environments (slope $\approx$ 0.85–0.91), and with the TPC and TPPC models in a normal environment (slope $\approx$ 0.85);
- The best R^2-lm values (R^2, Figure 7) were observed with the TPPrh for all moisture conditions ($R^2 \approx$ 0.57–0.73);
- The best MBE values (|MBE|, Figure 7) were observed with the TPPrh model in normal and wet environments (|MBE| $\approx$ 0.15–0.23), and with the TPPC model in a dry environment (|MBE| $\approx$ 0.23);
- The best RMSE values (RMSE, Figure 7) were observed with the TPPrh for all moisture conditions (RMSE $\approx$ 0.53–0.73).

From the conducted analysis, we see that the *SOC* and water-holding abilities are critical for the choice of optimal SR models. The TPPrh model with the quadratic dependency on the temperature becomes more optimal in most of the environmental conditions—normal, dry, and wet—for Haplic Luvisol having the finer texture (siltic), meaning lower permittivity to gasses [36], and the ability to hold larger amounts of water in comparison to Entic Podzol (Table 1) and for a longer time period. On the other hand, the TPPC model looks more optimal for sandy Entic Podzol, for which the weak water-holding ability leads to lack of water in the dry periods and brings forward the presence of *SOC*—as the substrate for the microbial community—to support SR when precipitation occurs.

These conclusions are also supported by the comparison of the mean annual SR measurements with modeled values (Figure 8). While the TPPrh model shows a better annual performance for T_{soil}, for T_{air}, the TPPC model becomes slightly better for Entic Podzol—the respective SR measured and modeled values stay within the standard deviation ranges of each other. However, it should be noted that these conclusions based on the annual means can be biased toward the larger summer SR values, underestimating the effects of the smaller winter SR. The underestimation of the mean annual SR is also in agreement with the fact that the model-comparison slopes are smaller than unity for both Entic Podzol (Figure 5) and Haplic Luvisol (Figure 6), causing a possible underestimation of the larger summer-time SR values. In the next section, we will see that the lower winter-time SR values are actually adequately modeled by our procedure and should not influence the respective model behavior showed in Figure 8.

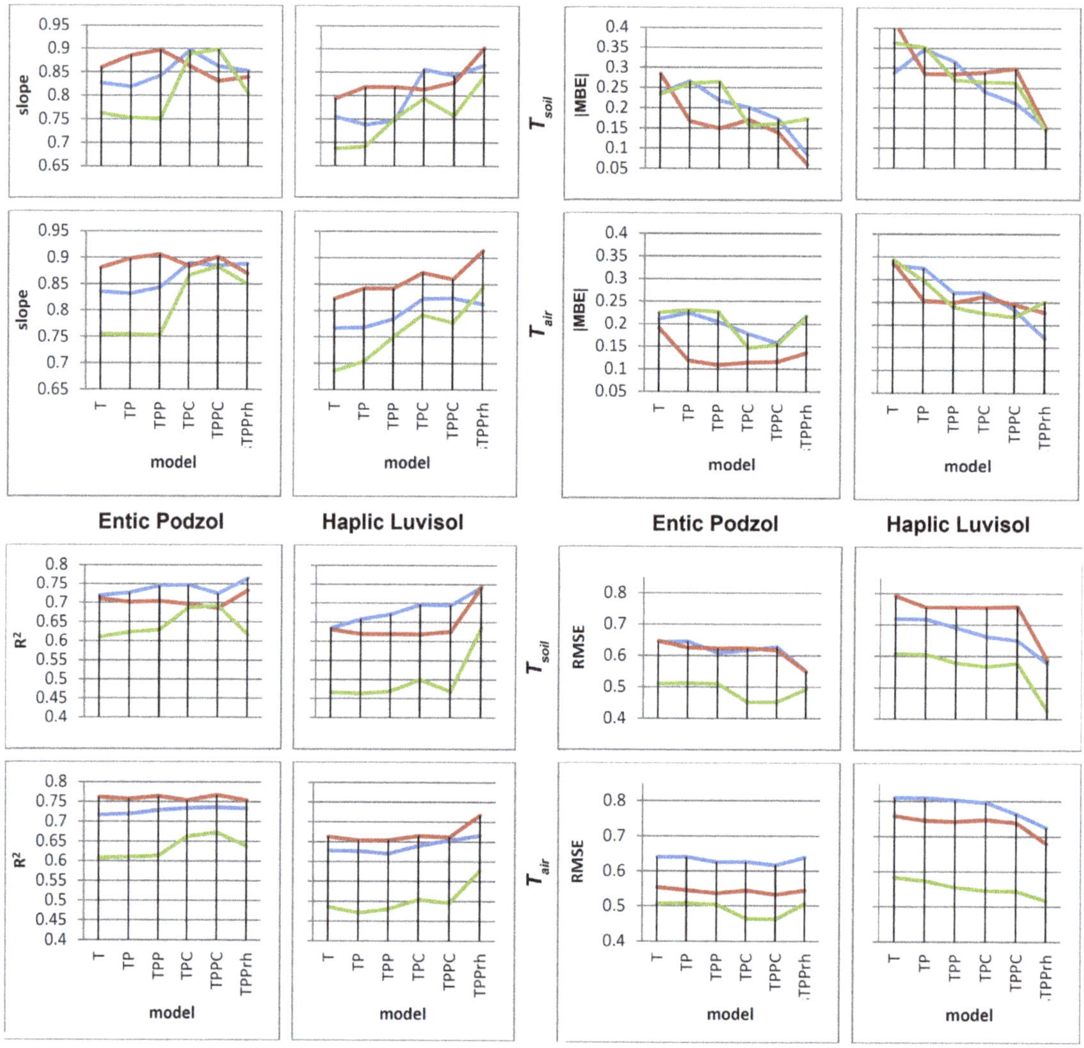

Figure 7. The comparison-statistics: slope—the slope of the lm; R^2—the determination coefficient of the lm; |MBE|—the absolute mean-bias error; and RMSE—the root-mean-square error between the modeled and measures soil respiration values for the normal (**blue**), wet (**red**), and dry (**green**) environmental conditions; in each panel: top—T_{soil} (local), bottom—T_{air} (monitoring station) used for parameterizations; in each panel: left—for Entic Podzol, right—for Haplic Luvisol.

3.3. An Optimal-Model Selection and the Winter Soil Respiration Control

The obtained above conclusions on the quality of the SR models parameterized with the different temperature sources (T_{soil} and T_{air}) for Entic Podzol (Figure 9) and Haplic Luvisol (Figure 10) are well illustrated by the time series generated from the TPPC (red) and TPPrh (blue) models. It should be noted that the TPPC model generates higher values than the TPPrh model and by this, the TPPC model better approximates the winter extremes but overestimates the summer SR pikes. It can be seen that an application of T_{soil} for the model parameterization allows more accurate estimations of the winter SR (both models) and reduces overestimations in summer (both models), improving SR estimates in the warm periods.

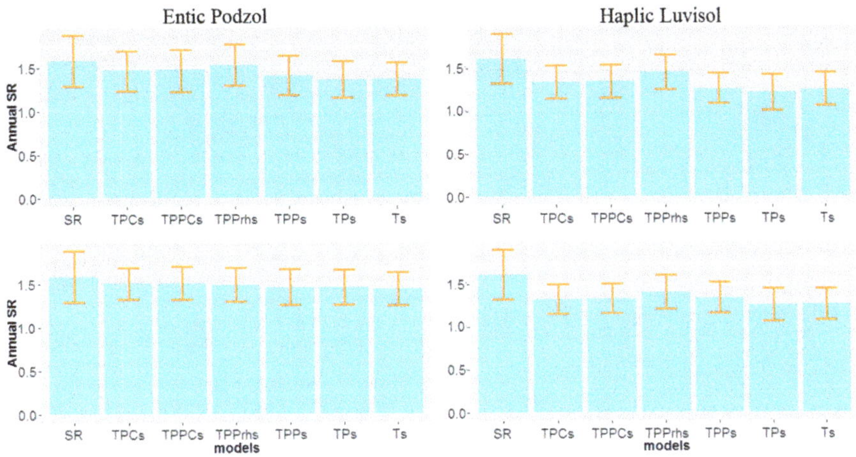

Figure 8. The mean annual SR (gC m^{-2} day^{-1}) over 25-year-long periods—blue bars—measured (SR) and modeled (labels on the x-axis) with the standard deviation ranges (orange bars) of the individual-year distributions of the mean annual SR; top—T_{soil} (local), bottom—T_{air} (monitoring station).

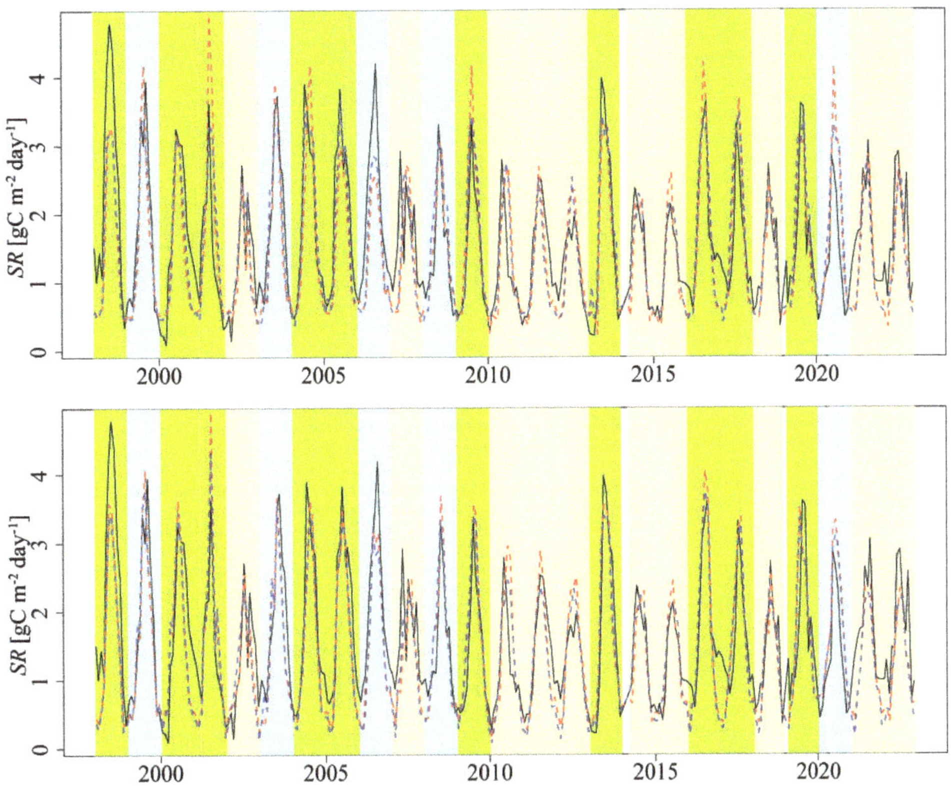

Figure 9. The SR time series modeling for Entic Podzol: SR measurements—black line, TPPC model—dashed red line, and TPPrh model—dashed blue line; top—for T_{soil} (local) and bottom—for T_{air} (monitoring station); green—normal, blue—wet, beige—dry years.

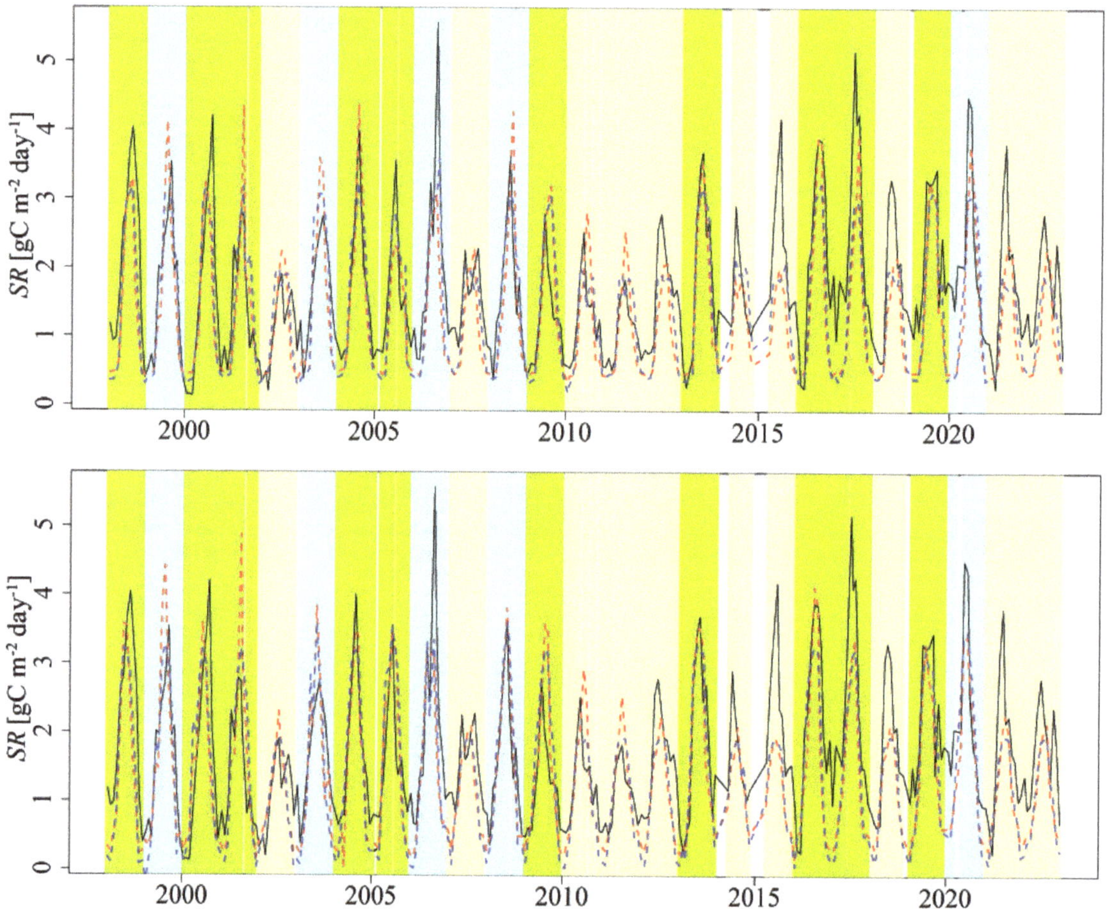

Figure 10. The SR time series modeling for Haplic Luvisol: SR measurements—black line, TPPC model—dashed red line, and TPPrh model—dashed blue line; top—for T_{soil} (local) and bottom—for T_{air} (monitoring station); green—normal, blue—wet, beige—dry years.

As an explanation of such model behavior, we can point to the soil condition difference between the winter and summer periods: when there is enough water in winters, the SR becomes more dependent on the T_{soil} interconnected with microbial activity and *SOC* presence as a substrate for the microbial community [24], while in dryer summer periods, the presence of precipitation and the respective temperature and humidity changes though the evapotranspiration of vegetation [60,61] start playing an important role in the soil water balance and SR regulation. All those mentioned above are illustrated by Figure 4 (Section 3.1) highlighting different regimes of dependency between the T_{soil} and T_{air} in the cold and warm times of the year—blue and red lines, respectively.

We combine the TPPC and TPPrh models by the regime-change condition ($T_{soil} \approx 2$ °C) from Figure 4, comparing the soil and air temperatures:

- with the $T_{soil} > 2$ °C—choose the TPPrh model;
- with the $T_{soil} \leq 2$ °C—choose the TPPC model.

For the comparison analysis of the combined TPPC, the TPPrh models (Table 2) show the best statistical values for R^2, MBE, and RMSE for the following model parameterizations:

All data: for the combined TPPC[T_{soil}], the TPPrh[T_{soil}] model parameterized by the soil temperature;

Cold period: for the TPPC[T_{soil}] model parameterized by the soil temperature;

Warm period: for the TPPrh[T_{soil}] or TPPrh[T_{air}] model parameterized by the soil or the air temperature.

Table 2. Quality control of the modeling by the combined TPPC: TPPrh models with the parameterization by different temperature sources ([T_{soil}] or [T_{air}]) conducted on the 25-year monitoring for Entic Podzol and Haplic Luvisol; top—all data, middle—cold periods ($T_{soil} \leq 2$ °C), bottom—warm periods ($T_{soil} > 2$ °C).

	Entic Podzol			Haplic Luvisol		
Model	R^2	**MBE**	**RMSE**	R^2	**MBE**	**RMSE**
		(all data)				
TPPC[T_{soil}]:TPPrh[T_{air}]	**0.734**	−0.150	**0.527**	0.624	−0.348	0.716
TPPC[T_{air}]:TPPrh[T_{air}]	0.731	−0.156	0.536	0.623	−0.357	0.723
TPPC[T_{soil}]:TPPrh[T_{soil}]	**0.735**	**−0.115**	**0.524**	**0.674**	**−0.287**	**0.651**
		$T_{soil} \leq 2$ (cold periods)				
TPPC[T_{soil}]	**0.116**	**−0.225**	**0.397**	**0.054**	**−0.376**	**0.553**
TPPC[T_{air}]	0.110	−0.241	0.428	0.047	−0.402	0.580
TPPrh[T_{soil}]	0.032	−0.224	0.411	0.110	−0.425	0.581
TPPrh[T_{air}]	0.070	−0.288	0.480	0.040	−0.456	0.643
		$T_{soil} > 2$ (warm periods)				
TPPC[T_{soil}]	0.583	−0.124	0.638	0.465	−0.413	0.852
TPPC[T_{air}]	0.616	−0.094	0.599	0.431	−0.412	0.856
TPPrh[T_{soil}]	**0.604**	**−0.051**	**0.584**	**0.512**	**−0.239**	**0.698**
TPPrh[T_{air}]	**0.604**	**−0.106**	**0.589**	0.431	−0.333	0.790

It should be noted that the (intercept → 0+) approach which we developed guarantees an adequate estimation of the cold-period SR and a good magnitude resolution of the model results (confirmed by Figure 8) in comparison to the measurements—an often observed inefficiency of the standard parameterization approaches [20,31,42].

For Haplic Luvisol (Figure 10), the SR measurements are well represented by the modeling values before the 2015 year; however, in the later period, the winter-time SR values are too large due to changes in the forest structure (tree fall down) that year, making soil more accessible to elements [62].

The low R^2 values for the cold period (Table 2) are directly associated with the high variability of the observations (see Figure 3) which typically occur during the winter-time measurements due to snow presence on the ground and freezing–thawing cycles and also due to changes in precipitation causing CO_2 accumulation in soil and interrupting gas exchange.

4. Conclusions

We have demonstrated an importance to account for the *SOC* storage and water-holding ability for the selection of the proper SR models. For the sandy Entic Podzol with a coarse texture and good drainage, it is better to use the models accounting for the *SOC* storage (TPC and TPPC), while for the loamy Haplic Luvisol, having a finer texture and high water-holding ability, it is better to apply Raich–Hashimoto-type models (TPPrh) with the quadratic temperature dependency being connected to the water presence in the soil and the reaction of the microbial biomass on temperature change.

Both in the dry and in the normal years, accounting for *SOC* storage significantly improves the modeling results (TPC and TPPC models) in comparison to the more simple model results (T, TP, and TPP models). In the dry years, the TPPC model is better than the TPC model, but in the normal years, the TPC model is better than the TPPC model, highlighting an importance of the prolonged presence of soil humidity in dry periods.

An effect of humidity change becomes the most important in the wet years (TPP и TPPC models). Optimal values for the parameter α included into the TPP and TPPC models become close to zero or negative in the dry years, highlighting an importance of the (continuous) moisture and precipitation presence from the previous month (PP).

We found that TPPC parameterized by the soil temperature adequately describes the SR measured during the cold periods ($T_{soil} \leq 2$ °C), whereas TPPrh parameterized by the soil or air temperature is better for describing the SR measured during the warm periods ($T_{soil} > 2$ °C).

The parameterization of the models with the soil temperature is shown to be an important factor for adequate SR estimates. With this parameterization, the TPPC model can be applied for the control of the winter-time SR measurements conducted at the sites.

Author Contributions: Conceptualization, I.K. and S.K.; methodology, I.K. and S.K.; formal analysis, S.K.; investigation, S.K.; resources, V.L.d.G., D.K., T.M. and D.S.; data curation, V.L.d.G., D.K. and S.K.; writing—original draft preparation, S.K.; writing—review and editing, I.K., K.I. and S.K.; visualization, D.K., V.L.d.G. and S.K.; supervision, I.K. and K.I.; project administration, I.K. and K.I.; funding acquisition, I.K. and K.I. All authors have read and agreed to the published version of the manuscript.

Funding: The data collection and preparation were carried out as part of the most important innovative project of national importance, "Development of a system for ground-based and remote monitoring of carbon pools and greenhouse gas fluxes in the territory of the Russian Federation, ensuring the creation of recording data systems on the fluxes of climate-active substances and the carbon budget in forests and other terrestrial ecological systems" (Registration number: 123030300031-6); the data analysis and modeling works were supported by state assignment No. 122111000095-8.

Data Availability Statement: The data presented in this study are available on reasonable request from the authors.

Acknowledgments: We give thanks to Vera Ableeva (Station of Background Monitoring, Roshydromet) who provides us the meteorological data set (1973 to 2021) for the study area. We especially thank Daniela Dalmonech for her comments and suggestions improving the objectivity of the better model selection process and anonymous reviewers for their constructive comments.

Conflicts of Interest: The authors declare no conflict of interest.

Appendix A

The results of parameterizations of the models and the comparisons between the soil respiration measurements and modeling results for the Entic Podzol (Tables A1 and A3) and Haplic Luvisol (Tables A2 and A4); T_{soil} or T_{air} was used for the model parameterizations.

Table A1. Parameters (R_0, Q, Q_2, K, α, ψ) of the models (T, TP, TPP, TPC, TPPC, TPPrh) and the comparisons with the measurements (|MBE|, RMSE, slope, intercept, R^2) for the Entic Podzol and forest ecosystem in normal (n), wet (w), and dry (d) years; T_{soil} was used for the model parameterizations.

Model	Wetness	R_0	Q	Q_2	K	α	ψ	Slope	Intercept	|MBE|	RMSE	R^2
T	n	0.545	0.118	-	-	-	-	0.827	0.063	0.237	0.643	0.720
TP	n	0.545	0.118	-	0.901	-	-	0.819	0.049	0.266	0.644	0.726
TPP	n	0.545	0.118	-	−0.941	1.137	-	0.843	0.054	0.219	0.609	0.745

Table A1. *Cont.*

Model	Wetness	R_0	Q	Q_2	K	α	ψ	Slope	Intercept	|MBE|	RMSE	R^2
TPC	n	0.545	0.118	-	6.838	-	−0.179	0.897	−0.022	0.200	0.617	0.747
TPPC	n	0.545	0.118	-	−0.571	1.753	−0.043	0.863	0.066	0.172	0.628	0.724
TPPrh	n	0.545	0.197	0.005	−1.694	2.266	-	0.853	0.172	0.083	0.549	0.765
T	w	0.508	0.121	-	-	-	-	0.860	−0.044	0.284	0.645	0.711
TP	w	0.508	0.121	-	−4.341	-	-	0.885	0.029	0.167	0.625	0.702
TPP	w	0.508	0.121	-	−4.954	0.072	-	0.897	0.026	0.149	0.623	0.704
TPC	w	0.508	0.121	-	−5.869	-	0.042	0.863	0.063	0.171	0.623	0.697
TPPC	w	0.508	0.121	-	−11.137	0.330	0.148	0.831	0.152	0.138	0.617	0.686
TPPrh	w	0.508	0.188	0.005	−5.843	0.119	-	0.840	0.214	0.060	0.548	0.732
T	d	0.526	0.094	-	-	-	-	0.763	0.095	0.233	0.511	0.610
TP	d	0.526	0.094	-	0.864	-	-	0.753	0.082	0.261	0.512	0.623
TPP	d	0.526	0.094	-	0.734	1.157	-	0.751	0.081	0.265	0.510	0.629
TPC	d	0.526	0.094	-	12.374	-	−0.353	0.889	−0.003	0.157	0.451	0.688
TPPC	d	0.526	0.094	-	20.012	0.870	−0.440	0.898	−0.019	0.161	0.452	0.692
TPPrh	d	0.526	0.094	0.005	0.734	1.157	-	0.805	0.097	0.174	0.493	0.617

Table A2. Parameters (R_0, Q, Q_2, K, α, ψ) of the models (T, TP, TPP, TPC, TPPC, TPPrh) and the comparisons with the measurements (|MBE|, RMSE, slope, intercept, R^2) for the Haplic Luvisol and forest ecosystem in normal (n), wet (w), and dry (d) years; T_{soil} was used for the model parameterizations.

Model	Wetness	R_0	Q	Q_2	K	α	ψ	Slope	Intercept	|MBE|	RMSE	R^2
T	n	0.448	0.119	-	-	-	-	0.755	0.112	0.287	0.718	0.635
TP	n	0.448	0.119	-	2.179	-	-	0.738	0.079	0.347	0.717	0.658
TPP	n	0.448	0.119	-	0.181	1.129	-	0.747	0.097	0.315	0.690	0.671
TPC	n	0.448	0.119	-	10.960	-	−1.050	0.856	−0.006	0.239	0.662	0.696
TPPC	n	0.448	0.119	-	4.537	1.099	−0.767	0.843	0.044	0.212	0.650	0.694
TPPrh	n	0.448	0.238	0.007	7.495	1.036	-	0.865	0.070	0.150	0.577	0.742
T	w	0.432	0.122	-	-	-	-	0.794	−0.055	0.416	0.792	0.631
TP	w	0.432	0.122	-	−5.128	-	-	0.818	0.035	0.285	0.756	0.619
TPP	w	0.432	0.122	-	−5.151	0.995	-	0.818	0.035	0.285	0.756	0.619
TPC	w	0.432	0.122	-	−5.316	-	0.030	0.813	0.039	0.288	0.756	0.618
TPPC	w	0.432	0.122	-	−2.683	1.298	−0.124	0.828	0.004	0.298	0.757	0.626
TPPrh	w	0.432	0.239	0.007	−0.030	1.537	-	0.902	0.022	0.150	0.587	0.742
T	d	0.408	0.093	-	-	-	-	0.687	0.057	0.363	0.607	0.467
TP	d	0.408	0.093	-	−0.356	-	-	0.691	0.061	0.352	0.605	0.463
TPP	d	0.408	0.093	-	−2.952	0.037	-	0.751	0.064	0.269	0.578	0.469
TPC	d	0.408	0.093	-	6.617	-	−1.147	0.794	0.012	0.264	0.567	0.500
TPPC	d	0.408	0.093	-	−1.954	0.009	−0.180	0.757	0.064	0.262	0.577	0.469
TPPrh	d	0.408	0.221	0.008	−3.074	0.384	-	0.842	0.069	0.143	0.426	0.635

Table A3. Parameters (R_0, Q, Q_2, K, α, ψ) of the models (T, TP, TPP, TPC, TPPC, TPPrh) and the comparisons with the measurements (|MBE|, RMSE, slope, intercept, R^2) for the Entic Podzol and forest ecosystem in normal (n), wet (w), and dry (d) years; T_{air} (meteostation) was used for parameterization.

Model	Wetness	R_0	Q	Q_2	K	α	ψ	Slope	Intercept	|MBE|	RMSE	R^2
T	n	0.686	0.087	-	-	-	-	0.836	0.074	0.211	0.640	0.717
TP	n	0.686	0.087	-	0.384	-	-	0.832	0.067	0.224	0.640	0.720
TPP	n	0.686	0.087	-	−0.473	1.135	-	0.844	0.066	0.205	0.625	0.729
TPC	n	0.686	0.087	-	4.362	-	−0.128	0.890	0.013	0.177	0.626	0.734
TPPC	n	0.686	0.087	-	1.001	1.122	−0.079	0.885	0.042	0.157	0.616	0.736
TPPrh	n	0.618	0.121	0.002	0.273	1.128	-	0.889	−0.025	0.218	0.639	0.734
T	w	0.724	0.082	-	-	-	-	0.881	0.014	0.191	0.553	0.762
TP	w	0.724	0.082	-	−2.618	-	-	0.898	0.056	0.118	0.544	0.757
TPP	w	0.724	0.082	-	−2.171	1.331	-	0.906	0.052	0.109	0.536	0.764
TPC	w	0.724	0.082	-	−4.089	-	0.032	0.884	0.083	0.114	0.544	0.753
TPPC	w	0.724	0.082	-	−1.114	−2.042	−0.055	0.901	0.054	0.116	0.532	0.766
TPPrh	w	0.651	0.095	0.001	−4.949	1.249	-	0.870	0.087	0.136	0.544	0.753
T	d	0.649	0.063	-	-	-	-	0.755	0.114	0.225	0.507	0.608
TP	d	0.649	0.063	-	0.167	-	-	0.754	0.111	0.231	0.507	0.611
TPP	d	0.649	0.063	-	−0.750	2.586	-	0.753	0.116	0.227	0.503	0.613
TPC	d	0.649	0.063	-	8.757	-	−0.291	0.867	0.038	0.147	0.464	0.662
TPPC	d	0.649	0.063	-	20.364	0.763	−0.435	0.883	0.009	0.153	0.462	0.672
TPPrh	d	0.584	0.110	0.002	0.084	1.234	-	0.850	−0.008	0.216	0.506	0.637

Table A4. Parameters (R_0, Q, Q_2, K, α, ψ) of the models (T, TP, TPP, TPC, TPPC, TPPrh) and the comparisons with the measurements (|MBE|, RMSE, slope, intercept, R^2) for the Haplic Luvisol and forest ecosystem in normal (n), wet (w), and dry (d) years; T_{air} (meteostation) was used for parameterization.

Model	Wetness	R_0	Q	Q_2	K	α	ψ	Slope	Intercept	|MBE|	RMSE	R^2
T	n	0.538	0.100	-	-	-	-	0.767	0.103	0.331	0.810	0.628
TP	n	0.538	0.100	-	−0.204	-	-	0.768	0.107	0.325	0.809	0.627
TPP	n	0.538	0.100	-	−2.136	0.440	-	0.784	0.133	0.270	0.803	0.621
TPC	n	0.538	0.100	-	3.588	-	−0.522	0.823	0.058	0.272	0.795	0.640
TPPC	n	0.538	0.100	-	−0.946	−0.933	−0.202	0.824	0.094	0.233	0.764	0.654
TPPrh	n	0.538	0.175	0.004	−0.346	2.364	-	0.813	0.178	0.170	0.724	0.667
T	w	0.611	0.093	-	-	-	-	0.823	−0.007	0.338	0.758	0.662
TP	w	0.611	0.093	-	−3.108	-	-	0.842	0.041	0.254	0.745	0.654
TPP	w	0.611	0.093	-	−2.787	1.268	-	0.842	0.046	0.249	0.742	0.655
TPC	w	0.611	0.093	-	1.107	-	−0.318	0.872	−0.023	0.263	0.747	0.665
TPPC	w	0.611	0.093	-	−0.242	2.004	−0.232	0.860	0.018	0.245	0.739	0.662
TPPrh	w	0.611	0.158	0.003	3.576	−0.618	-	0.914	−0.066	0.227	0.680	0.718

Table A4. *Cont.*

Model	Wetness	R_0	Q	Q_2	K	α	ψ	Slope	Intercept	|MBE|	RMSE	R^2
T	d	0.530	0.065	-	-	-	-	0.686	0.078	0.343	0.584	0.485
TP	d	0.530	0.065	-	−1.484	-	-	0.704	0.098	0.298	0.574	0.470
TPP	d	0.530	0.065	-	−3.552	0.156	-	0.750	0.096	0.239	0.555	0.480
TPC	d	0.530	0.065	-	3.968	-	−0.969	0.792	0.053	0.225	0.545	0.504
TPPC	d	0.530	0.065	-	−0.013	−0.432	−0.595	0.777	0.081	0.218	0.543	0.495
TPPrh	d	0.530	0.126	0.003	−0.011	−0.432	-	0.844	−0.043	0.251	0.516	0.578

References

1. Ryan, M.; Law, B. Interpreting, measuring, and modeling soil respiration. *Biogeochemistry* **2005**, *73*, 3–27. [CrossRef]
2. Le Quéré, C.; Moriarty, R.; Andrew, R.M.; Peters, G.P.; Ciais, P.; Friedlingstein, P.; Jones, S.D.; Sitch, S.; Tans, P.; Arneth, A.; et al. Global carbon budget 2014. *Earth Syst. Sci. Data* **2015**, *7*, 47–85. [CrossRef]
3. Valentini, R.; Matteucci, G.; Dolman, A.J.; Schulze, E.D.; Rebmann, C.; Moors, E.J.; Granier, A.; Gross, P.; Jensen, N.O.; Pilegaard, K.; et al. Respiration as the main determinant of carbon balance in European forests. *Nature* **2000**, *404*, 861–865. [CrossRef] [PubMed]
4. Peltoniemi, M.; Thürig, E.; Ogle, S.; Palosuo, T.; Schrumpf, M.; Wutzler, T.; Butterbach-Bahl, K.; Chertov, O.; Komarov, A.; Mikhailov, A.; et al. Models in country scale carbon accounting of forest soils. *Silva Fenn.* **2007**, *41*, 575–602. [CrossRef]
5. Lovett, G.M.; Cole, J.J.; Pace, M.L. Is Net Ecosystem Production Equal to Ecosystem Carbon Accumulation? *Ecosystems* **2006**, *9*, 152–155. [CrossRef]
6. Chapin, F.S., III; Woodwell, G.M.; Randerson, J.T.; Rastetter, E.B.; Lovett, G.M.; Baldocchi, D.D.; Clark, D.A.; Harmon, M.E.; Schimel, D.S.; Valentini, R.; et al. Reconciling Carbon-Cycle Concepts, Terminology, and Methods. *Ecosystems* **2006**, *9*, 1041–1050. [CrossRef]
7. UNFCCC. National Inventory Submissions. 2006. Available online: http://unfccc.int/national_reports/annex_i_ghg_inventories/national_inventories_submissions/items/3734.php (accessed on 1 June 2023).
8. Singh, B.K. (Ed.) *Soil Carbon Storage: Modulators, Mechanisms and Modeling*; Academic Press: London, UK, 2018.
9. Kirschbaum, M.U.F.; Eamus, D.; Gifford, R.M.; Roxburgh, G.H.; Sands, P.J. Definitions of Some Ecological Terms Commonly Used in Carbon Accounting. In *Net Ecosystem Exchange Workshop Proceedings CRC for Greenhouse Accounting*; Kirschbaum, M.U.F., Mueller, R., Eds.; Canberra Act: Canbera, Australia, 2001; pp. 2–5.
10. Aubinet, M.; Vesala, T.; Papale, D. (Eds.) *Eddy Covariance*; Springer: Dordrecht, The Netherlands, 2012.
11. Burba, G. *Eddy Covariance Method for Scientific, Industrial, Agricultural, and Regulatory Applications: A Field Book on Measuring Ecosystem Gas Exchange and Areal Emission Rates*; LI-COR Biosciences: Lincoln, NE, USA, 2013.
12. Baldocchi, D.D. Measuring fluxes of trace gases and energy between ecosystems and the atmosphere—The state and future of the eddy covariance method. *Glob. Chang. Biol.* **2014**, *20*, 3600–3609. [CrossRef]
13. Baldocchi, D.D.; Chu, H.; Reichstein, M. Inter-annual variability of net and gross ecosystem carbon fluxes: A review. *Agric. For. Meteorol.* **2017**, *249*, 520–533. [CrossRef]
14. Reichstein, M.; Falge, E.; Baldocchi, D.; Papale, D.; Aubient, M.; Berbiger, P.; Bernhofer, C.; Buchmann, N.; Gilmanov, T.; Garnier, A.; et al. On the separation of net ecosystem exchange into assimilation and ecosystem respiration: Review and improved algorithm. *Glob. Chang. Biol.* **2005**, *11*, 1424–1439. [CrossRef]
15. Kutsch, W.; Bahn, M.; Heinemeyer, A. *Soil Carbon Dynamics*; Cambridge University Press: Cambridge, UK, 2009.
16. Vekuri, H.; Tuovinen, J.P.; Kulmala, L.; Papale, D.; Kolari, P.; Aurela, M.; Laurila, T.; Liski, J.; Lohila, A. A widely-used eddy covariance gap-filling method creates systematic bias in carbon balance estimates. *Sci. Rep.* **2023**, *13*, 1720. [CrossRef]
17. Hanson, P.J.; Edwards, N.T.; Garten, C.T.; Andrews, J.A. Separating root and soil microbial contributions to soil respiration: A review of methods and observations. *Biogeochemistry* **2000**, *48*, 115–146. [CrossRef]
18. Lopes De Gerenyu, V.O.; Kurganova, I.N.; Rozanova, L.N.; Kudeyarov, V.N. Annual emission of carbon dioxide from soils of the southern taiga zone of Russia. *Eurasian Soil Sci.* **2001**, *34*, 931–944.
19. Kurganova, I.; Lopes de Gerenyu, V.; Rozanova, L.; Sapronov, D.; Myakshina, T.; Kudeyarov, V. Annual and seasonal CO_2 fluxes from Russian southern taiga soils. *Tellus B Chem. Phys. Meteorol.* **2003**, *55*, 338–344. [CrossRef]
20. Kurganova, I.N.; Lopes de Gerenyu, V.O.; Myakshina, T.N.; Sapronov, D.V.; Romashkin, I.V.; Zhmurin, V.A.; Kudeyarov, V.N. Native and model assessment of Respiration of forest sod-podzolic soil in Prioksko-Terrasny Biospheric Reserve. *Contemp. Probl. Ecol.* **2020**, *13*, 813–824. [CrossRef]
21. Kurganova, I.N.; Lopes de Gerenyu, V.O.; Khoroshaev, D.A.; Blagodatskaya, E. Effect of snowpack pattern on cold-season CO_2 efflux from soils under temperate continental climate. *Geoderma* **2017**, *304*, 28–39. [CrossRef]

22. Kurganova, I.N.; Lopes de Gerenyu, V.O.; Khoroshaev, D.A.; Myakshina, T.N.; Sapronov, D.V.; Zhmurin, V.A.; Kudeyarov, V.N. Analysis of the Long-Term Soil Respiration Dynamics in the Forest and Meadow Cenoses of the Prioksko-Terrasny Biosphere Reserve in the Perspective of Current Climate Trends. *Eurasian Soil Sci.* **2020**, *53*, 1421–1436. [CrossRef]
23. Reichstein, M.; Rey, A.; Freibauer, A.; Tenhunen, J.; Valnetini, R.; Banza, J.; Caslas, P.; Cheng, Y.; Grunzweig, J.M.; Irvine, J.; et al. Modeling temporal and large-scale spatial variability of soil respiration from soil water availability, temperature and vegetation productivity indices. *Glob. Biogeochem. Cycles* **2003**, *17*, 1104. [CrossRef]
24. Raich, J.W.; Potter, C.S. Global patterns of carbon dioxide emission from soils. *Glob. Biogeochem. Cycles* **1995**, *9*, 23–36. [CrossRef]
25. Hashimoto, S.; Carvalhais, N.; Ito, A.; Migliavacca, M.; Nishina, K.; Reichstein, M. Global spatiotemporal distribution of soil respiration modeled using a global database. *Biogeosciences* **2015**, *12*, 4121–4132. [CrossRef]
26. Pavelka, M.; Darenova, E.; Dusek, J. Modeling of soil CO_2 efflux during water table fluctuation based on in situ measured data from a sedge-grass marsh. *Appl. Ecol. Environ. Res.* **2016**, *14*, 423–437. [CrossRef]
27. Komarov, A.S.; Chertov, O.G.; Nadporozhskaya, M.A.; Priputina, I.V. *Models of Forest Soil Organic Matter Dynamics*; Kudeyarov, V.N., Ed.; Nauka: Moscow, Russia, 2007; 380p. (In Russian)
28. Chertov, O.G.; Nadporozhskaya, M.A. Models of soil organic matter dynamics: Problems and perspectives. *Comput. Res. Model.* **2016**, *8*, 391–399. (In Russian) [CrossRef]
29. Priputina, I.V.; Bykhovets, S.S.; Frolov, P.V.; Chertov, O.G.; Kurganova, I.N.; Lopes de Gerenyu, V.O.; Sapronov, D.V.; Myakshina, T.N. Application of Mathematical Models ROMUL and Romul_Hum for Estimating CO_2 Emissions and Dynamics of Organic Matter in Albic Luvisol under Deciduous Forest in Southern Moscow Region. *Eurasian Soil Sci.* **2020**, *53*, 1480–1491. [CrossRef]
30. Lopes De Gerenyu, V.O.; Kurganova, I.N.; Rozanova, L.N.; Kudeyarov, V.N. Effect of temperature and moisture on CO_2 evolution rate of cultivated Phaeozem: Analyses of long-term field experiment. *Plant Soil Environ.* **2005**, *51*, 213–219. [CrossRef]
31. Juhász, C.; Huzsvai, L.; Kovács, E.; Kovács, G.; Tuba, G.; Sinka, L.; Zsembeli, J. Carbon Dioxide Efflux of Bare Soil as a Function of Soil Temperature and Moisture Content under Weather Conditions of Warm, Temperate, Dry Climate Zone. *Agronomy* **2022**, *12*, 3050. [CrossRef]
32. Howard, P.J.A.; Howard, D.M. Respiration of decomposing litter in relation to temperature and moisture. *Oikos* **1979**, *33*, 457465. [CrossRef]
33. O'Connell, A.M. Microbial Decomposition (Respiration) of Litter in Eucalypt Forests of South-Western Australia: An Empirical Model Based on Laboratory Incubations. *Soil Biol. Biochem.* **1990**, *22*, 153–160. [CrossRef]
34. Orchard, V.A.; Cook, F.J. Relationship between soil respiration and soil moisture. *Soil Biol. Biochem.* **1983**, *15*, 447–453. [CrossRef]
35. Jarvis, P.; Rey, A.; Petsikos, C.; Wingate, L.; Rayment, M.; Pereira, J.; Banza, J.; David, J.; Miglietta, F.; Borghetti, M.; et al. Drying and wetting of Mediterranean soils stimulates decomposition and carbon dioxide emission: The "Birch effect". *Tree Physiol.* **2007**, *27*, 929–940. [CrossRef]
36. Maier, M.; Schack-Kirchner, H.; Hildebrand, E.E.; Holst, J. Pore-space CO_2 dynamics in a deep, well-aerated soil. *Eur. J. Soil Sci.* **2010**, *61*, 877–887. [CrossRef]
37. Chamindu Deepagoda, T.K.K.; Elberling, B. Characterization of diffusivity-based oxygen transport in Arctic organic soil. *Eur. J. Soil Sci.* **2015**, *66*, 983–991. [CrossRef]
38. Vygodskaya, N.N.; Varlagin, A.V.; Kurbatova, Y.A.; Ol'chev, A.V.; Panferov, O.I.; Tatarinov, F.A.; Shalukhina, N.V. Response of taiga ecosystems to extreme weather conditions and climate anomalies. *Dokl. Biol. Sci.* **2009**, *429*, 571–574. [CrossRef] [PubMed]
39. Yuste, J.C.; Janssens, I.A.; Carrara, A.; Meiresonne, L.; Ceulemans, R. Interactive effects of temperature and precipitation on soil respiration in a temperate maritime pine forest. *Tree Physiol.* **2003**, *23*, 1263–1270. [CrossRef] [PubMed]
40. Kurganova, I.N.; Lopes De Gerenyu, V.O.; Myakshina, T.N.; Sapronov, D.V.; Savin, I.Y.; Shorohova, E.V. Carbon balance in forest ecosystems of southern part of Moscow region under a rising aridity of climate. *Contemp. Probl. Ecol.* **2017**, *10*, 748–760. [CrossRef]
41. Karelin, D.V.; Zamolodchikov, D.G.; Isaev, A.S. Unconsidered sporadic sources of carbon dioxide emission from soils in taiga forests. *Dokl. Biol. Sci.* **2017**, *475*, 165–168. [CrossRef]
42. Sukhoveeva, O.E.; Karelin, D.V. Assessment of Soil Respiration with the Raich–Hashimoto Model: Parameterisation and Prediction. Izvestia RAN. *Ser. Geogr.* **2022**, *86*, 519–527.
43. Kayranli, B.; Scholz, M.; Mustafa, A.; Hedmark, Å. Carbon Storage and Fluxes within Freshwater Wetlands: A Critical Review. *Wetlands* **2010**, *30*, 111–124. [CrossRef]
44. Makhnykina, A.V.; Polosukhina, D.A.; Kolosov, R.A.; Prokushkin, A.S. Seasonal Dynamics of CO_2 Emission from the Surface of the Raised Bog in Central Siberia. *Geosph. Res.* **2021**, *4*, 85–93. (In Russian) [CrossRef]
45. Xu, X.; Inubushi, K.; Sakamoto, K. Effect of vegetations and temperature on microbial biomass carbon and metabolic quotients of temperate volcanic forest soils. *Geoderma* **2006**, *136*, 310–319. [CrossRef]
46. Del Grosso, S.; Parton, W.; Mosier, A.; Holland, E.A. Modeling soil CO_2 emissions from ecosystems. *Biogeochemistry* **2005**, *73*, 71–91. [CrossRef]
47. Lal, R. Forest soils and carbon sequestration. *For. Ecol. Manag.* **2005**, *220*, 242–258. [CrossRef]
48. Chen, S.; Huang, Y.; Zou, J.; Shen, Q.; Hu, Z.; Qin, Y.; Chen, H.; Pan, G. Modeling interannual variability of global soil respiration from climate and soil properties. *Agric. For. Meteorol.* **2010**, *150*, 590–605. [CrossRef]
49. Ilyasov, D.V.; Molchanov, A.G.; Glagolev, M.V.; Suvorov, G.G.; Sirin, A.A. Modelling of carbon dioxide net ecosystem exchange of hayfield on drained peat soil: Land use scenario analysis. *Comput. Res. Model.* **2020**, *12*, 1427–1449. (In Russian) [CrossRef]

50. WRB 2014. 2015. Available online: https://www.fao.org/soils-portal/data-hub/soil-classification/world-reference-base/en/ (accessed on 1 June 2023).
51. Reddy, K.R.; DeLaune, R.D. *Biogeochemistry of Wetlands: Science and Applications*; CRC Press: Boca Raton, FL, USA, 2008.
52. Brown, R.B. *Soil Texture*; Agronomy Fact Sheet Series: Fact Sheet SL-29; Department of Crop and Soil Sciences, Cornell University: Ithaca, NY, USA, 2007.
53. Kurganova, I.; Lopes de Gerenyu, V.; Khoroshaev, D.; Myakshina, T.; Sapronov, D.; Zhmurin, V. Temperature Sensitivity of Soil Respiration in Two Temperate Forest Ecosystems: The Synthesis of a 24-Year Continuous Observation. *Forests* **2022**, *13*, 1374. [CrossRef]
54. Liu, Y.; Shang, Q.; Wang, Z.; Zhang, K.; Zhao, C. Spatial Heterogeneity of Soil Respiration Response to Precipitation Pulse in A Temperate Mixed Forest in Central China. *J. Plant Anim. Ecol.* **2017**, *1*, 1–13. [CrossRef]
55. Wu, Z.; Dijkstra, P.; Koch, G.W.; Penuelas, J.; Hungate, B.A. Responses of terrestrial ecosystems to temperature and 696 precipitation change: A meta-analysis of experimental manipulation. *Glob. Chang. Biol* **2011**, *17*, 927–942. [CrossRef]
56. Deng, Q.; Hui, D.; Zhang, D.; Zhou, G.; Liu, J.; Liu, S.; Chu, G.; Li, J. Effects of precipitation increase on soil respiration: A 698 three-year field experiment in subtropical forests in China. *PLoS ONE* **2012**, *7*, e41493. [CrossRef]
57. Schindlbacher, A.; Wunderlich, S.; Borken, W.; Kitzler, B.; Zechmeister-Boltenstern, S.; Jandl, R. Soil respiration under climate 656 change: Prolonged summer drought offsets soil warming effects. *Glob. Chang. Biol.* **2012**, *18*, 2270–2279. [CrossRef]
58. Lundquist, E.J.; Scow, K.M.; Jackson, L.E.; Uesugi, S.L.; Johnson, C.R. Rapid response of soil microbial communities from conventional, low input, and organic farming systems to a wet/dry cycle. *Soil Biol. Biochem.* **1999**, *12*, 1661–1675. [CrossRef]
59. Xu, X.; Luo, X. Effect of wetting intensity on soil GHG fluxes and microbial biomass under a temperate forest floor during dry season. *Geoderma* **2012**, *170*, 118–126. [CrossRef]
60. Monteith, J.L.; Unsworth, M.H. Chapter 13—Steady-State Heat Balance: (i) Water Surfaces, Soil, and Vegetation. In *Principles of Environmental Physics*, 4th ed.; Monteith, J.L., Unsworth, M.H., Eds.; Academic Press: Cambridge, MA, USA, 2013; pp. 217–247.
61. Cuxart, J.; Boone, A.A. Evapotranspiration over Land from a Boundary-Layer Meteorology Perspective. *Bound.-Layer Meteorol.* **2020**, *177*, 427–459. [CrossRef]
62. Karelin, D.V.; Zamolodchikov, D.G.; Shilkin, A.V.; Popov, S. The effect of tree mortality on CO_2 fluxes in an old-growth spruce forest. *Eur. J. Forest Res.* **2021**, *140*, 287–305. [CrossRef]

forests

MDPI

Article

Environmental Response of Tree Species Distribution in Northeast China with the Joint Species Distribution Model

Juan Yong [1,2], Guangshuang Duan [3], Shaozhi Chen [1,*] and Xiangdong Lei [4,*]

1 Research Institute of Forestry Policy and Information, Chinese Academy of Forestry, Beijing 100091, China; yj@cfgc.cn
2 China Forestry Group Corporation, Beijing 100036, China
3 College of Mathematics and Statistics, Xinyang Normal University, Xinyang 464000, China; duans@xynu.edu.cn
4 State Key Laboratory of Efficient Production of Forest Resources, Key Laboratory of Forest Management and Growth Modelling, National Forestry and Grassland Administration, Institute of Forest Resource Information Techniques, Chinese Academy of Forestry, Beijing 100091, China
* Correspondence: szchen@caf.ac.cn (S.C.); xdlei@ifrit.ac.cn (X.L.)

Abstract: The composition, distribution, and growth of native natural forests are important references for the restoration, structural adjustment, and close-to-nature transformation of artificial forests. The joint species distribution model is a powerful tool for analyzing community structure and interspecific relationships. It has been widely used in biogeography, community ecology, and animal ecology, but it has not been extended to natural forest conservation and restoration in China. Therefore, based on the 9th National Forest Inventory data in Jilin Province, combined with environmental factors and functional traits of tree species, this study adopted the joint species distribution model—including a model with all variables (model FULL), a model with environmental factors (model ENV), and a model with spatial factors (model SPACE)—to examine the distribution of multiple tree species. The results show that, in models FULL and ENV, the environmental factors explaining the model variation were ranked as follows, climate > site > soil. The explanatory power was as follows: model FULL (AUC = 0.8325, Tjur R^2 = 0.2326) > model ENV (AUC = 0.7664, Tjur R^2 = 0.1454) > model SPACE (AUC = 0.7297, Tjur R^2 = 0.1346). Tree species niches in model ENV were similar to those in model FULL. Compared to predictive power, we found that the information transmitted by environmental and spatial predictors overlaps, so the choice between model FULL and ENV should be based on the purpose of the model, rather than the difference in predictive ability. Both models can be used to study the adaptive distribution of multiple tree species in northeast China.

Keywords: joint species distribution model; niche; environmental factors; Tjur R^2

Citation: Yong, J.; Duan, G.; Chen, S.; Lei, X. Environmental Response of Tree Species Distribution in Northeast China with the Joint Species Distribution Model. *Forests* **2024**, *15*, 1026. https://doi.org/10.3390/f15061026

Academic Editors: Daniela Dalmonech, Alessio Collalti and Gina Marano

Received: 28 April 2024
Revised: 4 June 2024
Accepted: 10 June 2024
Published: 13 June 2024

1. Introduction

Natural forests are the most stable, diverse, and structurally complex terrestrial ecosystems in nature; they play an irreplaceable role in responding to climate change, protecting biodiversity, and maintaining ecological balance [1,2]. China's natural forest area reaches 138.68 million hectares, accounting for 63.55% of national forest area [3], making it a crucial strategic resource. The stock volume per unit area of China's natural forests is 113.36 m^3/ha, which is significantly lower compared to European countries with similar site conditions (where the stock volume exceeds 200 m^3/ha) [4]. In July 2019, the General Office of the Central Committee of the Communist Party of China and the General Office of the State Council issued the "Natural Forest Conservation and Restoration System Scheme", proposing the conservation and improvement of natural forest structure, focusing on cultivating native tree species, and enhancing forest quality. Therefore, protecting natural forest resources and enhancing the quality of natural forests has become a focus of forest management now and in the future.

The structure of stands, especially tree species composition, is a critical factor to consider in the restoration of natural forests. For a long time, the selection of target tree species and the determination of target forests in natural forest management have been mainly based on expert experience, lacking quantitative research. The composition, distribution, and growth conditions of tree species in virgin natural forests are important references for natural forest restoration, structural adjustment, and the near-natural management of plantations [5]. Currently, species distribution studies in Jilin Province are mainly limited to a few tree species and forest types, such as *Betula platyphylla*, *Betula ermanii*, *Quercus mongolica*, *Larix gmelinii*, and the broad-leaved forests of *Pinus koraiensis* [6–8]. Liu et al. conducted more systematic research using the MaxEnt model, but these studies have overlooked the interactions between species and cannot meet the needs of natural forest restoration and quality improvement in the region [9]. Therefore, it is necessary to adopt new methods to study the potential distribution suitability of the main forest types in northeast China's natural forests.

Species in nature do not exist independently in the environment but coexist and interact within communities, also influenced by surrounding environmental factors [10]. Thus, species distribution is determined by both abiotic environmental factors (such as climate, soil, and topography) and biotic interactions between different species (such as predation, competition, and mutualism) [11,12]. With the development of computer technology and statistical methods, a species distribution modeling approach that combines environmental variables with interactions among multiple species (i.e., joint species distribution models) has been used in studies on the simulation and prediction of multi-species distributions [13–15]. Joint species distribution models, by using species correlation information and latent variables to predict missing environmental factors, can clearly select models and assess the model's performance within the model framework, thereby simulating multi-species species–environment relationships and predicting the intensity and type of interactions between different species [16]. Not only does this enhance the interpretative power over ecological questions, but it also improves the flexibility and effectiveness of predicting species' distribution suitability. Therefore, joint species distribution models are powerful tools for analyzing the structure of biological communities and interspecific relationships, widely applied in biogeography, community ecology, and animal ecology [17–22], but these models have not yet been widely applied in the conservation and restoration of China's natural forests. Thus, understanding the distribution of multiple tree species and their interrelationships in natural forests has significant theoretical and practical significance for forest restoration.

Jilin province is located in the temperate zone, which has various types of natural forests, including coniferous forest, broad-leaved forest, mixed forest, etc. The distribution and combination of different tree species in the forest form a unique forest ecosystem. By studying the natural distribution of tree species, we can reveal the structure, function, and succession law of the forest ecosystem, and provide a scientific basis for the protection, restoration, and management of forest ecology. Therefore, the objectives of this study were as follows: (1) to construct a hierarchical model of species communities (HMSC) in combination with environmental factors and species functional traits for the distribution of multiple species; (2) to interpret tree species niches through the establishment of the best fitting joint species distribution model; and (3) to analyze effects of tree species traits and phylogeny on the HMSC model.

2. Materials and Methods

2.1. Study Area

The area involved in this study is the entirety of Jilin Province, located in the central part of northeast China, spanning from 121°38′ E to 131°19′ E longitude and 40°52′ N to 46°18′ N latitude. The terrain of Jilin Province exhibits a characteristic of being higher in the southeast and lower in the northwest, divided by the central Greater Khingan Range into eastern mountainous areas and central–western plains. There are 19 soil types within the province, with dark brown earth being predominant. The natural environment belongs to a temperate continental monsoon climate, with distinct seasonal changes and regional differences, forming a gradient of vegetation types from southeast to northwest characterized by moist, semi-moist, and semi-arid climates—from the moist forest climate in the east to the semi-moist forest-steppe climate in the center and the semi-arid steppe climate in the west. The average winter temperature does not exceed −11 °C, with summer temperatures generally above 23 °C. The annual average precipitation is between 400 and 600 mm, with annual average sunshine hours ranging from 2259 to 3016 h, and the frost-free period lasts between 100 and 160 days [9].

2.2. Data Collection

The data for the study area primarily revolve the ninth (2014) National Forest Inventory database in Jilin Province, encompassing five types of data needed: community data (species presence or abundance), environmental factors (site, soil, or climate), tree species functional traits (maximum tree height, wood density, and leaf area index, etc.), phylogenetic relationships of tree species, and plot spatial data.

2.2.1. Plot Data

The construction of the joint species distribution model primarily uses sample plot data on tree species composition by basal area, site factors, and latitude and longitude coordinates, focused on the central Changbai Mountain subregion. The species' basal area compositions are extracted to form community data, with the basal area per hectare for each species calculated to form the community abundance data. Further, the basal area per hectare data are used to form presence–absence data for the community. Site factor data mainly include elevation, slope, aspect, position, soil (type and thickness), humus layer thickness, and litter thickness, from which site factor data in the environmental covariates can be obtained. Plot latitude and longitude coordinates form the spatial data. A total of 3309 permanent sample plot data points with each 600 m^2 from the ninth National Forest Inventory were collected for this study; their distributions are shown in Figure 1.

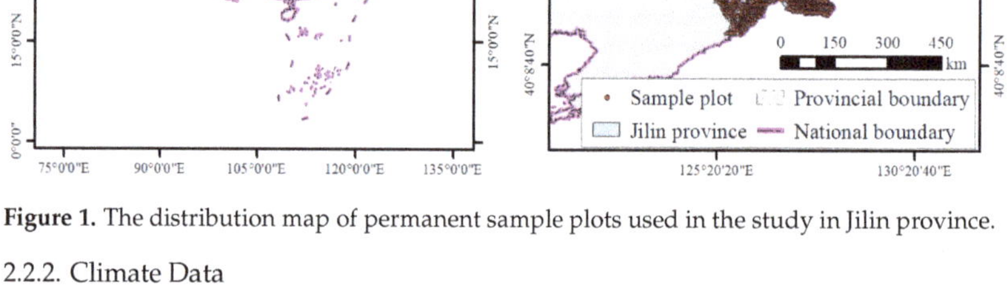

Figure 1. The distribution map of permanent sample plots used in the study in Jilin province.

2.2.2. Climate Data

The climate data were sourced from the software ClimateAP V3.10, which is specifically designed for extracting climate factors in the East Asia–Pacific region (from https://web.climateap.net, accessed on 5 September 2023). ClimateAP allows users to extract and downscale gridded climate data to site-specific, scale-free climate data through a dynamic local downscaling method. ClimateAP calculates and exports many biologically relevant climate variables at monthly, seasonal, and annual timesteps [23]. The study area's climate data were formed by registering and clipping with the basic geographic data of Jilin Province. For this study area, the climate data of 3309 sample plots were acquired, including 10 annual variables (MAT, MWMT, MCMT, TD, MAP, AHM, NFFD, EMT, Eref, and CMD), 8 seasonal variables (Tave_DJF, Tave_JJA, Tmax_DJF, Tmax_JJA, Tmin_DJF, Tmin_JJA, PPT_DJF, and PPT_JJA), and 48 monthly variables (Tave01–Tave12, Tmin01–Tmin12, Tmax01–Tmax12, and PPT01–PPT12). These abbreviations refer to Wang et al. [23].

2.2.3. Soil Data

Soil data were obtained from the SoilGrids system (version updated in June 2016), with a spatial resolution of 250 m. Based on the World Reference Base (WRB) and the United States Department of Agriculture (USDA) classification systems, totaling approximately 280 raster layers, the SoilGrids system provides global predictions of standard soil properties (organic carbon, bulk density, cation exchange capacity (CEC), pH value, soil structure ratio, and proportion of coarse fragments), bedrock depth, and soil type distributions at seven standard depths (0, 5, 15, 30, 60, 100, and 200 cm) [24]. Under the Open Database License (ODbL), the 250 m maps from SoilGrids V2.0 can be downloaded from www.SoilGrids.org (accessed on 15 September 2023). For this study area, 11 soil properties were acquired.

2.2.4. Tree Species Trait Factors Data

To assess the relationship between tree species niches and functional traits, it was necessary to collect functional trait data for the tree species appearing in the community data. Among the 3309 permanent sample plot data collected, a total of 75 tree species were involved. Due to the very low frequency of occurrence of many species, this study selected 31 tree species for further research, based on the criterion that the species appeared in more than 3% of all permanent sample plots. The species selected include the following: *Pinus koraiensis Siebold et Zuccarini* (sp1), *Picea koraiensis* Nakai (sp2), *Picea jezoensis Carr.* var. microsperma (Lindl.) Cheng et L.K.Fu (sp3), *Abies nephrolepis* (Trautv.) Maxim. (sp4), *Abies holophylla* Maxim. (sp5), *Larix olgensis* Henry (sp6), *Pinus sylvestris* Linn. var. mongolica Litv. (sp7), *Quercus mongolica* Fischer ex Ledebour (sp8), *Tilia mandshurica* Rmpr.et Maxim. (sp9), *Tilia amurensis Rupr.* (sp10), *Ulmus davidiana* Planch var. japonica (Rehd.) Nakai (sp11), *Carpinus cordata Bl.* (sp12), *Ulmus laciniata* (Trautv.) Mayr (sp13), *Betula dahurica* Pall. (sp14), *Betula platyphylla* Suk. (sp15), *Betula costata* Trautv. (sp16), *Fraxinus mandschurica* Rupr. (sp17), *Juglans mandshurica* Maxim. (sp18), *Phellodendron amurense* Rupr. (sp19), *Acer mono* Maxim. (sp20), *Acer tegmentosum* Maxim. (sp21), *Acer mandshuricum* Maxim. (sp22), *Acer ukurunduense* Trautv. et Mey. (sp23), *Acer triflorum* Komarov (sp24), *Acer pseudo-sieboldianum* (Pax) Komarov (sp25), *Sorbus alnifolia* (Sieb. et Zucc.) K. Koch (sp26), *Fraxinus rhynchophylla* Hance (sp27), *Populus davidiana* Dode (sp28), *Populus ussuriensis* Kom. (sp29), *Populus simonii* Carr. (sp30), and *Salix matsudana* Koidz. (sp31).

The trait factors for these tree species were obtained from the literature. Due to limitations in data acquisition, the trait factors compiled in this study include the following: wood density (WD), maximum height (H), leaf area (LA), specific leaf area (SLA), leaf dry matter density (LMA), leaf dry matter content (LDMC), leaf carbon concentration (Cmass), leaf nitrogen concentration (Nmass), leaf phosphorus concentration (Pmass), leaf potassium concentration (Kmass), area-based nitrogen content (Narea), area-based phosphorus content (Parea), and area-based potassium content (Karea).

2.2.5. Tree Species Phylogenetic Data

To evaluate the extent to which tree species niches reflect phylogenetic relationships, phylogenetic data for the species were required. These data typically take the form of a phylogenetic tree, usually constructed by running genomic sequence data through phylogenetic analysis software and then generating a phylogenetic correlation matrix to describe the phylogenetic relationships among the species.

The procedure for generating the phylogenetic tree for the 31 tree species in this study began with obtaining protein molecular sequences for the species from the National Center for Biotechnology Information (www.ncbi.nlm.nih.gov, accessed on 12 December 2023).

Due to the limitation of protein molecular sequence data for different organs of these 31 tree species, the chosen organ was the mature enzyme K (maturase K [chloroplast]) in tree leaves. Then, the Molecular Evolutionary Genetics Analysis software (MEGA_X_10.2.4) was used to perform phylogenetic analysis on the protein molecular sequence information of these tree species and generate usable phylogenetic tree files in NWK format. The finalized phylogenetic tree for these 31 tree species is shown in Figure 2.

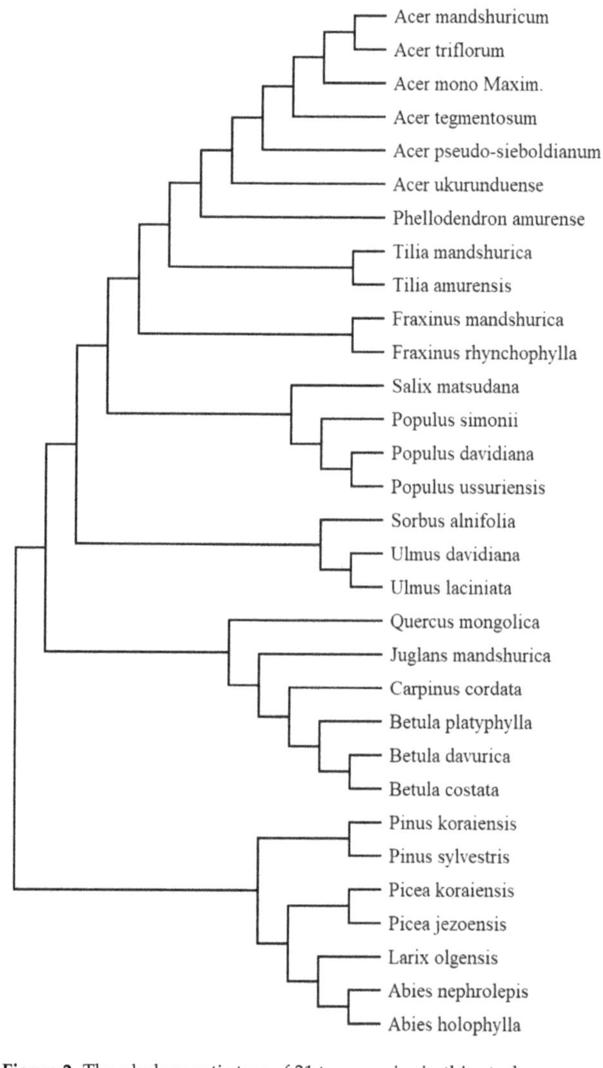

Figure 2. The phylogenetic tree of 31 tree species in this study.

2.3. Methods

A multivariate hierarchical generalized linear mixed model for 31 tree species in Jilin Province was constructed to study the relationship between environmental variables and the distribution of tree species, i.e., the niches of tree species, and further to elucidate the relationship between tree species niches, tree species traits, and phylogenetic trees.

2.3.1. Model Structure Setup and Fitting

Given that the structure of multiple tree species community data is of a presence–absence type, meaning the dependent variable is a 0–1 variable, the error distribution type of the model was chosen to use a probit link function, considering spatial random effects based on the geographic coordinates of the plots. To more comprehensively study the relationship between environmental covariates and spatial random effects, three types of model forms were considered. The first type is the full model (FULL), which includes both environmental covariates and spatial random effects of the plots. The second type is the environmental factor model (ENV), which includes only environmental covariates without the spatial random effects of the plots. The third type is the spatial factor model (SPACE), which includes only spatial random effects without environmental covariates. All three model structures introduced tree species traits and phylogenetic relationships.

After setting up the HMSC model structure, the parameter estimation phase requires fitting the data using the sampleMcmc function. Two MCMC chains were set, i.e., nChains = 2, with a sampling step size of 10 (thin = 10), and sampling 1000 times for each parameter (samples = 1000), discarding the first 5000 estimates (transient = 5000). Considering the spatial random effects and the sample size exceeding 1000, the computational complexity increased sharply, making the computation infeasible. Therefore, when dealing with the spatial structure, the Nearest Neighbor Gaussian Process (NNGP) was used. In this case, the parameter Method was set to "NNGP" and the corresponding parameter neighbors were set to the standard number, 10 according to the literature [25].

2.3.2. Variable Selection

In terms of variable selection, it is hypothesized that several typical predictive variables could influence the occurrence or abundance of species. Among these variables, some might have no impact, or some might be collinear with other predictive variables, thus carrying redundant information. Increasing the number of variables in the model raises the risk of over-parameterization, thereby increasing the risk of overfitting. Therefore, it is generally recommended to pre-select as few variables as possible, ensuring that the information content of these variables is maximized, and they are mutually independent [26]. In summary, at the first step, the most ecologically meaningful predictive variables should be selected, and predictive variables closely related to other predictors should be excluded to eliminate the multicollinearity among predictive variables.

In this study, environmental variables include 66 climatic factors (10 annual variables, 8 seasonal variables, and 48 monthly variables), 11 soil factors, and 7 site factors. Based on the principles analyzed above, using the results of previous work and considering ecological significance, along with the importance ranking in machine learning's random forest algorithm, a selection is made. Liu found a significant response relationship between common tree species and forest types in Jilin Province and environmental factors, quantitatively measuring the impact of environmental variables on the distribution of tree species or forest types: climate > site > soil, indicating that climate is the most critical factor affecting vegetation distribution [4]. The specific environmental factors ranking (only top five selected) are as follows: the highest temperature of the hottest month, the average temperature of the hottest quarter, elevation, annual average temperature, and the average temperature of the coldest quarter. Soil factors have a smaller impact, but for conifer species, the significant ones are as follows: soil pH, bulk density, exchangeable hydrogen ions, exchangeable calcium ions, and available phosphorus. Machine learning with the data in the study is used to establish the relationship between tree species distribution and these environmental factors, employing the random forest algorithm. Based on the importance ranking, the top 10 influencing factors for each tree species are selected, and the frequency of each influencing factor is counted. The recommended results include climate factors NFFD, AHM, Eref, MCMT, TD, site factor ELE, soil factors bdod, soc, cfvo, and phh2o. Finally, to avoid variable collinearity, a correlation analysis is conducted for the shortlisted environmental variables, excluding variables with a correlation coefficient of 0.7 or above.

Therefore, from the environmental variables, seven climatic factors were selected: the average temperature of the hottest quarter (Tave_JJA), rainfall of the hottest quarter (PPT_JJA), number of frost-free days per year (NFFD), annual humidex (AHM), evapotranspiration (Eref), average temperature of the coldest month (MCMT), and the temperature difference between the hottest and coldest months (TD). One site factor was selected: elevation (ele). Four soil factors were selected: bulk density of soil particles (bdod), soil organic carbon content (soc), proportion of coarse fragments (cfvo), and soil pH value (phh2o). The specific value distributions are shown in Table 1.

Table 1. Summary statistics of environmental variables and tree species functional traits in this study.

Variable	Unit	Mean	Standard Deviation	Min	Max
Tave_JJA	°C	19.01	1.38	12.60	22.80
PPT_JJA	mm	448.73	66.99	277.00	690.00
NFFD	/	172.10	10.37	129.00	202.00
AHM	/	19.18	3.76	7.90	44.00
Eref	/	688.20	32.97	401.00	839.00
MCMT	°C	−15.89	0.94	−19.10	−10.50
TD	°C	36.22	1.29	32.00	40.50
ele	m	666.38	261.69	90.00	1860.00
bdod	kg/dm^3	132.57	3.43	118.84	144.05
soc	g/kg	269.52	56.43	65.96	516.38
cfvo	$cm^3/100\ cm^3$	235.78	37.38	81.78	407.60
phh2o	pH	60.25	2.22	52.23	83.14
H	m	24.68	9.56	8.00	50.00
WD	g/cm^3	0.50	0.11	0.32	0.71
LA	m^2	0.00	0.01	0.00	0.04
LMA	kg/m^2	0.08	0.08	0.02	0.38
Cmass	g/kg	407.84	82.73	240.20	512.77
Nmass	g/kg	17.67	7.01	2.20	30.86
Pmass	g/kg	1.65	0.65	0.67	3.86
Kmass	g/kg	12.42	6.31	5.02	30.25

This study includes thirteen tree species functional traits: wood density (WD), maximum tree height (H), leaf area (LA), specific leaf area (SLA), leaf dry matter density (LMA), leaf dry matter content (LDMC), leaf carbon concentration (Cmass), leaf nitrogen concentration (Nmass), leaf phosphorus concentration (Pmass), leaf potassium concentration (Kmass), area-based nitrogen content (Narea), area-based phosphorus content (Parea), and area-based potassium content (Karea). Based on pair scatter plots and correlation analysis, it was found that SLA and LMA are inversely related, LDMC is related to Cmass, Nmass, and Kmass, and the relationship between Nmass and Narea is related to LA. Considering the meanings of these indicators, there were eight tree species trait factors selected for this study: H, WD, LA, LMA, Cmass, Nmass, Pmass, and Kmass.

2.3.3. Model Evaluation Metrics

Changing the structure of the HMSC model will produce different joint species distribution models, requiring the evaluation of the fitting effectiveness and predictive capability of different HMSC models. The predictive capability can be assessed through cross-validation. The chosen evaluation metrics are the Area Under the Receiver Operating Characteristic (ROC) Curve (AUC) and Tjur R^2 [27].

AUC is a performance metric that measures the excellence of a learner, quantifying the classification capability expressed by the ROC curve. The larger the AUC, the better the classification capability, the more reasonable the output probabilities, and the more sensible the order of results. The AUC value has a range of [0, 1], with an AUC greater than 0.5 indicating that the model's fit is superior to random guessing [28]. Tjur R^2 is primarily used to evaluate logistic regression models and is similar to the coefficient of determination R^2 used in ordinary regression models [29,30].

$$Tjur\ R^2 = \frac{1}{2}\left(\frac{\sum_{i=1}^{n}(\hat{\pi}_i - \overline{y})^2 - \sum_{i=1}^{n}(y_i - \hat{\pi}_i)^2}{\sum_{i=1}^{n}(y_i - \overline{y})^2} + 1\right)$$

In the formula, y_i represents a binary response variable, in which $y_i = 1$ means presence and $y_i = 0$ means absence. $\overline{y}$ represents the average value of y_i, and $\hat{\pi}_i$ represents the predictive probability value of occurrence for species i. n represents the total number of data records.

2.3.4. Data Analysis Tools

The software or dataset used for data processing and analysis in this study includes R 4.3.2, ClimateAP V3.10, SoilGrids V2.0, and MEGA_X_10.2.4. Details are as follows: R packages, with dplyr for data processing [31], ggplot2 for plotting [32], Hmsc for analyzing joint species distribution data [33], ClimateAP for obtaining climate data for permanent sample plots, SoilGrids for obtaining soil data for permanent sample plots, and MEGA_X_10.2.4 for generating phylogenetic trees of tree species.

3. Results

3.1. Tree Species Distribution Patterns

The richness (frequency of occurrence of tree species within permanent sample plots) and prevalence (proportion of plots in which a given tree species appears out of the total surveyed plots) distributions of these 31 tree species are shown in Figure 3. The richness of tree species varies significantly across plots, ranging from 1 to 18 (plots with ≤3 species account for 18.77%, and those with >10 species account for 16.50%). The prevalence of tree species ranges from 5% to 60%, reflecting that the tree species selected in this study include both common and rare species (top three prevalent common species—*Acer mono* 59.78%, *Quercus mongolica* 55.03%, *Tilia amurensis* 54.37%; top three prevalent rare species—*Salix matsudana* 3.35%, *Populus tremula* 3.93%, *Pinus sibirica* 4.08%).

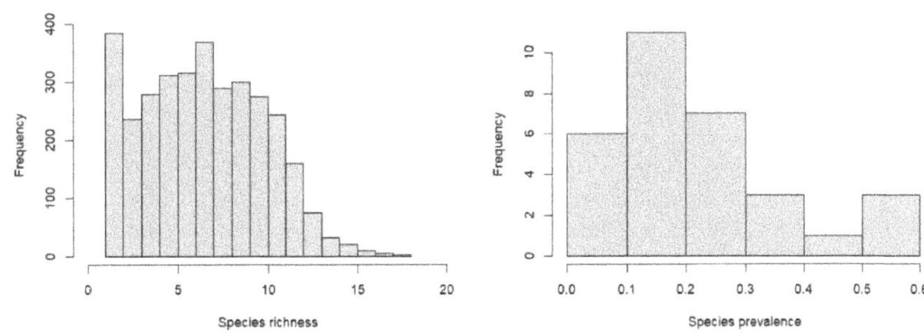

Figure 3. Tree species distribution frequency of the community data (for species richness (**left**), the y-axis (frequency) corresponds to the number of sampling plots, and the x-axis to the number of species in each sampling plot; for species prevalence (**right**), the y-axis (frequency) corresponds to the number of species, and the x-axis to the fraction of sampling plots in which the species is present).

3.2. Model Interpretability and Predictive Power

3.2.1. Overall Evaluation of Interpretability and Predictive Power of the Three Models

For the models FULL, ENV, and SPACE, their interpretability was evaluated by calculating two indicators, AUC and Tjur R^2. Cross-validation was conducted based on plot numbers using a two-fold sampling method, and both AUC and Tjur R^2 were calculated to evaluate the models' predictive power (Table 2).

Table 2. The evaluation of HMSC model.

Models	Fitting		Cross Validation	
	AUC	**Tjur R^2**	**AUC**	**Tjur R^2**
Model FULL	0.8325	0.2326	0.6940	0.1412
Model ENV	0.7664	0.1454	0.7528	0.1399
Model SPACE	0.7297	0.1346	0.6719	0.0705

Table 2 indicates that, in terms of model fitting, the AUC and Tjur R^2 metrics show consistent results, reflecting that, in terms of interpretability, model FULL > model ENV > model SPACE. However, in the cross-validation, the performance of model SPACE was slightly worse than the other two models. The AUC metric shows that model ENV performed better than model FULL, but the Tjur R^2 metric shows that its performance was slightly worse than model FULL. By comparing the evaluation results of cross-validation and model fitting, it can be seen that spatial random effects significantly improve the model's fitting effectiveness (AUC increased by 8.62%, Tjur R^2 increased by 60.04%), but its predictive effect is not ideal (AUC actually decreased by 7.81%, Tjur R^2 increased by 0.97%). The inclusion of spatial random effects in model FULL led to a significant drop in both AUC and Tjur R^2 metrics in cross-validation (AUC decreased by 16.64%, Tjur R^2 decreased by 39.30%). The cross-validation indicators for model ENV were close to those of the model fitting (AUC decreased by 1.78%, Tjur R^2 decreased by 3.79%), indicating that this model maintains consistency in both interpretability and predictive power. At first glance, the interpretability of model SPACE being greater than zero is questionable since it does not contain environmental covariates, but it does have a spatial random effects part, which contributes to its interpretability. However, the random effects part has limited help in predicting new plots (AUC decreased by 7.92%, Tjur R^2 decreased by 47.62%), as shown by the predictive power of model SPACE based on cross-validation (Tjur R^2 value is 0.0705), suggesting there is overlap in information between environmental covariates and spatial coordinates. In summary, whether considering interpretability or predictive power, and whether evaluated by AUC or Tjur R^2 for model fitting effectiveness, model FULL and ENV perform as expected without overfitting, while model SPACE might have overfitting issues.

3.2.2. Evaluation of Interpretability and Predictive Power by Tree Species

Combining the model forms—model FULL (model 1), model ENV (model 2), and model SPACE (model 3)—with the type of model prediction, i.e., interpretability (MF) and predictive power (MFCV), six combination schemes are formed. The evaluation of these 31 tree species using AUC and Tjur R^2 metrics is displayed in Figure 4. The results reflected by the AUC and Tjur R^2 metrics for the distribution of the 31 tree species are consistent across all models, indicating that, within the same model, interpretability exceeds predictive power. This outcome is expected since interpretability uncovers all the information within the data. The interpretability and predictive power of the 31 tree species in model ENV are almost identical, resembling the predictive power of model FULL, suggesting that model ENV has stable extrapolation capabilities.

There are significant differences in the HMSC model fits among the 31 tree species in the study area, with AUC and Tjur R^2 metrics showing a consistent pattern of differences between interpretability and predictive power: model FULL > model SPACE > model ENV. For example, considering AUC values, the top three species with the largest difference in model FULL are as follows: *Pinus sibirica* (sp7, 0.48), *Pinus sylvestris* (sp5, 0.29), and *Ulmus pumila* (sp12, 0.28). The top three species with the largest difference in model SPACE are as follows: *Acer ukurunduense* (sp21, 0.14), *Betula ermanii* (sp16, 0.13), and *Acer ginnala* (sp23, 0.12). The top three species with the largest difference in model ENV are as follows: *Populus tremula* (sp30, 0.05), *Pinus sibirica* (sp7, 0.05), and *Salix matsudana* (sp31, 0.04). Furthermore, in terms of consistency and good fit between model interpretability and predictive power, the results for tree species in model FULL and ENV are consistent, including *Picea jezoensis* (sp3), *Abies nephrolepis* (sp4), and *Acer ginnala* (sp23). Notably, in model SPACE, *Picea jezoensis* (sp3), *Abies nephrolepis* (sp4), *Betula ermanii* (sp16), and *Acer ginnala* (sp23) all exhibit the best interpretability and predictive power, even exceeding model FULL. Among the more prevalent tree species, those with good interpretability and predictive power in model FULL include the following: *Pinus koraiensis* (sp1, MF: 0.83, MFCV: 0.76), *Tilia amurensis* (sp10, MF: 0.89, MFCV: 0.73), *Ulmus laciniata* (sp13, MF: 0.92, MFCV: 0.73), *Betula ermanii* (sp16, MF: 0.91, MFCV: 0.81), *Acer ukurunduense* (sp21, MF: 0.92, MFCV: 0.76), and *Acer pseudosieboldianum* (sp25, MF: 0.89, MFCV: 0.77).

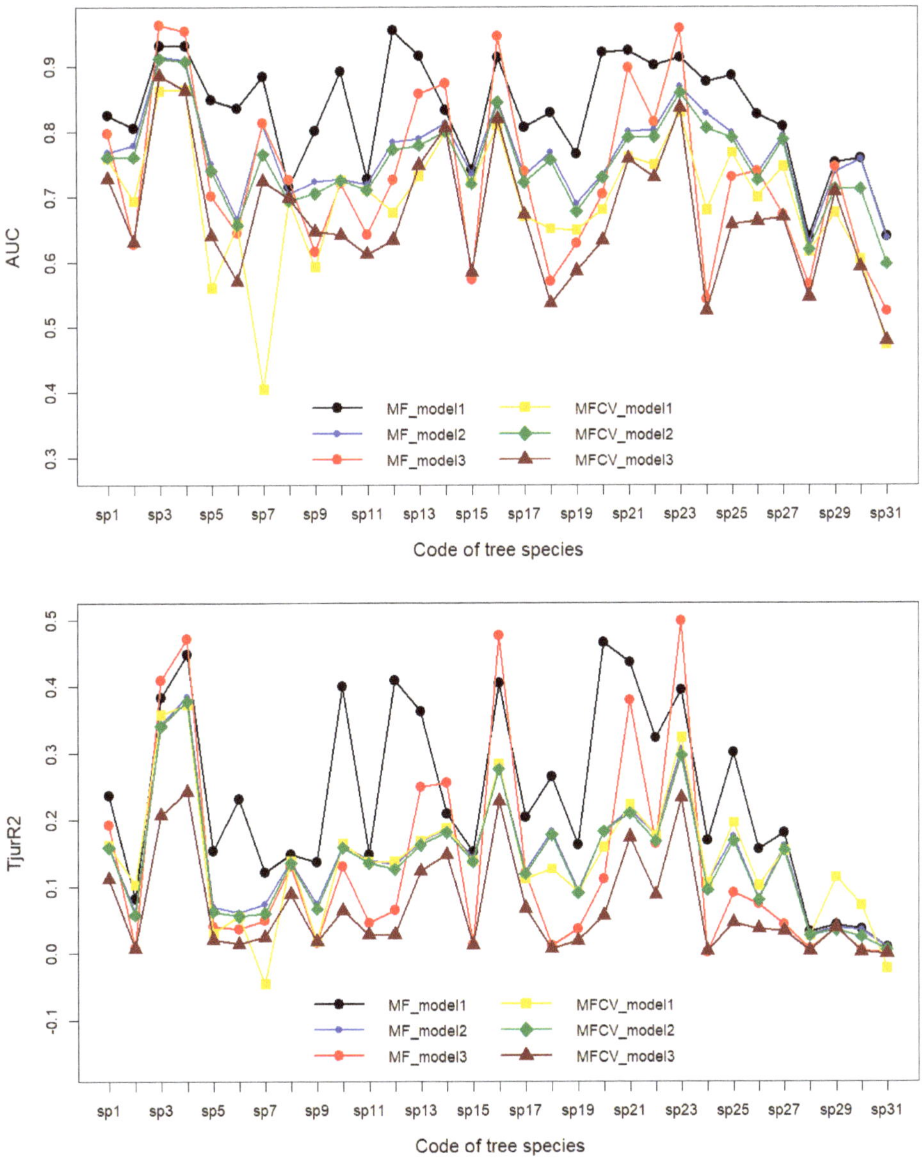

Figure 4. The evaluation of 31 tree species for HMSC model.

3.3. Variance Contribution of Environmental Covariates

The study first investigates the variance contribution of environmental covariates to the models, i.e., variance partitioning for these three model types. Since model SPACE only contains spatial random effects, all explained variance is attributed to random effects. Therefore, comparisons between model FULL and ENV will more clearly demonstrate the importance of environmental covariates and spatial random effects in explaining community ecosystems.

The environmental covariates are grouped as follows: average temperature of the hottest quarter (Tave_JJA), rainfall of the hottest quarter (PPT_JJA), number of frost-free days per year (NFFD), annual humidex (AHM), evapotranspiration (Eref), average temperature of the coldest month (MCMT), and the temperature difference between the hottest and coldest months (TD) are grouped as climate variables; elevation (ELE) as a site variable; and bulk density of soil particles (bdod), soil organic carbon content (soc), proportion of coarse fragments (cfvo), and soil pH value (phh2o) as soil variables. For simplicity, the intercept is allocated to climate variables. The variance contribution rates of the model components are shown in Table 3. It is observed that the ranking of environmental factors explaining model variance is as follows: climate > site > soil. The two components of environmental change (climate and site) explain a significant portion of the model variance, with soil factors being the weakest. The variance contribution of spatial random effects is also significant, with the importance of specific environmental factors varying by tree species.

Table 3. The variance partition of models FULL, ENV, and SPACE.

Models	Climate	Site	Soil	Random
Model FULL	0.5742	0.1275	0.0914	0.2069
Model ENV	0.7322	0.1617	0.1061	0.0000
Model SPACE	0.0000	0.0000	0.0000	1.0000

The variance contribution results of model ENV also indicate that the ranking of environmental factors explaining model variance is as follows: climate > site > soil, consistent with what is shown by model FULL. Two species are dominated by site factors, each with a variance contribution rate of over 45%, specifically *Ulmus japonica* (sp11, 0.62) and *Tilia amurensis* (sp9, 0.47), with the remaining 29 species being dominated by climate factors. The results of the variance partitioning indicate that model FULL has a substantial dependency on spatial random effects for some species, which also explains the performance differences between it and model ENV. Moreover, climate change explains nearly five times the model variability than site changes.

3.4. Tree Species Niches

To examine the differences between the niches of tree species, this study utilized a visualization method to display the response relationships of the 31 tree species to environmental covariates in model FULL, representing the niches of tree species (Figure 5). Figure 5 provides strong statistical support for the posterior distribution, showing either positive (red) or negative (blue) relationships, indicating clear environmental filtering signals. For most tree species, there is a significant parabolic relationship between their presence and the average temperature of the hottest quarter. Due to the negative coefficient of the quadratic term, there is an optimal temperature for their existence. Their responses to environmental factors (rainfall in the hottest quarter PPT_JJA, annual humidex AHM, average temperature of the coldest month MCMT, the temperature difference between the hottest and coldest months TD, and soil pH value phh2o) are negative, while their preferences for the number of frost-free days per year NFFD, evapotranspiration Eref, elevation ele, bulk density of soil particles bdod, soil organic carbon content soc, and the proportion of coarse fragments cfvo vary. In terms of the relationship between various tree species and environmental covariates, *Larix olgensis* (sp6), *Ulmus japonica* (sp11), *Betula platyphylla* (sp15), and *Acer triflorum* (sp24) show a more positive response.

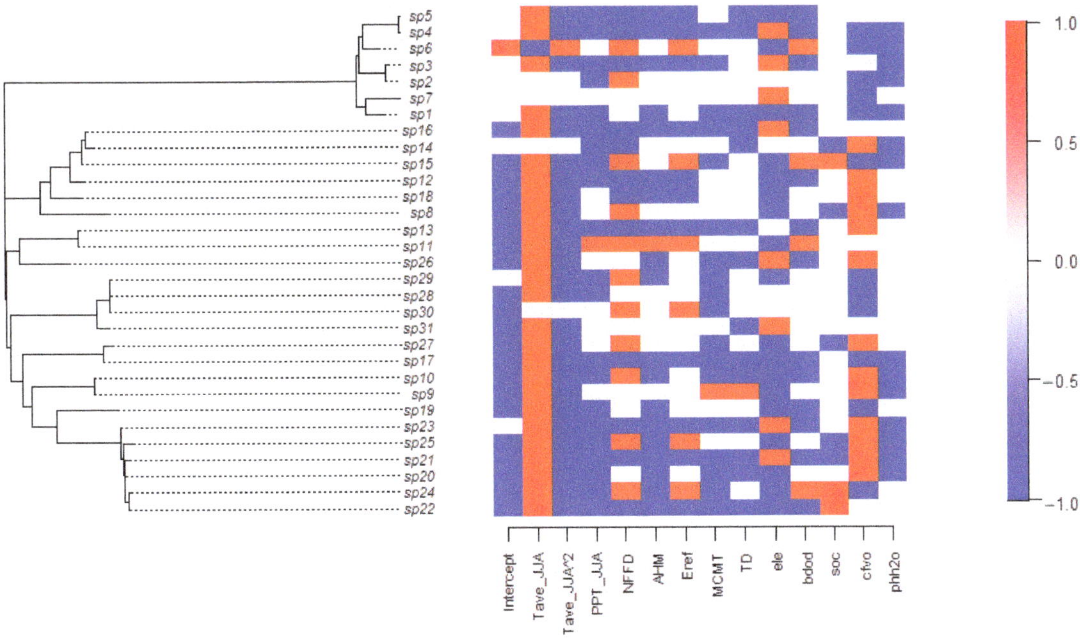

Figure 5. Heatmap of estimated parameters β of model FULL, i.e., species niches. Red and blue colors show parameters that are estimated to be positive and negative, respectively (with at least 0.95 posterior probability in Model FULL).

The visualization results of tree species niches in model ENV are similar to those in model FULL, with the difference being that the response of some tree species to environmental covariates shifts from positive to non-correlated. For instance, in model ENV, the response relationship to the number of frost-free days per year (NFFD) for *Tilia amurensis* (sp10), *Acer triflorum* (sp24), and *Fraxinus rhynchophylla* (sp27) show non-correlation. *Ulmus japonica* (sp11)'s response to the annual humidex AHM and *Acer triflorum* (sp24)'s response to evapotranspiration Eref are non-correlated.

4. Discussion

4.1. Relationship between Tree Species Niches and Traits—Phylogeny

The HMSC models tree species niches (parameters β) as a function of tree species traits (regression parameters γ) and phylogeny (phylogenetic signal parameters ρ); thus, these connections can be explored by plotting parameter estimates for tree species communities [16]. First, we investigate whether there is a correlated relationship between tree species niches and their traits and phylogeny in model FULL (Figure 6). At a posterior distribution statistical support level of 0.95, it indicates that the response of tree species to environmental covariates (average temperature of the coldest month MCMT and the temperature difference between the hottest and coldest months TD) is positively correlated with tree species traits (leaf dry matter density, LMA, and leaf nitrogen concentration, Nmass), while elevation, ele, is negatively correlated with leaf nitrogen concentration, Nmass, and the proportion of coarse fragments, cfvo, is negatively correlated with maximum tree height, H (Figure 6 left). When reducing the statistical support level to 0.85, a richer relationship between tree species niches and traits is observed (Figure 6 right). The relationships revealed by model ENV are mostly consistent, for example, at the 0.95 level, elevation, ele, is not correlated with leaf nitrogen concentration, Nmass.

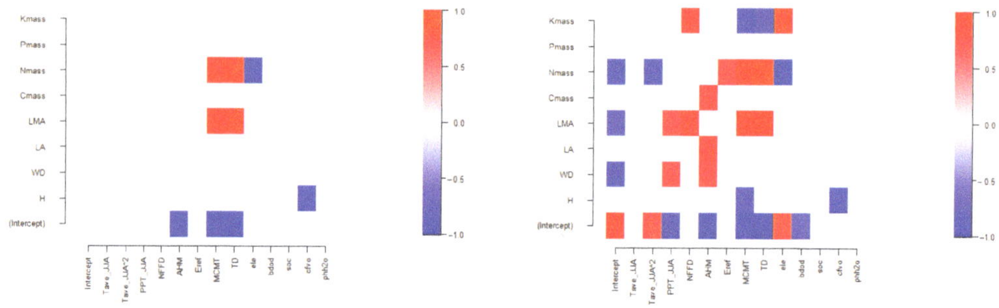

Figure 6. Heatmap of estimated parameters γ linking species traits to species niches of model FULL. Red and blue colors show parameters that are estimated to be positive and negative, respectively (with at least 0.95 posterior probability in the **left**, and 0.85 in the **right**).

Another method of examining the impact of tree species traits is to evaluate how much variation they explain in the response of tree species to their environmental covariates (Table 4). Leaf dry matter density, LMA, leaf nitrogen concentration, Nmass, and maximum tree height, H, explain a significant portion of the environmental covariate variation, while the remaining tree species traits only explain a small part of the environmental covariate variation; this is consistent with the patterns presented in Figure 5. Further quantification of the variance in the occurrence dependent variable of tree species explained by tree species traits reveals the same pattern, i.e., the contribution to explained variance is not high (0.1832 for model FULL and 0.1797 for model ENV). Comparing models FULL and ENV, the former shows a slight improvement in the degree to which tree species traits explain the variance of environmental covariates and the occurrence dependent variable, which may be due in part to the contribution of spatial random effects in model FULL.

Table 4. The explanatory power of species traits.

Models		Model FULL	Model ENV
Environmental covariates	Intercept	0.3291	0.2521
	Tave_JJA	0.0881	0.0659
	Tave_JJA^2	0.1521	0.0935
	PPT_JJA	0.1458	0.1254
	NFFD	0.0684	0.0839
	AHM	0.0541	0.0471
	Eref	0.0405	0.0494
	MCMT	0.4355	0.4569
	TD	0.4393	0.4522
	ele	0.3504	0.3229
	bdod	0.0418	0.0444
	soc	0.0394	0.0406
	cfvo	0.2119	0.1812
	phh2o	0.0677	0.0515
Species occurrence		0.1832	0.1797

Further, examining the strength of the phylogenetic signal in tree species niches, the percentiles of the parameter for the phylogenetic signal are statistically obtained from the posterior distribution. (Model FULL: 2.5%, 0.24; 50%, 0.48; 97.5%, 0.67. Model ENV, 2.5%, 0.31; 50%, 0.55; 97.5%, 0.71.) This clearly indicates the presence of a phylogenetic signal within the tree species niches. This suggests that the trait factors affecting tree species niches that are missing in the data have a phylogenetic structure [16,34]. Additionally, the average value of the parameter representing the strength of the phylogenetic signal in model FULL is 0.47, slightly lower than 0.54 in model ENV, indicating that spatial random effects reduce the impact of missing tree species trait factors on tree species niches to some extent [16].

4.2. Impact of Introducing Tree Species Traits and Phylogenetic Trees on Prediction

The HMSC model uses tree species trait data to estimate parameters reflecting the response of tree species traits on their niches and uses tree species phylogenetic information to estimate parameters indicating the strength of the phylogenetic signal in tree species niches. These two parameters integrate tree-level information into community-level parameters, hence incorporating tree species traits and phylogeny into the model helps to synthesize information extracted from community data. However, whether the introduction of tree species traits and phylogeny can enhance the predictive ability of the HMSC model still needs to be assessed [16]. To this end, a new model, NTP, which does not include information on tree species traits and phylogeny, was set up on the basis of model FULL and applied to the same data. The predictive evaluation metrics AUC and Tjur R^2 of both models were then compared in cross-validation (Figure 7).

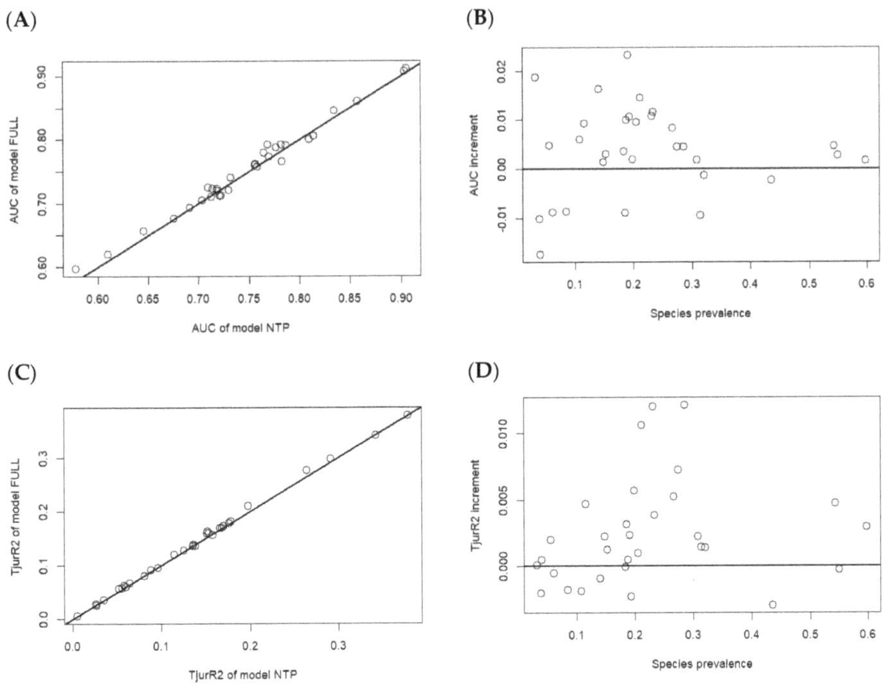

Figure 7. Difference in predictive power between models that include traits and phylogeny versus those that do not. In four panels, each dot corresponds to one tree species. Panel (**A**) shows the AUC statistic for the models that do (y-axis) and do not (x-axis) include traits and phylogeny. Panel (**B**) shows the difference in AUC between the two models as a function of species prevalence. Positive values of AUC increment indicate that the model with traits and phylogeny performs better in cross-validation. The same as panels (**C**,**D**) for index Tjur R^2, respectively.

The comparison between model FULL and NTP (Figure 7) shows that the patterns reflected by the indicators AUC and Tjur R^2 are consistent. The predictive ability gap between model FULL and NTP is very small, and this difference diminishes as the prevalence of tree species increases. On average, the model including tree species traits and phylogeny performs better, especially for rare tree species that are typically difficult to predict accurately. This is because the introduction of traits and phylogeny allows the model to borrow information from other tree species, especially those with similar traits and closely related phylogenies. For tree species that occur infrequently in the data, this minor information can make a significant difference [34]. In contrast, tree species with sufficient data do not need to borrow information from others, which explains why the difference between the two models decreases when the species prevalence approaches 0.3.

As mentioned above, the introduction of tree species traits and phylogeny in the HMSC model can enhance the predictive ability for evaluating tree species occurrence. However, the contribution of tree species traits to explain the variation of tree species occurrence is not high, and there are phylogenetic signals in tree species niches, which means that the relevant tree species traits affecting tree species niches are missing in the data. Therefore, it is necessary to find suitable tree species traits for improving the prediction of tree species occurrence, especially for rare tree species in further study [34]. Additionally, we can use tree species traits and phylogeny in the HMSC model to study the biotic relationship between tree species in following research [22,35].

5. Conclusions

This study employed the widely used hierarchical modeling of species communities (HMSC) to fit the distribution data of multiple tree species in Jilin Province. The data types included the presence or abundance of tree species from the ninth National Forest Inventory in 2014, environmental factors corresponding to the plots (site, soil, or climate data), tree species functional traits (maximum tree height, wood density, and leaf area index, etc.), phylogenetic relationships of tree species, and the geographic coordinates of the plots. A joint species distribution model for multiple tree species was constructed, including model structure design, selection of predictive factors, MCMC convergence test of model parameters, and model evaluation and comparison. The HMSC fitting determined environmental factors or tree species trait factors, establishing the joint species distribution model with the best fitting effect. Finally, the application and extension of the constructed joint species distribution model included interpreting tree species niches, studying the relationship between tree species niches and tree species traits and phylogeny, and the results of model comparison indicating that the information conveyed by environmental and spatial predictors overlaps to some extent. Therefore, the choice between model FULL and ENV should be based on the purpose of the model's use, not the difference in predictive ability.

Author Contributions: Conceptualization, S.C. and X.L.; Methodology, J.Y.; Software, G.D.; Data curation, X.L.; Writing—original draft, J.Y.; Writing—review & editing, J.Y., G.D., S.C. and X.L.; Supervision, S.C.; Project administration, X.L. All authors have read and agreed to the published version of the manuscript.

Funding: This research was funded by National Key R&D Program of China (Grant No. 2022YFD2200501).

Informed Consent Statement: Informed consent was obtained from all subjects involved in c study.

Data Availability Statement: The data are not publicly available due to proprietary rights.

Conflicts of Interest: Author Juan Yong was employed by the company China Forestry Group Corporation. The remaining authors declare that the research was conducted in the absence of any commercial or financial relationships that could be construed as a potential conflict of interest.

References

1. Duan, G.; Lei, X.; Zhang, X.; Liu, X. Site index modeling of larch using a mixed-effects model across regional site types in northern China. *Forests* **2022**, *13*, 815. [CrossRef]
2. Hu, X.F.; Duan, G.S.; Zhang, H.R. Modelling individual tree diameter growth of Quercus mongolica secondary forest in the northeast of China. *Sustainability* **2021**, *13*, 4533. [CrossRef]
3. National Forestry and Grassland Administration. *China Forest Resources Report 2014–2018*; China Forestry Publishing House: Beijing, China, 2019. (In Chinese)

4. Liu, D. *Research on Quantitative Place Suitability and Tree Suitability Based on Distribution Suitability and Potential Productivity*; Chinese Academy of Forestry: Beijing, China, 2018. (In Chinese)
5. Lan, J.; Lei, X.; He, X.; Gao, W.; Guo, H. Stand density, climate and biodiversity jointly regulate the multifunctionality of natural forest ecosystems in northeast China. *Eur. J. For. Res.* **2023**, *142*, 493–507. [CrossRef]
6. Leng, W.; He, H.S.; Bu, R.; Dai, L.; Hu, Y.; Wang, X. Predicting the distributions of suitable habitat for three larch species under climate warming in northeastern China. *For. Ecol. Manag.* **2008**, *254*, 420–428. [CrossRef]
7. Wang, X.; Fang, J.; Sanders, N.J.; White, P.S.; Tang, Z. Relative importance of climate vs local factors in shaping the regional patterns of forest plant richness across northeast China. *Ecography* **2009**, *32*, 133–142. [CrossRef]
8. Yin, X.J.; Zhou, G.S.; Sui, X.H.; He, Q.J.; Li, R.P. Dominant climatic factors and their thresholds in the geographical distribution of *Quercus mongolica*. *Acta Ecol. Sin.* **2013**, *33*, 103–109. (In Chinese)
9. Liu, D.; Li, Y.T.; Hong, L.X.; Guo, H.; Xie, Y.S.; Zhang, Z.L.; Lei, X.D.; Tang, S.Z. Potential distribution suitability of major natural forests in Jilin Province based on maximum entropy model. *Sci. Silvernica Sin.* **2018**, *54*, 1–15. (In Chinese)
10. Pugnaire, F.I. *Positive Plant Interactions and Community Dynamics*; CRC Press: Boca Raton, FL, USA, 2010.
11. Wisz, M.S.; Pottier, J.; Kissling, W.D.; Pellissier, L.; Lenoir, J.; Damgaard, C.F.; Dormann, C.F.; Forchhammer, M.C.; Grytnes, J.-A.; Guisan, A.; et al. The role of biotic interactions in shaping distributions and realised assemblages of species: Implications for species distribution modelling. *Biol. Rev.* **2013**, *88*, 15–30. [CrossRef]
12. Barrero, A.; Ovaskainen, O.; Traba, J.; Gómez-Catasús, J. Co-occurrence patterns in a steppe bird community: Insights into the role of dominance and competition. *Oikos* **2023**, *2023*, e09780. [CrossRef]
13. Ovaskainen, O.; Soininen, J. Making more out of sparse data: Hierarchical modeling of species communities. *Ecology* **2011**, *92*, 289–295. [CrossRef]
14. Randin, C.F.; Ashcroft, M.B.; Bolliger, J.; Cavender-Bares, J.; Coops, N.C.; Dullinger, S.; Dirnböck, T.; Eckert, S.; Ellis, E.; Fernández, N.; et al. Monitoring biodiversity in the Anthropocene using remote sensing in species distribution models. *Remote Sens. Environ.* **2020**, *239*, 111626. [CrossRef]
15. Chardon, N.I.; Pironon, S.; Peterson, M.L.; Doak, D.F. Incorporating intraspecific variation into species distribution models improves distribution predictions, but cannot predict species traits for a wide-spread plant species. *Ecography* **2020**, *43*, 60–74. [CrossRef]
16. Ovaskainen, O.; Abrego, N. *Joint Species Distribution Modelling with Applications in R*; Cambridge University Press: Cambridge, UK, 2020.
17. Clark, J.S.; Gelfand, A.E.; Woodall, C.W.; Zhu, K. More than the sum of the parts: Forest climate response from joint species distribution models. *Ecol. Appl.* **2014**, *24*, 990–999. [CrossRef] [PubMed]
18. Warton, D.I.; Blanchet, F.G.; O'hara, R.B.; Ovaskainen, O.; Taskinen, S.; Walker, S.C.; Hui, F.K. So many variables: Joint modeling in community ecology. *Trends Ecol. Evol.* **2015**, *30*, 766–779. [CrossRef] [PubMed]
19. Harris, D.J. Generating realistic assemblages with a joint species distribution model. *Methods Ecol. Evol.* **2015**, *6*, 465–473. [CrossRef]
20. Hui, F.K.C. Boral-Bayesian ordination and regression analysis of multivariate abundance data in r. *Methods Ecol. Evol.* **2016**, *7*, 744–750. [CrossRef]
21. Ovaskainen, O.; Tikhonov, G.; Norberg, A.; Blanchet, F.G.; Duan, L.; Dunson, D.; Roslin, T.; Abrego, N. How to make more out of community data? A conceptual framework and its implementation as models and software. *Ecol. Lett.* **2017**, *20*, 561–576. [CrossRef] [PubMed]
22. Zurell, D.; Pollock, L.J.; Thuiller, W. Do joint species distribution models reliably detect interspecific interactions from co-occurrence data in homogenous environments. *Ecography* **2018**, *41*, 1812–1819. [CrossRef]
23. Wang, T.; Wang, G.; Innes, J.L.; Seely, B.; Chen, B. ClimateAP: An application for dynamic local downscaling of historical and future climate data in Asia Pacific. *Front. Agric. Sci. Eng.* **2017**, *4*, 448–458. [CrossRef]
24. Poggio, L.; de Sousa, L.M.; Batjes, N.H.; Heuvelink, G.B.M.; Kempen, B.; Ribeiro, E.; Rossiter, D. SoilGrids 2.0: Producing soil information for the globe with quantified spatial uncertainty. *Soil* **2021**, *7*, 217–240. [CrossRef]
25. Tikhonov, G.; Duan, L.; Abrego, N.; Newell, G.; White, M.; Dunson, D.; Ovaskainen, O. Computationally efficient joint species distribution modeling of big spatial data. *Ecology* **2019**, *101*, e02929. [CrossRef] [PubMed]
26. Norberg, A.; Abrego, N.; Blanchet, F.G.; Adler, F.R.; Anderson, B.J.; Anttila, J.; Araújo, M.B.; Dallas, T.; Dunson, D.; Elith, J.; et al. A comprehensive evaluation of predictive performance of 33 species distribution models at species and community levels. *Ecol. Monogr.* **2019**, *89*, e01370. [CrossRef]
27. Tjur, T. Coefficients of determination in logistic regression models a new proposal: The coefficient of discrimination. *Am. Stat.* **2009**, *63*, 366–372. [CrossRef]
28. Valavi, R.; Guillera-Arroita, G.; Lahoz-Monfort, J.J.; Elith, J. Predictive performance of presence-only species distribution models: A benchmark study with reproducible code. *Ecol. Monogr.* **2022**, *92*, e01486. [CrossRef]
29. Tikhonov, G.; Opedal, H.; Abrego, N.; Lehikoinen, A.; de Jonge, M.M.J.; Oksanen, J.; Ovaskainen, O. Joint species distribution modelling with the R-package Hmsc. *Methods Ecol. Evol.* **2020**, *11*, 442–447. [CrossRef] [PubMed]
30. Abrego, N.; Ovaskainen, O. Evaluating the predictive performance of presence–absence models: Why can the same model appear excellent or poor? *Ecol. Evol.* **2023**, *13*, e10784. [CrossRef] [PubMed]

31. Wickham, H.; François, R.; Henry, L.; Müller, K. dplyr: A Grammar of Data Manipulation. R Package Version 1.1.2. 2023. Available online: https://CRAN.R-project.org/package=dplyr (accessed on 15 October 2023).
32. Wickham, H. *ggplot2: Elegant Graphics for Data Analysis*; Springer: New York, NY, USA, 2016.
33. Tikhonov, G.; Ovaskainen, O.; Oksanen, J.; De Jonge, M.; Opedal, O.; Dallas, T. Hmsc: Hierarchical Model of Species Communities. R Package Version 3.0-13. 2022. Available online: https://CRAN.R-project.org/package=Hmsc (accessed on 15 October 2023).
34. Zhang, C.; Chen, Y.; Xu, B.; Xue, Y.; Ren, Y. Improving prediction of rare species' distribution from community data. *Sci. Rep.* **2020**, *10*, 12230. [CrossRef]
35. Blanco-Cano, L.; Navarro-Cerrillo, R.M.; González-Moreno, P. Biotic and abiotic effects determining the resilience of conifer mountain forests: The case study of the endangered Spanish fir. *For. Ecol. Manag.* **2022**, *520*, 120356. [CrossRef]

forests

MDPI

Article

Effects of Environment Change Scenarios on the Potential Geographical Distribution of *Cunninghamia lanceolata* (Lamb.) Hook. in China

Jiajie Feng †, Yiwei Cao †, Teja Manda, Delight Hwarari, Jinhui Chen and Liming Yang *

State Key Laboratory of Tree Genetics and Breeding, College of Life Sciences, Nanjing Forestry University, Nanjing 210037, China; jiajiefengonly@163.com (J.F.); caoyiwei123@outlook.com (Y.C.); teja.manda27@gmail.com (T.M.); tondehwarr@njfu.edu.cn (D.H.); chenjh@njfu.edu.cn (J.C.)

* Correspondence: yangliming@njfu.edu.cn

† These authors contributed equally to this work.

Abstract: Changes in climate and environmental conditions have aggravated the severity and unpredictability of plant survival and growth. *Cunninghamia lanceolata* (Lamb.) Hook. is an economically important timber tree. Exploring its potential distribution and dynamic changes and identifying the leading environmental variables affecting it will help to adjust the planting range reasonably according to the habits and climate change, thus contributing to its survival and growth. Based on the MaxEnt model and ArcGIS tool, climate, soil, terrain, human activities, variable environment layers, and 395 *C. lanceolata* distribution points were used to simulate and analyze the geographical distribution characteristics of *C. lanceolata* in the current and future periods (the 2050s and 2070s) under RCP2.6, RCP4.5, RCP6.0, and RCP8.5. The results showed that *C. lanceolata* was suitable to grow in a subtropical monsoon climate with warm, humid, abundant rainfall and a relatively gentle topography. Additionally, using percent contribution, permutation importance, and the knife-cutting test, we noted that the annual precipitation (Bio12), human activities (Hfp), minimum temperature of the coldest month (Bio6), mean temperature of the coldest quarter (Bio11), precipitation of coldest quarter (Bio19), annual temperature range (Bio7), and elevation were the leading environmental factors affecting the geographical distribution of *C. lanceolata*. Among them, it should be noted that the impact of human activities was negatively correlated with suitable habitat areas of *C. lanceolata* and led to the degeneration of suitable habitats and fragmentized distribution. In addition, predictions have shown that the areas of habitats under other scenarios will be characterized by an increasing and then decreasing trend by the 2050s and 2070s, except for the RCP2.6 scenario, under which the suitable habitats area of *C. lanceolata* will increase continuously. The core distributional shifts showed that the suitable habitats of *C. lanceolata* will gradually shift and migrate to high-latitude areas due to global warming. This study focused on the characteristics of suitable habitats of *C. lanceolata* under different climatic scenarios using more environmental factors and scenarios than before, aiming to provide a theoretical basis and guidance for the management and utilization of forest resources, the planning of suitable planting areas, and germplasm protection.

Keywords: *Cunninghamia lanceolata* (Lamb.) Hook.; MaxEnt; geographical distribution; environmental factors; human activities; carbon emission level

Citation: Feng, J.; Cao, Y.; Manda, T.; Hwarari, D.; Chen, J.; Yang, L. Effects of Environment Change Scenarios on the Potential Geographical Distribution of *Cunninghamia lanceolata* (Lamb.) Hook. in China. *Forests* **2024**, *15*, 830. https://doi.org/10.3390/f15050830

Academic Editors: Daniela Dalmonech, Alessio Collalti and Gina Marano

Received: 13 March 2024
Revised: 5 May 2024
Accepted: 6 May 2024
Published: 9 May 2024

1. Introduction

Cunninghamia lanceolata (Lamb.) Hook. belongs to Cupressaceae, and it is popular for its moderate intensity, lightweight, and easy processing characteristics, making it a commercially important tree that is useful for landscaping, etc. [1,2]. The origin of *C. lanceolata* is in the northeast and surroundingFF areas of north China and Inner Mongolia as early as the late Jurassic period, based on the early fossil records. In light of the mass extinction during the Quaternary Ice Age, only some remnant areas remained, which were gradually excavated, utilized, and cultivated by people. The cultivation history can be traced back to the end of the Warring States period [3]. According to 2010 statistics, the planting area of *C. lanceolata* in 10 provinces in southern China was approximately 11.26 million hectares [4], and the vegetation types are mainly cultivated vegetation, coniferous forest, shrub, broad-leaved forest, grass, coniferous, and broad-leaved mixed forest, accounting for about 25% of the plantations in the subtropical region of China [5] and providing up to 30% of the harvested logs in the Chinese timber industry [6]. At present, *C. lanceolata* has become the main timber tree species in China. Moreover, the utilization of *C. lanceolata* can be more diversified, in-depth, and efficient. For instance, Yao et al. developed and proved the melamine–formaldehyde-modified furfurylation to improve the dimensional stability and mechanical properties of *C. lanceolata* and to overcome the high costs of traditional reaction methods [7]; Yan and Chang also developed a waterborne thermochromic topcoat film with color-changing microcapsules, which exhibits different colors in different seasons, as well as enhanced stability and aging resistance, consequently increasing the availability of *C. lanceolata* raw materials [8]. Sun et al. demonstrated that the volatiles of *C. lanceolata* in the natural state contain a variety of terpenoid, aliphatic, and aromatic compounds [9]. In particular, the terpenoid possesses antibacterial, anti-inflammatory, expectorant, and antitussive effects.

The increase in industrialization, greenhouse gas (GHG) emissions, and other human activities has proportionally increased global warming and its effects. This is visible through extended frost-free seasons, seasonal droughts, heat waves, extreme precipitation, etc. [10,11], resulting in disordered phenology [12], crop productivity reductions [13], increased tree mortality [14,15], loss of species diversity, and even the extinction of cold-tolerant plants [16]. Reducing GHG emissions is a top priority to increase the resilience of natural systems [17]. Therefore, in light of global warming, there is a great need to investigate the environmental factors affecting the geographical distribution and to predict the geographical distribution characteristics of plants. It is of great significance for plant protection, introduction, and development, as well as resource distribution [18]. Species distribution models (SDMs) mainly use existing species distribution and environmental data to estimate the species niches using specific algorithms. In addition, the models reflect the species' preference for habitat as a probability, which in turn predicts the range of suitable habitats for plant species well [18]. At present, the commonly used models for predicting the potential distribution of species include GARP, MaxEnt, CLIMEX, and BIOCLIM. Among them, MaxEnt is widely used due to its good prediction effect, stability, and simple and rapid operation [19]. Liu et al. analyzed the impact of human activities on the environment and the impact of climate change on the distribution of three Cypripedium species in northeast China using the MaxEnt model combined with ArcGIS technology [20]. Likewise, the MaxEnt model was also used by Garah and Bentouati to predict the potential geographical distribution of Aurasian Aleppo pine (*Pinus halepensis* Mill.) [21]; Wang et al. compared four niche models: GARP, BIOCLIM, DOMAIN, and MaxEnt., and concluded that the prediction result of MaxEnt was comparably the most accurate [22].

Previous research has mainly reported the impacts of climate factors on suitable habitats of *C. lanceolata* in China and found that annual average precipitation and minimum temperature of the coldest month were the main factors affecting the distribution of *C. lanceolata* [23,24]. Given the importance of human activities, soil, and topography on distributions of species populations, this research collected all of relevant information, such as the existing investigation and research data of *C. lanceolata*, specimen data, as well as digital network information of climate, ecological, and environmental factors, under different carbon emission scenarios to characterize the geographical distributions of *C. lanceolata* at present and in the future.

The objectives of this study were: (1) to identify the potential distribution of *C. lanceolata* in China under the current climatic conditions and the leading environmental factors affecting the potential distribution of *C. lanceolata*; (2) to analyze the response curves of environmental factors and predict the environmental characteristics of suitable habitats of *C. lanceolata*; and (3) to predict the potential distribution and dynamic change characteristics of *C. lanceolata* in China under future climate change scenarios. This study will provide a scientific theoretical basis for studying the distribution pattern of *C. lanceolata* and preventing the decrease in *C. lanceolata* resources.

2. Materials and Methods

2.1. Collection and Processing of Species Distribution Data and Environmental Factors

Distribution data: The geographical information data of *C. lanceolata* used in this paper were collected from National Specimen Information Infrastructure (http://www.nsii.org.cn/, accessed on 1 June 2022), National Plant Specimen Resource Center (https://www.cvh.ac.cn/index.php, accessed on 1 June 2022) and Global Biodiversity Information Facility (https://www.gbif.org/, accessed on 3 June 2022). All the specimen data since the 19th century were selected, and duplicate samples, fuzzy records, and artificial cultivation records were screened out and deleted, only keeping specimens of wild communities. The spatial resolution adopted in this paper was 2.5 arc-minutes (about 4.5 km), and the buffer diameter was set at 3 km; when two distribution points were in the same buffer, a single point was reserved [25]. A total of 389 *C. lanceolata* specimen records (Figure 1) were obtained, which were processed by EXCEL 2016 into CSV. Format files contained only species name, longitude, and latitude to make them convenient for the subsequent construction of the MaxEnt model.

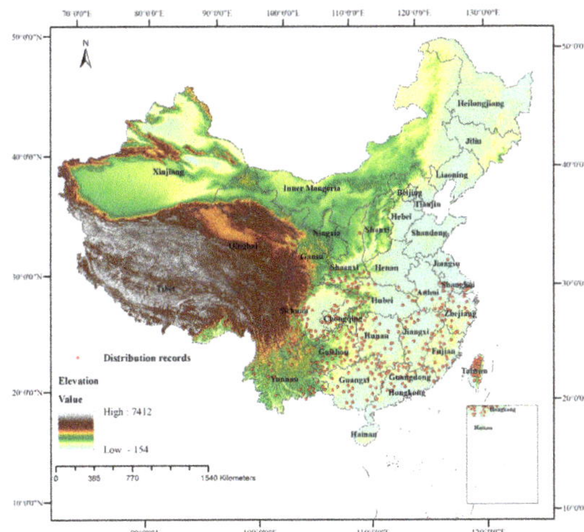

Figure 1. The geographical location of distribution points of *Cunninghamia lanceolata* (Lamb.) Hook. in China.

Climate factors: 19 present and future bioclimatic factors were obtained from the World Climate Data (http://www.worldclim.org/, accessed on 15 June 2022). Future climate variables were based on the prediction of global future climate change, which was made in the fifth assessment report published by IPCC. The atmospheric circulation model and the BCC_CSM model developed by China National Climate Center were adopted, in which RCP2.6, RCP4.5, RCP6.0, and RCP8.5 represented various scenario assumptions about future climate under four different carbon emission scenarios, respectively. Additionally, the latter figures indicated that the radiation forcing level will be 2.6 W m^{-2} to 8.5 W m^{-2} by 2100. In total, there were 8 combinations of climate scenarios considered by this article: RCP2.6—2050s, RCP2.6—2070s, RCP4.5—2050s, RCP4.5—2070s, RCP6.0—2050s, RCP6.0—2070s, RCP8.5—2050s, and RCP8.5—2070s.

Terrain factor: elevation data were obtained from the Shuttle Radar Topography Mission (SRTM), which could be obtained from the world climate database (http://www.worldclim.org/, accessed on 20 June 2022). Global terrain slope and aspect data were obtained from the World Soil Database (https://www.fao.org/soils-portal/data-hub/soil-maps-and-databases/harmonized-world-soil-database-v12/en/, accessed on 20 June 2022).

Soil factors: 16 soil data points were obtained from the World Soil Database (https://www.fao.org/soils-portal/data-hub/soil-maps-and-databases/harmonized-world-soil-database-v12/en/, accessed on 20 June 2022).

Human activity factor: the human activity intensity data were obtained from the V2 (1995–2004) data set of the Global Human Footprint of Socioeconomic Data and Applications Center (https://sedac.ciesin.columbia.edu/data/set/wildareas-v2-human-footprint-geographic, accessed on 1 July 2022). See Table S1 for details of all environmental factors.

2.2. Preprocessing of Environmental Factor Data

Firstly, all the data were processed according to the vector map of the 1:1,000,000 administrative division of China and converted into ASC file format using "Toolbox/Conversion Tools/From Raster/Raster to ASCII" in ArcGIS. In addition, Chinese soil and the HWSD DATA files were imported into ArcGIS to establish a connection; then, 16 preliminarily selected raster layers of surface soil factors in the MU_GLOBAL layer were extracted and converted into ASC format files. Secondly, further processing of environmental factors was conducted. All environmental variables were extracted using the mask tool in ArcGIS10.6, "toolbox/spatial analyst tools/extraction/extract by mask", and further resampled with "toolbox/data management tools/raster/raster processing/Resample". Finally, the environmental variables of data grid image range and pixel size were completely consistent, with a unified resolution of 2.5 min and a geographic coordinate system GCS_WGS_1984. To avoid over-fitting among environmental variables and to ensure the accuracy of the model's operation, 50 environmental variables and data of *C. lanceolata* distribution points were imported into MaxEnt 3.4.1 for pre-simulation, and the contribution rate of each environmental variable was preliminarily obtained. The "Multivariate" tool in ArcGIS 10.6 was used to conduct a multiple linear analysis of 50 environmental variables. In addition, when the correlation of two environmental variables was $\geq$0.8, the environmental variable with a larger contribution rate in the pre-simulation was retained, and then the subsequent secondary simulation was carried out [26].

2.3. Model Building and Accuracy Evaluation

Environmental variables and *C. lanceolata* distribution point data were imported into MaxEnt 3.4.1; then, "Basic" was set to 25% of the distribution data as a test set to verify the accuracy and 75% of the distribution data was used as a training set to drive the model. Then, "Random seed" was checked, followed by "Subsample" selection; for the repetition type, the operation was repeated 10 times, and output distribution values were set in "logistic" format. Afterwards, "create response curves" was selected to draw curves of how climate factors determine the predicted occurrence probability value, and "do jackknife" was selected to output the contribution rate of each climate factor. Other parameters were set as default, and the file was output as ASCII type.

The contribution rate of each environmental variable to the distribution of *C. lanceolata* habitats was calculated using the Jackknife method. The receiver operating characteristic curve (ROC curve) was drawn with specificity as the abscissa and sensitivity as the ordinate. The quality of the model was determined by the area under the ROC curve (AUC value). The numerical range of AUC was set at 0~1, and the larger the numerical value, the higher the accuracy of the model. Theoretically, when the AUC value was 0.5~0.6, the model had no prediction ability; at 0.6~0.7, the prediction ability was poor; at 0.7~0.8, the prediction ability was medium; at 0.8~0.9, the prediction ability was good; and the accuracy of the model was extremely high when the AUC value was >0.9 [27]. Furthermore, the prediction results were imported into ArcMap 10.6, and the classification tool was used to divide the layer into four grades using the natural breaks (Jenks) method [23]. Various suitable habitats were obtained based on *p*-value, as follows: unsuitable habitat ($p < 0.07$), low suitable habitat ($0.07 < p < 0.2$), moderate suitable habitat ($0.2 < p < 0.5$), and high suitable habitat ($p > 0.5$). Finally, the distribution maps of current and future suitable habitats of *C. lanceolata* were made.

2.4. Dynamic Changes in the Distribution of Suitable Habitats under Different Climatic Scenarios in the Future and Core Distributional Shifts under Different Climatic Scenarios

According to the calculation result of MaxEnt, the potential distribution map of species was obtained by loading the result and visualizing the suitable habitat grades in ArcGIS 10.6. The change in the area of potential distribution habitats and core distributional shifts of *C. lanceolata* in the future were analyzed using the SDM toolbox v2.5, written in Python language in ArcGIS 10.6 [28]. The statistics of the area mainly showed three types of changes: expansion, unchanged, and contraction. In this research, this tool was used to investigate the changes in the distribution of *C. lanceolata* habitats in different scenarios compared with current suitable habitats.

As mentioned above, the SDM toolbox (Tool of ArcGIS based on Python 2.7.14) was also used to treat the suitable habitats of *C. lanceolata* as a whole and to reduce them to a vector particle. Then, the changing trend of the suitable habitats and the distributional core position of suitable habitats in four scenarios in the present and future were calculated. The geometric core represents the overall spatial position of the suitable habitats of *C. lanceolata*, and the change reflects the overall spatial migration trend of suitable habitats [21].

3. Results

3.1. Accuracy Test of MaxEnt Model and the Leading Environmental Factors Affecting the Distribution of C. lanceolata

The AUC value of the area under the ROC curve was used to test the accuracy of the model. After 10 calculations, the mean test AUC was 0.947 (Figure 2), showing that the model could accurately predict suitable habitats.

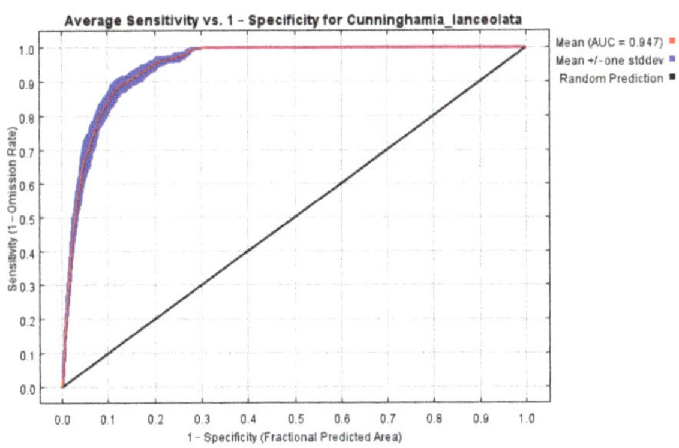

Figure 2. The receiver operating characteristic (ROC) curve of the MaxEnt model.

According to the contribution percentage in the first simulation and the correlation among factors (Figure 3), 16 environmental factors were selected, including 8 climatic factors, 5 topographic factors, 2 soil factors, and human activities (Hfp). Then, the 16 factors were simulated for the second time, and the leading environmental variable factors that limited the potential geographical distribution of *C. lanceolata* in China were selected based on the percent contribution, permutation importance, and knife-cutting test. The percent contribution was used to indicate the contribution degree of the variable to the model. The permutation importance indicated the dependence of the model on this variable. The knife-cutting method was used to analyze the relationship between different environmental factors and the distribution of suitable habitats for *C. lanceolata*. Criteria: The blue band represents "With only variable", and the longer the band, the higher the score, indicating that this variable has a strong predictive ability for species distribution. The cyan band represents the training score of "Without variable", that is, the sum of the contributions of the remaining variables except for this variable. If the score of "Without variable" is low, it means that this variable contains some unique information, which is also important for species distribution. The red stripe represents the cumulative contribution rate of all environmental variables to the established model.

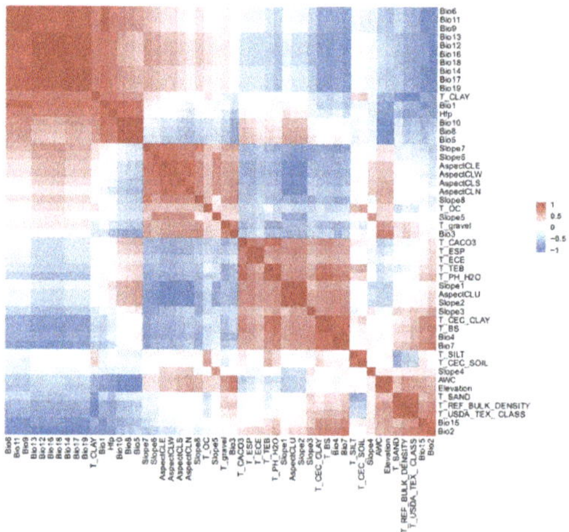

Figure 3. Correlation analysis of environmental factors.

According to the contribution percent (Figure S1), the environmental factors that reached more than 10% were Bio12 (41.3%), Hfp (20%), Bio19 (13.1%), and Bio6 (11.1%). The cumulative contribution rate of these four factors reached 85.5%, and they were the leading variables affecting the distribution of *C. lanceolata*.

According to the permutation importance (Figure S2), the leading variables reaching more than 5% were Bio12 (40.1%), Hfp (19.2%), Bio7 (17.4%), Bio6 (6.5%), and Elevation (5.3%).

According to the analysis of the normalized training gain obtained by the knife-cutting test, under the condition of "With only variable", the scores of Bio3, SlopeCl1, SlopeCl3, AspectCIU, and AspectCIE were lower than 0.2, suggesting little influence on the distribution of *C. lanceolata*. Interestingly, most terrain factors had little influence on the prediction results. However, the scores of Bio11, Bio12, and Bio6 were over 1.2, which showed that precipitation and temperature were the leading environmental factors. In addition, the "Without variable" score of human activities was low, indicating the importance of this variable to the geographical distribution of *C. lanceolata*, which should not be ignored (Figure 4). Therefore, Bio11, Bio12, Bio6, and human activity factors (Hfp) were determined to be the leading variables for the knife-cutting test.

Combined with the percent contribution, permutation importance, and knife-cutting test, Bio12, Hfp, Bio6, Bio11, Bio19, Bio7, and Elevation were eventually selected as the leading environmental factors.

The above content is the importance test of the influence of each factor on the distribution of *C. lanceolata* habitats at present, the results of eight future scenarios were roughly the same as those of current times. Bio12, Hfp, Bio6, and Bio19 still accounted for more than 80% of the contribution percent; Bio12, Hfp, and Bio7 still accounted for more than 80% of the permutation importance; and the results of knife-cutting were not different. (For details, see Table S2).

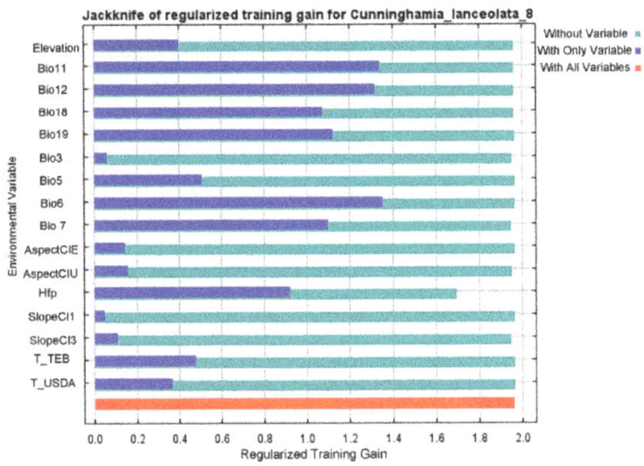

Figure 4. Results of knife-cutting of environmental factors.

To further clarify the relationships between the distribution habitats of *C. lanceolata* and leading environmental factors, the response curves of the single environmental factor (Bio12, Hfp, Bio6, Bio11, Bio19, Bio7, and Elevation) were obtained based on the MaxEnt model (Figure 5). These curves show how the predicted probability of species distribution changed with each environmental variable.

As shown in Figure 5, the presence probability of *C. lanceolata* fluctuated greatly with the change in environmental variables. The presence probability response curve of *C. lanceolata* to seven leading environmental factors was divided by 0.5 as the dividing line, and the range with a presence probability greater than 0.5 was considered as the suitable environmental interval conducive for the growth of *C. lanceolata*, as shown in Figure 5. The thresholds for the leading environmental parameters were obtained at a presence probability of >0.5. Regarding elevation, its influence on the presence probability of *C. lanceolata* was stable and remained unchanged below 0 m. However, the increase in the presence probability of *C. lanceolata* was proportionally linked to the increase in elevation. In detail, the presence probability started to rapidly increase at −100 m, peaked at 100 m, and then plummeted. Regarding the minimum temperature of the coldest month (Bio6), when the temperature was below −10 °C, it was unfavorable for the growth of *C. lanceolata*. As the temperature gradually increased, the presence probability also rapidly increased and peaked when the temperature reached about 12 °C, and then decreased rapidly. The influence of the temperature annual range (Bio7) on the presence probability was stable at first, then increased when the temperature range rose from 13 °C, and reached its peak when the range was 16 °C. The presence probability was slightly decreased after that, then slightly increased when the temperature range increased to 22 °C, and finally dropped sharply.

As for the mean temperature of coldest quarter (Bio11), *C. lanceolata* showed no growth below −5 °C, suggesting that this is not suitable for its growth. In addition, its presence probability increased with the increase in the mean temperature of the coldest quarter until the temperature reached about 14 °C, and then the presence probability of *C. lanceolata* reached its peak and later had a steep decrease. Concerning the annual precipitation (Bio12), the presence probability gradually increased with its increase from 500 mm, then stabilized at 1100 mm~1800 mm and increased sharply after that. When the precipitation reached about 3250 mm, the presence probability reached a maximum value and then decreased, but remained above 0.5. Moreover, as for precipitation of coldest quarter (Bio19), the presence probability sharply increased when Bio19 was from 10 mm to 50 mm, and increased constantly later, peaking at about 0.9 when Bio19 stabilized at 450 mm. The presence probability of *C. lanceolata* increased layer by layer with the increase in the intensity of human activities. Generally speaking, the presence probability of *C. lanceolata* increased rapidly when human activity factors were between 10 and 50. After 50, the presence probability rose slowly and dropped to 70, then rose rapidly until about 91, and after that, the probability reached a maximum value. Wholly, the elevation ranged from 0 to 600 m, the minimum temperature of the coldest month (Bio6) ranged from 1 to 12 °C, the temperature annual range (Bio7) ranged from 14 to 30 °C, the mean temperature of coldest quarter (Bio11) ranged from 6 to 15 °C, the annual precipitation (Bio12) ranged from 1100 to 4700 mm, the precipitation of the coldest quarter (Bio19) ranged from 120 to 510 mm, and the human activity index ranged from 40 to 105 (Table 1).

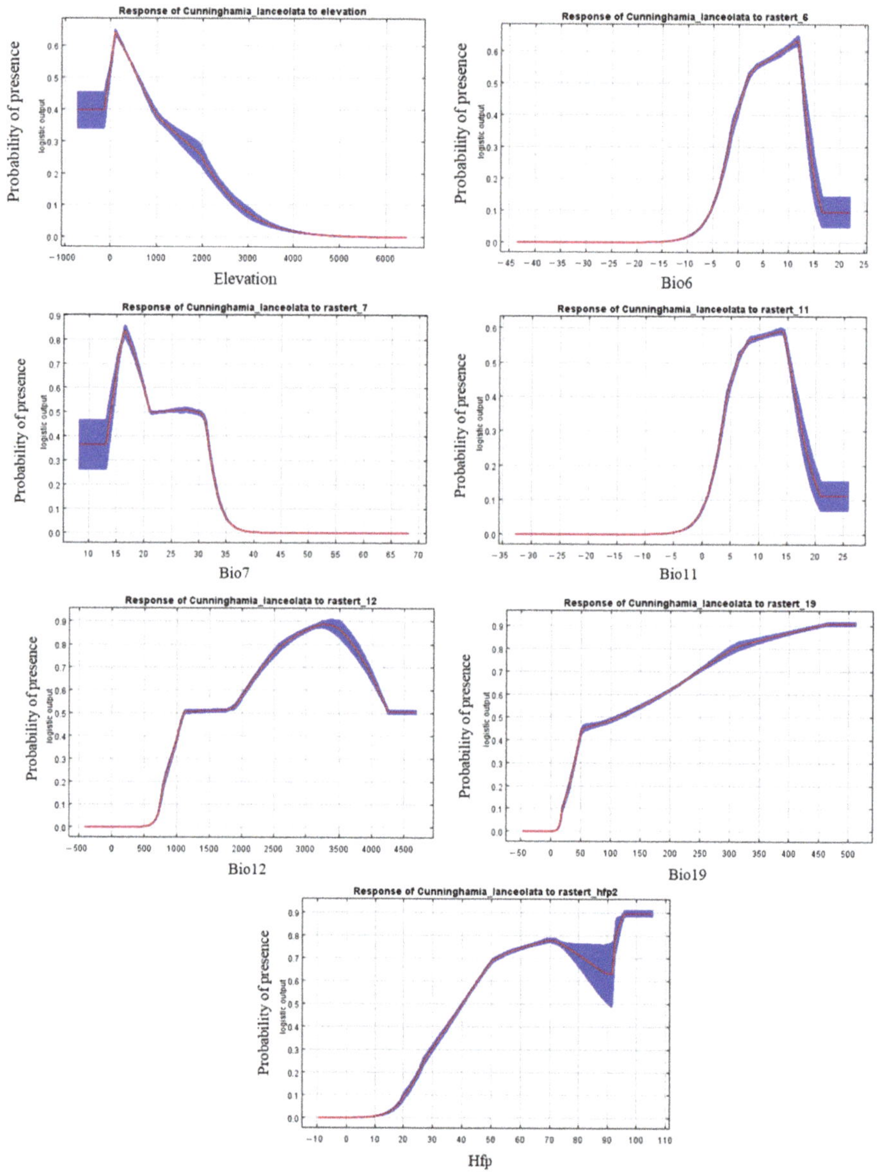

Figure 5. Response curve of leading environmental factors. The curves show the mean response of the 10 replicate Maxent runs (red) and the mean ± one standard deviation (blue, two shades for categorical variables).

Table 1. The suitable range of the leading variable environmental factors.

Environmental Factors	Elevation (m)	Bio6 (°C)	Bio7 (°C)	Bio11 (°C)	Bio12 (mm)	Bio19 (mm)	Hfp
Suitable minimum value	0	1	14	6	1100	120	40
Suitable maximum value	600	12	30	15	4700	510	105

3.2. Potential Geographical Distribution of Suitable Habitats in Different Periods

3.2.1. Current Geographical Distribution of Suitable Habitats

According to the division of natural breaks (Jenks) method, the currently suitable habitats of *C. lanceolata* were divided into four types: unsuitable habitat, low suitable habitat, moderate suitable habitat, and high suitable habitat.

As can be seen from Figure 6, suitable habitats were mainly distributed in the south of the Yangtze River, with the main distribution range between 97.59°~121.24° E and 20.303°~34.575° N. According to the area calculation, the total geographical distribution habitat area of *C. lanceolata* was 193.78 × 10^4 km^2, accounting for 20.19% of China's land area, among which the high suitable habitat accounted for 2.62%, the moderate suitable habitat accounted for 6.10%, and the low suitable habitat accounted for 11.47%.

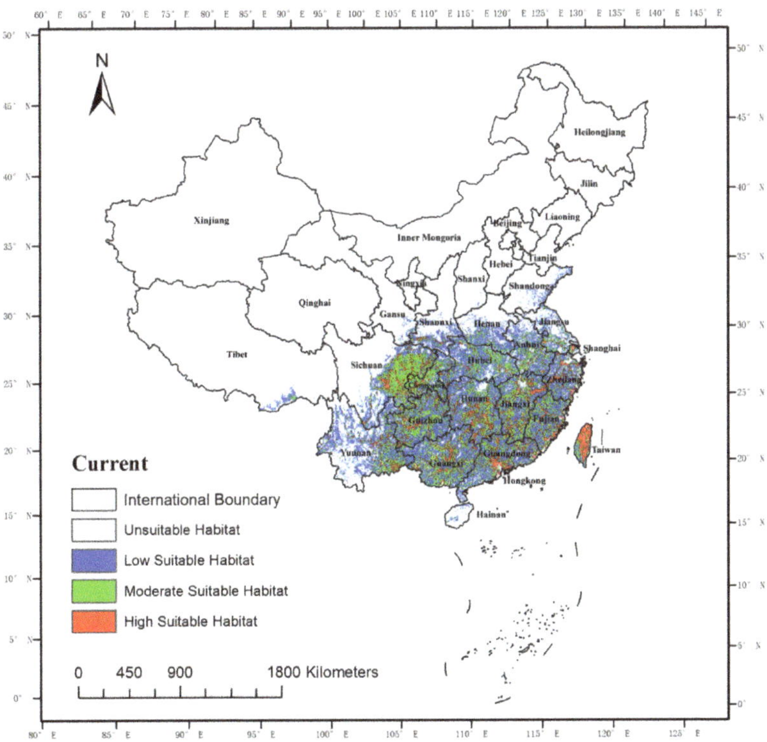

Figure 6. Current distribution of suitable habitats of *C. lanceolata*.

To explore the influence of human activities (high percent contribution) on the distribution of *C. lanceolata*, human activities were removed and the simulation was conducted again. It was found that the distribution of suitable habitats of *C. lanceolata* was concentrated and fragmented rather than scattered, characterized by an increase in the moderate and high suitable habitats (Figure 7).

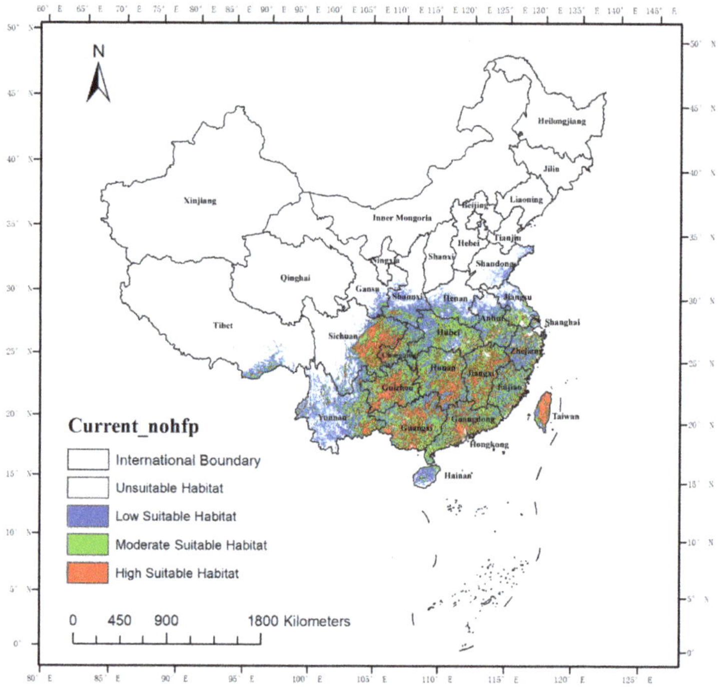

Figure 7. Current distribution of suitable habitats of *C. lanceolata* without human activities.

Furthermore, the distribution area range did not change much, except in the expansion of Hainan Island. The distribution of suitable habitats without interference of human activities was upgraded to a low suitable habitat in the original unsuitable habitat, the low suitable habitat was upgraded to a moderate suitable habitat, and the moderate suitable habitat was upgraded to a high suitable habitat. In the area calculations without human activity factors, the total geographical distribution area of *C. lanceolata* was 218.65 $\times$ 10^4 km^2, accounting for 22.78% of China's land area and being 2.60% higher than the original suitable habitat area. The high suitable habitat accounted for 4.93%, the moderate suitable area accounted for 8.67%, and the low suitable area accounted for 9.18% of the total suitable habitat area. Without the impact of human activities, high suitable habitats were mainly distributed in eastern Sichuan, western Chongqing, Guizhou, Taiwan, Guangxi, southern Guangdong, Hunan, and Jiangxi. Moderate suitable habitat was widely distributed in the south of the Yangtze River, inlaid with the high suitable habitat. Generally, subtropical monsoon climate areas and tropical monsoon climate areas in the south of the Yangtze River were more suitable for the growth of *C. lanceolata*. The precipitation conditions were relatively abundant and the temperature conditions were relatively appropriate. Not considering the influence of human activities, these two kinds of areas were mainly high and moderate suitable habitats for *C. lanceolata*.

Considering the influence of human activities, the distribution of *C. lanceolata* was fragmented, and we speculated that human activities interrupted the distribution of the original suitable habitats and led to its eastward and southward reduction. Hong et al. also showed that the natural distribution habitats of *C. lanceolata* were continuous at first, and then discontinuous later due to the influence of climate and human factors [29]. The influence of human activities on the distribution of other species was similar [20]. In terms of the total area, the suitable habitats area was reduced by 24.87 $\times$ 10^4 km^2 due to human activities, which indicated that there was a negative correlation between human activities and the distribution of *C. lanceolata*.

3.2.2. Potential Geographical Distribution of Suitable Habitats in the Future

In this paper, four RCP scenarios and two future periods were selected. Compared with the current times, the total suitable habitat area of *C. lanceolata* increased and migrated by different degrees in each period (Figure S3). Areas of potential distribution under different climate scenarios of *C. lanceolata* are shown in Table S3.

In the 2050s, the proportions of the total suitable habitat area of *C. lanceolata* in China's land area under the RCP2.6, RCP4.5, RCP6.0, and RCP8.5 scenarios were predicted to be 21.49%, 21.60%, 22.34%, and 21.70%, respectively, which are higher than that in current times (20.18%), with the area under RCP6.0 being the largest. In contrast to the potential total suitable habitats in the 2050s (except for RCP2.6), the areas under other scenarios were greatly reduced. By the 2070s, their area proportions are predicted to account for 21.80%, 20.59%, 21.73%, and 20.80%, respectively; of these four scenarios, the area under RCP2.6 will be the largest.

In addition, the area of habitat under RCP8.5 will be the largest by the 2050s, and by the 2070s, the largest will be RCP2.6, not accounting for the influence of human activities. The proportions of the total suitable habitat area of *C. lanceolata* in China's land area under the RCP2.6, RCP4.5, RCP6.0, and RCP8.5 scenarios will be 23.35%, 22.43%, 23.30%, and 23.63% in the 2050s, respectively. By the 2070s, the proportion of habitat area under four scenarios will be 23.82%, 23.56%, 23.42%, and 22.57%, respectively. This result once again shows that human activities have a considerable impact on the potential distribution of *C. lanceolata*, so this factor is additionally considered and analyzed below.

Considering different GHG emissions in the same period, the 2050s, the areas of low suitable habitat under different GHG emission scenarios were RCP8.5 > RCP6.0 > RCP4.5 > RCP2.6. In the moderate suitable habitats, the areas under different scenarios were RCP6.0 > RCP2.6 > RCP8.5 > RCP4.5, while the areas for high suitable habitats were RCP6.0 > RCP4.5 > RCP2.6 > RCP8.5. On the whole, RCP6.0 was the most ideal based on the area of each suitable habitat under different scenarios. Under this scenario, not only was total suitable habitat area was the largest, but the middle and high suitable habitat areas were also the largest. In the 2070s, the area of low suitable habitat of *C. lanceolata* under different GHG emission scenarios was ordered as RCP2.6 > RCP6.0 > RCP4.5 > RCP8.5. The area condition of the moderate suitable habitat under different GHG emission scenarios was RCP2.6 > RCP6.0 > RCP8.5 > RCP4.5. Finally, the high suitable habitat areas were ordered as RCP8.5 > RCP2.6 > RCP6.0 > RCP4.5. Noteworthily, the RCP2.6 scenario was the highest in this period based on the distribution areas of moderate and high suitable habitats.

Considering different periods and the same GHG emission scenarios under the four carbon emission scenarios, the low suitable habitat of *C. lanceolata* in the 2070s was reduced compared with the 2050s, and the largest contraction was 12.08×10^4 km^2 under RCP8.5, while the smallest was 0.34×10^4 km^2 under RCP2.6. Likewise, in the moderate suitable habitat, the area of the 2070s under RCP2.6 climate scenarios increased greatly compared with that of the 2050s. In addition, the situation in high suitable habitat was similar to that of the moderate suitable habitat. It should be noted that, under the RCP6.0 scenario, the area cover fluctuated greatly. Wholly, the general trend of suitable habitats was that the total habitat area increased continuously and reached the highest value of 209.27×10^4 km^2 under RCP2.6, but under RCP4.5, RCP6.0, and RCP8.5, it first increased in the 2050s and then decreased in the 2070s.

3.3. Dynamic Changes of Distribution Habitats under Different Climatic Scenarios in the Future

The distribution of *C. lanceolata* habitats in different scenarios in the future two periods is shown in Figure 8. Under RCP2.6—2050s, both expansion and contraction were predicted to be obvious. The expansion was mainly concentrated in Hainan province and the northern regions like Shaanxi, Henan, the west of Yunnan, and Sichuan. The reduction in the middle and lower reaches of the Yangtze River affected the performance of the overall suitable habitats, but in the 2070s, the contraction area was obviously reduced, and the expansion area was slightly increased. At this time, the total suitable area reached the maximum among all scenarios in the 2070s. Notably, the southern Tibet valley and the Ali area of the

Tibet Autonomous Region were also expanded; other studies have shown that these two places may have acted as shelter escapes for plants in the ice age [30,31].

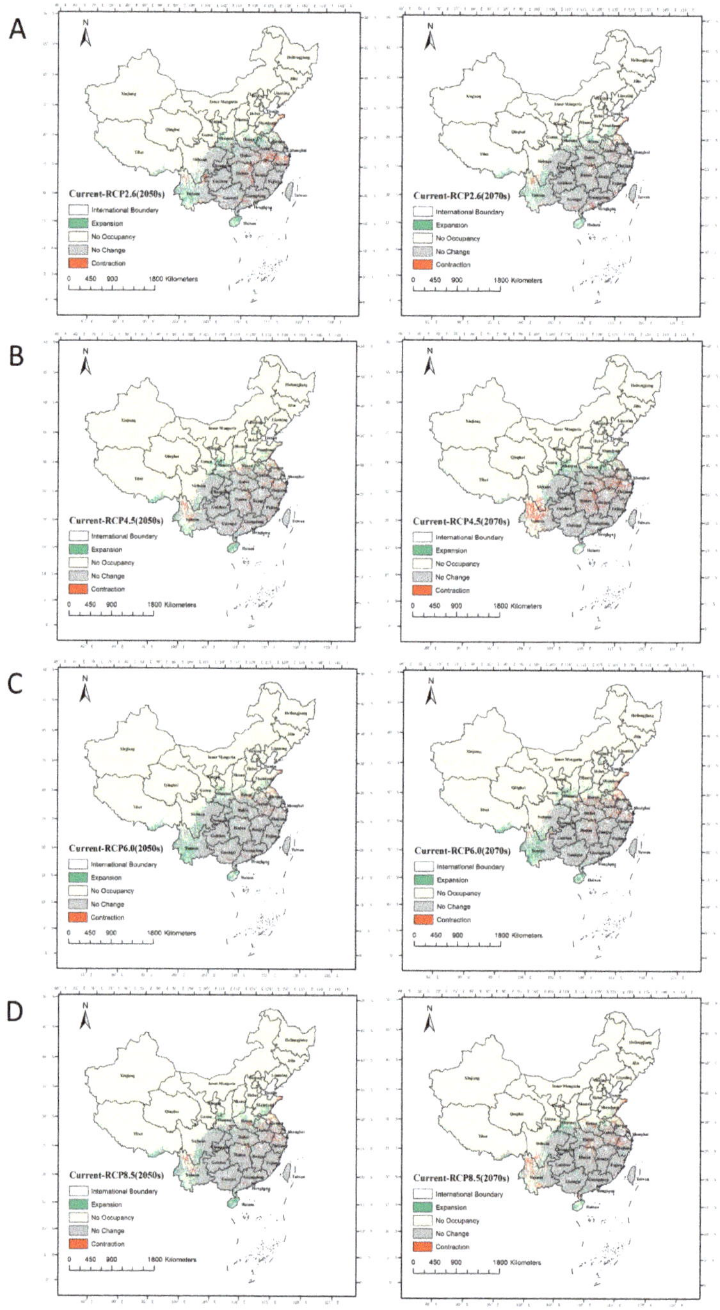

Figure 8. Change in the distribution of *C. lanceolata* habitats under different scenarios in the 2050s and 2070s. (**A**) RCP2.6, (**B**) RCP4.5, (**C**) RCP6.0, (**D**) RCP8.5.

RCP4.5—2050s was similar to RCP2.6—2050s, but by the 2070s, the contraction was predicted to be 13.37 × 10^4 km^2 and to reach the maximum. However, the expansion was lower, and it was the least suitable area under this scenario of the 2070s. Compared to the 2050s, the areas of the three kinds of suitable habitats were all reduced, especially the low suitable habitat. The contraction area was mainly concentrated in Yunnan and the middle and lower reaches of the Yangtze River.

Under RCP6.0, the situations in the 2050s and that in the 2070s had few differences. In the 2050s, the maximum expansion of the suitable area reached 27.33×10^4 km^2, especially in the west and north, and reduction was at its minimum, being 4.15×10^4 km^2. The suitable habitat condition in the 2070s was slightly worse than that in the 2050s. Compared with the 2050s, the areas of the three suitable habitats decreased, especially the high suitable habitat.

RCP8.5—2050s had the most obvious change in Yunnan, in contrast to RCP6.0—2050s, which can be observed in Figure 8. Most of the original extended regions disappeared. By the 2070s, the expansion was the least, especially in the western region. Expanded areas (mainly low suitable habitat) retracted around the range of current suitable habitats such as northeast Sichuan and southwest Shaanxi. The contraction of suitable areas was mainly reflected in Yunnan, Hunan, Hubei, Zhejiang, Anhui, Henan, and the coastal areas of Shandong and Jiangsu. All the areas of expansion and contraction under different scenarios are shown in Table 2.

Table 2. Changes in suitable habitat areas under different scenarios in the 2050s and 2070s.

	RCP2.6		**RCP4.5**		**RCP6.0**		**RCP8.5**	
	2050s	**2070s**	**2050s**	**2070s**	**2050s**	**2070s**	**2050s**	**2070s**
Expansion /×104 km^2	22.76	23.16	23.27	17.94	27.33	24.88	22.83	16.44
contraction /×104 km^2	8.46	5.74	8.04	13.37	4.15	7.81	6.26	9.14
No change /×104 km^2	205.96	208.68	206.39	201.05	210.28	206.61	208.16	205.28
No occupancy /×104 km^2	722.82	722.42	722.31	727.64	718.25	720.70	722.75	729.14

3.4. The Core Distributional Shifts under Different Climatic Scenarios

The centroid of the current habitats of *C. lanceolata* was located in Xinhua County, Loudi City, Hunan Province, China (27.71° N, 111.03° E). Under RCP2.6—2050s, the distribution centroid shifted to Xinxupu County, Huaihua City, Hunan Province (27.81° N, 110.56° E). Under RCP2.6—2070s and RCP4.5—2050s, the distribution centroids were the same as those of RCP2.6—2050s, while the coordinates were slightly different (27.82° N, 110.63° E and 27.84° N, 110.62° E). The centroid of RCP4.5—2070s shifted to Anhua County, Yiyang City, Hubei Province (28.03° N, 110.88° E). And that of RCP6.0—2050s was XinXupu County (27.81° N, 110.39° E), then shifted to Zhongfang County, Huaihua City, Hunan Province (27.64° N, 110.27° E) under RCP6.0—2070s. The distribution centroids under RCP8.5—2050s and RCP8.5—2070s were XinXupu County, Huaihua City, Hunan Province (27.80° N, 110.60° E) and Xinhua County, Loudi City, Hunan Province (27.81° N, 110.80° E), respectively.

In the 2050s, the centroid under RCP2.6 migrated 47.83 km to the northwest, 42.99 km to the northwest under RCP4.5, 64.49 km to the northwest under RCP6.0, and 43.59 km to the northwest under RCP8.5. In the 2070s, the centroid of RCP2.6 migrated 6.69 km northeast compared with that of 2050s, that of RCP4.5 moved 32.26 km northeast, that of RCP8.5 moved 20.22 km northeast and returned to the centroid in current times, that of RCP6.0 moved in a different direction from the others, the centroid moved 19.97 km southwest. However, the centroid under RCP6.0 kept going west and had a trend to the south by the 2070s, while the centroids under other scenarios all went northwest first and then northeast (Figure 9).

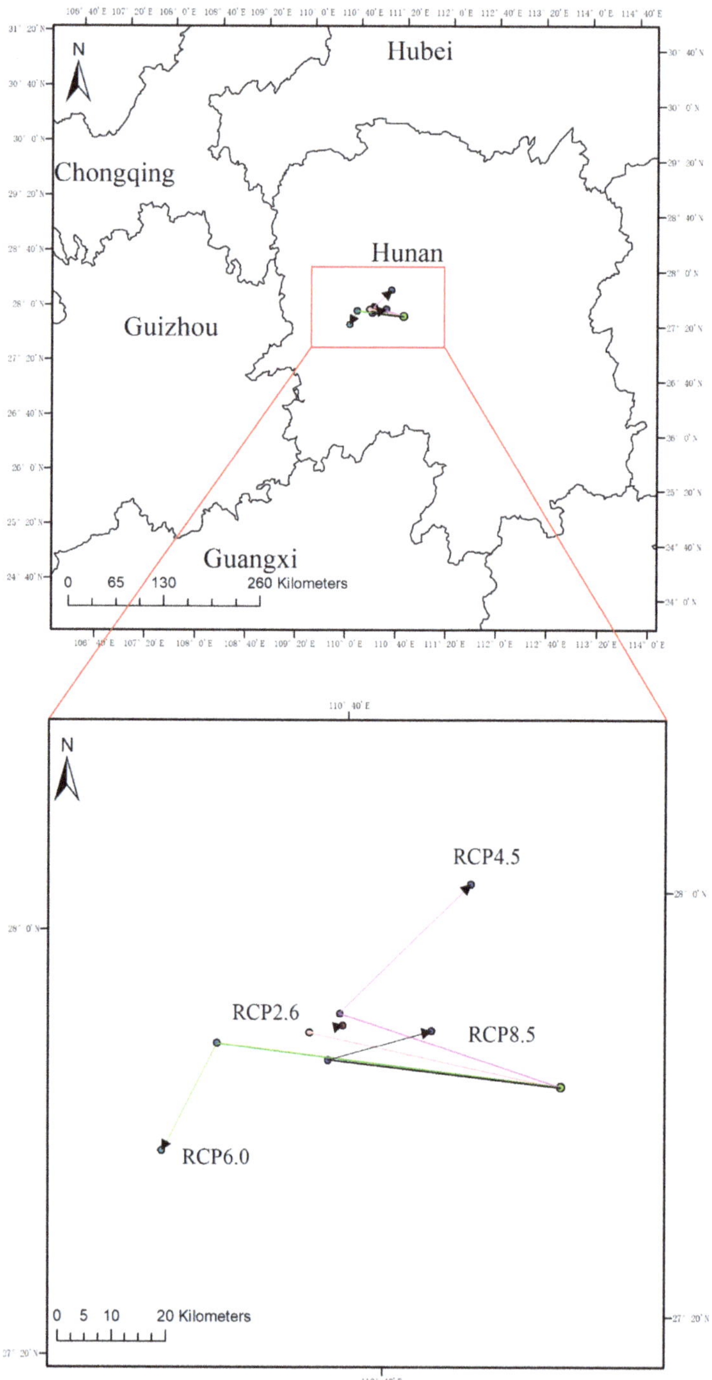

Figure 9. The core distributional shifts of *C. lanceolata* under different climate scenarios (the pink line represents RCP2.6, the purple line represents RCP4.5, the green line represents RCP6.0, and the black line represents RCP8.5).

4. Discussion

4.1. Changes in Suitable Habitats of C. lanceolata *in Different Periods*

Based on the MaxEnt model, this study predicted the potential suitable habitats of *C. lanceolata* in China, using climatic factors, human activity, soil factors, and terrain factors. The results showed that the current potential suitable habitats of *C. lanceolata* are mainly concentrated in the areas south of Qinling-Huai River, such as the Sichuan Basin and its surrounding areas, Yunnan Plateau, the middle and lower reaches of the Yangtze River and South China. The contemporary predicted results were consistent with the actual distribution of *C. lanceolata* in China [32], and the mean AUC value of the ROC curve was 0.947, which indicated that the prediction result was credible. By simulating and predicting the distribution range of *C. lanceolata* in different periods, the future change trend was investigated.

This paper studied the distribution of *C. lanceolata* habitats in current and future times, and its historical origin was also valued by many scholars. According to the research [3], the origin time of *C. lanceolata* was in the late Jurassic, and its origin centers were in the northeast of China, north China and Inner Mongolia, and the southeast of Siberia, Russia. However, due to the harsh climate during the Quaternary glacial period, the cold climate zone migrated to the middle and low latitudes, which led to the widespread development of ice sheets or glaciers at high latitudes and on mountains and made *C. lanceolata* migrate to the south; most origins also became extinct one after another. According to the fossil record of sporopollenin, *C. lanceolata* was mainly distributed in Lantian, Shaanxi Province; Yuanmou, Yunnan Province; Jiujiang and Nanchang, Jiangxi Province; coastal areas of Jiangsu and Zhejiang Provinces; Lixian and Changde, Hunan Province; Taihu Lake, Jiangsu Province; Sanshui, Zhongshan, and Dongguan, Guangdong Province; Hui'an Fujian Province; the Mianning and Anning River basins in Sichuan Province; Anqing, Anhui Province; Hangzhou Bay, Yuyao Plain, and Ningfeng Plain in Zhejiang Province, etc. [33–37]. Before current times, it formed its distribution south of the Qinling Mountains-Huaihe River, and these areas were also considered as the remaining land of *C. lanceolata*. Additionally, it was cultivated artificially as early as the pre-Qin period and the Spring and Autumn Warring States period, and was introduced to the Yellow River Basin in the Qin and Han Dynasties. Historically, its cultivated areas were mainly concentrated in the middle and lower reaches of the Yangtze River, Nanling Mountain, Wuyi Mountain, Xuefeng Mountain, and Sichuan Basin. In addition, there are also records of its introduction in Taiwan Province [38–40]. Furthermore, SSR and cpDNA molecular markers have been used to prove that *Cunninghamia konishii* Hayata was introduced and cultivated artificially across the Taiwan Strait [41]. Although, currently, the distribution of *C. lanceolata* habitats is mostly a result of years of cultivation, Wu regarded the current distribution areas as natural habitats, and these areas are worth studying [42].

With the continuous increase in GHG emissions in the future, the greenhouse effect will gradually intensify, leading to different changes in climate such as precipitation and temperature, etc., and the distribution of suitable habitats of *C. lanceolata* will change in terms of area and spatial pattern correspondingly. The future trend predicted in this research is that the expansion in the northern region will be more stable than that in the western region, and the central and eastern habitats will shrink to varying degrees; moreover, northern Yunnan will contract more. In several provinces, the suitable habitats will expand to slightly higher altitude areas. This expansion was noticed in previous studies, and it was concluded that to adapt to climate change, the distribution of border species will have to shift to higher-altitude areas gradually [43].

When the scenario was different and the period was the same, the area of total suitable habitats in the 2050s first increased and then decreased with the augmentation of GHGs, and reached its maximum value under the RCP6.0 scenario. The area change in the total suitable habitats in the 2070s fluctuated in a wave shape, reaching its maximum value under the RCP2.6 scenario. The expansion areas increased with the increase in the radiation forcing level in the 2050s, showing a positive correlation. In the 2070s, there was a downward trend, with the radiation forcing level rising, while the contraction areas showed the opposite. However, when the scenario was the same but in different periods, the area change in the total suitable habitats with time showed that it increased at first and then decreased. Except for the RCP2.6 scenario, which showed an increasing trend all the time, the expanded areas increased and the contraction areas decreased. Under the RCP2.6 scenario, it was predicted that the change of energy utilization types in the global scope would reduce the greenhouse effect significantly, and this scenario would have the largest increase in crop areas in the world; it is almost a suitable environment for all existing plants. Compared with RCP4.5, RCP6.0 had a lower GHG concentration and radiation intensity before the 2050s, which may be the reason why RCP6.0 was more suitable for the 2050s [44]. To sum up, RCP2.6 and RCP6.0 were more suitable for the future growth of *C. lanceolata*.

Under different scenarios in the future, the movement of habitat centroids was also slightly different. The centroid of RCP6.0 kept shifting to the west and also to the south by the 2070s. The centroids of other scenarios, especially RCP4.5, went northwest first and then northeast. In most cases, with the increase in the GHG concentration and the passage of time, the living environment deteriorated, and the centroids of *C. lanceolata* habitats migrated to high latitudes. Similarly, Shugart et al. found that with global warming, plants migrated northward at the end of the last glacier [45]. According to the latest research by Parmesan and Yohe [46], more than 1700 plant species migrated to the polar regions at an average speed of 6.1 m/10 a.

4.2. Ecological Characteristics of the Distribution of *C. lanceolata* Habitats

The results of this study showed that, among the 16 selected environmental factors, temperature, precipitation, human activities, and elevation had certain influences on the geographical distribution of *C. lanceolata*. According to the three test methods, precipitation was the most important factor, followed by human activities and temperature. On the whole, the influence of precipitation was stronger than that of temperature. The leading environmental factors affecting the growth of *C. lanceolata* were annual precipitation (Bio12), human activities (Hfp), minimum temperature of the coldest month (Bio6), mean temperature of the coldest quarter (Bio11), precipitation of the coldest quarter (Bio19), annual temperature range (Bio7) and elevation. The suitable ranges were 1100 mm~4700 mm (Bio12), 40~105 (Hfp), 1~12 °C (Bio6), 6~15 °C (Bio11), 120 mm~510 mm (Bio19), 14~30 °C (Bio7), and 0~600 m (elevation), respectively. Therefore, we concluded that a suitable environment for the growth of *C. lanceolata* should be warm, humid, rich in precipitation, and with a relatively flat terrain. As mentioned above, there was a negative correlation between human activities and the area of the distribution habitat of *C. lanceolata*, that is, the greater the intensity of human activities, the less suitable it was for *C. lanceolata*. According to the area calculation, the total geographical distribution habitat area of *C. lanceolata* was 193.78×10^4 km^2, accounting for 20.19% of China's land area. Without the influence of human activities, the proportion was about 22.78%; moreover, moderate and high suitable habitats increased significantly and the distribution of whole habitats was more continuous. Chen et al. pointed out that the high intensity of human activities led to the formation of acid rain and a decline in soil fertility, and large-scale logging also affected the nitrogen cycle in the ecological cycle, which is not conducive for plant growth [47]. The most suitable habitats of existing *C. lanceolata* were the subtropical monsoon climate, with ample rainfall and proper temperature conditions for growth, and mainly cultivated vegetation, coniferous forest, and shrubs. The high suitable habitat was predicted to be concentrated in the north and east of Taiwan Province; Taiwan's population is mainly concentrated in the west [48]. There are mountains in the east of the middle part of the land, which are less affected by human activities. This area is a subtropical monsoon climate, with high temperatures and much rain in summer and mild temperatures and little rain in winter. In addition, the undulating terrain weakens the winter wind, so it is more suitable for *C. lanceolata*. Another province worth noting is Sichuan, compared with the eastern coastal areas such as Jiangsu; the annual precipitation (Bio12), minimum temperature of the coldest month (Bio6), mean temperature of the coldest quarter (Bio11), and annual temperature range (Bio7), especially the first three environmental factors, were more suitable for the growth habits of *C. lanceolata*. Our prediction showed that with the increase in GHG emission concentration in the future, the moderate suitable habitat in the Sichuan Basin would be partially upgraded to a high suitable habitat and form a high suitable habitat to surround the low suitable habitat, which may be due to the following reasons: with the intensification of the greenhouse effect, climate change in Sichuan Basin is mainly expressed by an increase in temperature, rainfall (mainly concentrated in June-September), and even extreme rainfall. Additionally, the soil in the basin is loamy clay with a rich nitrogen content, and the surface and groundwater are abundant, which is more suitable for the growth of *C. lanceolata* [49,50]. The last area noteworthy is Hainan Province. Hainan was not found to be suitable for *C. lanceolata* growth in the current prediction due to the fact that mean temperature of the coldest quarter (Bio11) was more than 17 °C, the precipitation in the coldest quarter (Bio19) was less than 94 mm, and the minimum temperature of the coldest month (Bio6) was more than 13 °C. However, the greenhouse effect will advance the spring phenology of tropical plants, prolong the growing season of plants, advance the exhibition period of leaves, advance the initial flowering period, and delay the senescence of leaves, which may lead to the emergence of a low suitable habitat in this area [51].

Other factors, such as soil, slope, and aspect, had little influence on the distribution of *C. lanceolata*, among which AspectCIU and SlopeCL1-4 had slightly greater contributions, representing the position with a slope of 2%~10%. These two coefficients indicated that *C. lanceolata* is suitable for growing in an environment with a gentle slope, such as a flat plain, central basin, piedmont, piedmont inclined plain, valley bottom, platform, foothills, basin surroundings, hills, etc. The steeper the slope, the more serious the water and soil loss and the poorer soil water and fertilizer conservation became [52]. Among the soil factors, the soil texture (T_USDA_TEX_ CLASS) contributed a little to *C. lanceolata*, which mainly describes the relative proportion of mineral particles with different sizes in the soil. According to the response curve, the soil texture suitable for *C. lanceolata* growth was silty clay, clay, silty clay loam, clay loam, silt, and silty loam. Research has shown that loose soil without excessive rock obstacles is more suitable for plant growth [1].

Although *C. lanceolata* has been widely planted in recent years, many problems need to be solved urgently. Firstly, the utilization efficiency is low. Due to the prosperity of new building materials such as metal and plastic, the applicable scope of *C. lanceolata* wood has been greatly reduced, and the unsalable stock is overstocked. It is urgent to use *C. lanceolata* with high added value. Secondly, there are some problems, such as unreasonable continuous planting. Tian et al. found that continuous rotation will lead to a decline in the *C. lanceolata* forest yield, and properly prolonging the rotation period of *C. lanceolata* forest in the same place will be beneficial for maintaining high-quality wood [53]. While efficiently utilizing and rationally planting *C. lanceolata*, we should pay attention to the impact of global warming and try to take protective measures in areas with reduction tendencies. In fact, in this study, we found that in the cultivated vegetation in eastern China, there will be many contractions in the future, so it is time to carry out resource investigation and dynamic monitoring activities actively and to improve the adaptability of *C. lanceolata* to climate change. Meanwhile, we should also make sustainable land use plans in places with stable expansion and suitable environments, and reduce the intensity and frequency of human activities on the edge of the newly increased habitats so as to increase the populations of *C. lanceolata* forest resources, improve the ecological service function, and realize both economic and ecological benefits.

5. Conclusions

In this study, 39 environment factors, including climate, soil, terrain and human activities, were used to predict the potential distribution habitats of *C. lanceolata* in China based on the MaxEnt model under different scenarios in current and future periods (the 2050s and 2070s). The results showed that the annual precipitation (Bio12), human activities

(Hfp), minimum temperature of the coldest month (Bio6), mean temperature of the coldest quarter (Bio11), precipitation of the coldest quarter (Bio19), annual temperature range (Bio7), and elevation are predicted to be the leading environmental factors affecting distribution of *C. lanceolata*, and the suitable habitats are characterized by warmth, humidity, abundant rainfall, and relatively gentle topography. Human activities will cause the distribution of suitable habitats to degenerate and fragmentize. The areas of habitats under different scenarios were predicted to increase first and then decrease by the 2050s and 2070s, except for the RCP2.6 scenario, under which the suitable habitats area of *C. lanceolata* will increase continuously. Furthermore, the suitable habitats will gradually shift to high-latitude areas due to climate change.

Supplementary Materials: The following supporting information can be downloaded at: https://www.mdpi.com/article/10.3390/f15050830/s1, Table S1: Ecological factors used in simulation; Table S2: Contribution rate and permutation importance of environmental factors under current and other eight future scenarios; Table S3: Areas of potential distribution under different climate scenarios of *Cunninghamia lanceolata* (Lamb.) Hook.; Figure S1: Contribution rate of environmental factors; Figure S2: Permutation importance of environmental factors; Figure S3: Distribution of *C. lanceolata* habitats in the 2050s and 2070s in the future under different carbon emission concentration scenarios, (A) RCP2.6, (B) RCP4.5, (C) RCP6.0, (D) RCP8.5.

Author Contributions: L.Y. confirmed the topic and goals of this study; J.F. and Y.C. extracted and analyzed the data and wrote the manuscript; T.M., D.H. and J.C. reviewed and edited the manuscript; L.Y. sourced funding and supervised the progress of this manuscript. All authors have read and agreed to the published version of the manuscript.

Funding: This research was supported by the Research Start-up Fund for High-Level and High-Educated Talents of Nanjing Forestry University.

Data Availability Statement: The original data presented in the study are openly available in National Specimen Information Infrastructure (http://www.nsii.org.cn/, accessed on 1 June 2022), National Plant Specimen Resource Center (https://www.cvh.ac.cn/index.php, accessed on 1 June 2022), Global Biodiversity Information Facility (https://www.gbif.org/, accessed on 3 June 2022), World Climate Data (http://www.worldclim.org/, accessed on 20 June 2022), Food and Agriculture Organization of the United Nations (https://www.fao.org/soils-portal/data-hub/soil-maps-and-databases/harmonized-world-soil-database-v12/en/, accessed on 20 June 2022), and Global Human Footprint of Socioeconomic Data and Applications Center (https://sedac.ciesin.columbia.edu/data/set/wildareas-v2-human-footprint-geographic, accessed on 1 July 2022).

Conflicts of Interest: The authors declare no conflicts of interest.

References

1. Yang, H.F. Characteristics of *Cunninghamia lanceolata* and afforestation techniques of improved varieties. *Mod. Agric. Sci. Technol.* **2022**, *5*, 97–98+101. (In Chinese)
2. Shen, G.W.; Shen, G.H.; Fu, D. Cultivation techniques and application of urban roadside trees *Cunninghamia lanceolata*. *Xiandai Nongcun Keji* **2013**, *17*, 52. (In Chinese)
3. Hou, B.X. The origin and development history of Chinese fir. *Agric. Archaeol.* **1996**, *1*, 161–171. (In Chinese)
4. Xu, H.; Sun, Y.J.; Wang, X.J.; Wang, J.; Fu, Y. Linear mixed-effects models to describe individual tree crown width for China-fir in Fujian Province, southeast China. *PLoS ONE* **2015**, *10*, e0122257. [CrossRef]
5. Duan, H.J.; Cao, S.; Zheng, H.Q.; Hu, D.H.; Lin, J.; Lin, H.Z.; Hu, R.Y.; Sun, Y.H.; Li, Y. Variation in the growth traits and wood properties of Chinese fir from six provinces of southern China. *Forests* **2016**, *7*, 192. [CrossRef]
6. Li, M.; Chen, X.Z.; Huang, M.S.; Wu, P.F.; Ma, X.Q. Genetic diversity and relationships of ancient Chinese fir (*Cunninghamia lanceolata*) genotypes revealed by sequence-related amplified polymorphism markers. *Genet. Resour. Crop Evol.* **2017**, *64*, 1087–1099. [CrossRef]
7. Yao, M.M.; Yang, Y.Q.; Song, J.L.; Yu, Y.; Jin, Y.C. Melamine formaldehyde modified furfurylation to improve Chinese fir's dimensional stability and mechanical properties. *BioResources* **2017**, *12*, 3057–3066. [CrossRef]
8. Yan, X.X.; Chang, Y.J. Investigation of waterborne thermochromic topcoat film with color-changing microcapsules on Chinese fir surface. *Prog. Org. Coat.* **2019**, *136*, 105262. [CrossRef]
9. Sun, Q.X.; Peng, Z.H.; Zhang, Q.S. Volatiles of wood of Chinese fir in nature and its effect on human heaith. *J. Anhui Agric. Univ.* **2004**, *2*, 158–163. (In Chinese) [CrossRef]

10. Kurpis, J.; Serrato-Cruz, M.A.; Feria Arroyo, T.P. Modeling the effects of climate change on the distribution of *Tagetes lucida* Cav. (*Asteraceae*). *Glob. Ecol. Conserv.* **2019**, *20*, e00747. [CrossRef]
11. Gilani, H.; Arif Goheer, M.; Ahmad, H.; Hussain, K. Under predicted climate change: Distribution and ecological niche modelling of six native tree species in Gilgit-Baltistan, Pakistan. *Ecol. Indic.* **2020**, *111*, 106049. [CrossRef]
12. Walther, G.R.; Post, E.; Convey, P.; Menzel, A.; Parmesan, C.; Beebee, T.J.; Fromentin, J.M.; Hoegh-Guldberg, O.; Bairlein, F. Ecological responses to recent climate change. *Nature* **2002**, *416*, 389–395. [CrossRef]
13. Ulukan, H. Climate change and global warming effect(s) on wheat Landraces: A General Approach. In *Wheat Landraces*; Zencirci, N., Baloch, F.S., Habyarimana, E., Chung, G., Eds.; Springer International Publisher: Berlin, Germany, 2021; pp. 169–191. [CrossRef]
14. Steel, E.J.; Fontaine, J.B.; Ruthrof, K.X.; Burgess, T.I.; Hardy, G.E.S.J. Changes in structure of over- and midstory tree species in a Mediterranean-type forest after an extreme drought-associated heatwave. *Austral Ecol.* **2019**, *44*, 1438–1450. [CrossRef]
15. Matusick, G.; Ruthrof, K.X.; Brouwers, N.C.; Dell, B.; Hardy, G.S.J. Sudden forest canopy collapse corresponding with extreme drought and heat in a mediterranean-type eucalypt forest in southwestern Australia. *Eur. J. For. Res.* **2013**, *132*, 497–510. [CrossRef]
16. Thomas, C.D.; Cameron, A.; Green, R.E.; Bakkenes, M.; Beaumont, L.J.; Collingham, Y.C.; Erasmus, B.F.N.; de Siqueira, M.F.; Grainger, A.; Hannah, L.; et al. Extinction risk from climate change. *Nature* **2004**, *427*, 145–148. [CrossRef] [PubMed]
17. Moritz, C.; Agudo, R. The future of species under climate change: Resilience or decline? *Science* **2013**, *341*, 504–508. [CrossRef] [PubMed]
18. Phillips, S.J.; Anderson, R.P.; Schapire, R.E. Maximum entropy modeling of species geographic distributions. *Ecol. Model.* **2006**, *190*, 231–259. [CrossRef]
19. Phillips, S.J.; Dudík, M. Modeling of species distributions with Maxent: New extensions and a comprehensive evaluation. *Ecography* **2008**, *31*, 161–175. [CrossRef]
20. Liu, H.C.; Jacquemyn, H.; He, X.Y.; Chen, W.; Huang, Y.Q.; Yu, S.; Lu, Y.P.; Zhang, Y. The Impact of human pressure and climate change on the habitat availability and protection of *Cypripedium* (*Orchidaceae*) in Northeast China. *Plants* **2021**, *10*, 84. [CrossRef]
21. Garah, K.; Bentouati, A. Using the MaxEnt model for assessing the impact of climate change on the Aurasian Aleppo pine distribution in Algeria. *Afr. J. Ecol.* **2019**, *57*, 500–511. [CrossRef]
22. Wang, G.Z.; Geng, Q.F.; Xiao, M.Y.; Zhang, M.Y.; Zhang, Y.Y.; Wang, Z.S. Predicting *Pseudolarix amabilis* potential habitat based on four Niche models. *Sheng Tai Xue Bao* **2020**, *40*, 6096–6104. (In Chinese) [CrossRef]
23. Zhao, Y.; Deng, X.W.; Xiang, W.H.; Chen, L.; Ouyang, S. Predicting potential suitable habitats of Chinese fir under current and future climatic scenarios based on Maxent model. *Ecol. Inform.* **2021**, *64*, 101393. [CrossRef]
24. Chen, Y.G.; Yue, X.G.; Chen, Y.H.; Cheng, W.X.; Du, J.G.; Zhong, Q.L.; Cheng, D.L. Identification of potential distribution area of *Cunninghamia lanceolata* in China under climate change based on the MaxEnt model. *Ying Yong Sheng Tai Xue Bao* **2022**, *33*, 1207–1214. [CrossRef]
25. Wang, R.L.; Li, Q.; Feng, C.H.; Shi, Z.P. Predicting potential ecological distribution of *Locusta migratoria tibetensis* in China using MaxEnt ecological niche modeling. *Sheng Tai Xue Bao* **2017**, *37*, 8556–8566. [CrossRef]
26. Guo, J.; Liu, X.P.; Zhang, Q.; Zhang, D.F.; Xie, C.X.; Liu, X. Prediction for the potential distribution area of *Codonopsis pilosula* at global scale based on Maxent model. *Ying Yong Sheng Tai Xue Bao* **2017**, *28*, 992–1000. [CrossRef]
27. Zhang, S.; Liu, X.G.; Li, R.M.; Wang, X.L.; Cheng, J.H.; Yang, Q.L.; Kong, H. AHP-GIS and MaxEnt for delineation of potential distribution of Arabica coffee plantation under future climate in Yunnan, China. *Ecol. Indic.* **2021**, *132*, 108339. [CrossRef]
28. Etherington, T.R. Python based GIS tools for landscape genetics: Visualising genetic relatedness and measuring landscape connectivity. *Methods Ecol. Evol.* **2011**, *2*, 52–55. [CrossRef]
29. Hong, G.X.; Lv, S.Y.; Peng, J.Y.; Jiang, Z.Y. Genetic geography and conservation of *Cunninghamia lanceolata* and *Cunnighamia konishii*. *Nat. Conserv. Q.* **2000**, *31*, 33–36. (In Chinese)
30. Hu, Z.J.; Zhang, Y.L.; Yu, H.B. Simulation of Stipa purpurea distribution pattern on Tibetan Plateau based on MaxEnt model and GIS. *Ying Yong Sheng Tai Xue Bao* **2015**, *26*, 505–511. (In Chinese) [CrossRef]
31. Ding, W.N.; Ree, R.; Spicer, R.; Xing, Y. Ancient orogenic and monsoon-driven assembly of the world's richest temperate alpine flora. *Science* **2020**, *369*, 578–581. [CrossRef] [PubMed]
32. An, J. Study on Molecular Phylogeography of *Cunninghamia lanceolata* (Lamb.) Hook. Master's Thesis, Central South University of Forestry and Technology, Hunan, China, 2012. (In Chinese)
33. Wang, K.F.; Xu, S. *Quaternary Palynology*; Guizhou People's Publishing House: Guizhou, China, 1988; p. 333. (In Chinese)
34. Song, Z.C. Late Cenozoic palyno-flora from Zhaotong, Yunnan. *J. Nanjing Inst. Geol. Paleontol.* **1988**, *24*, 1–108. (In Chinese)
35. Liu, H.L.; Wang, D.Y. A study on the remains of ancient forest in Mianning, Sichuan. *Sci. Silvae Sin.* **1984**, *20*, 175–184. (In Chinese)
36. Huang, Z.G.; Li, P.R.; Zhang, Z.Y.; Li, K.H.; Qiao, P.N. *Formation, Development and Evolution of Pearl River Delta*; Science and Technology of China Press: Guangzhou, China, 1982; p. 274. (In Chinese)
37. Gu, H.B. Sporopollen Analysis and Paleoenvironment Discussion of Pengtoushan Site in Lixian County, Hunan Province. *Cult. Relics* **1990**, *8*, 30–32. (In Chinese)
38. Peng, Z.H. On the origin of Chinese fir from historical documents. *J. Anhui Agric. Univ.* **1984**, *2*, 23–32. (In Chinese) [CrossRef]
39. Huang, B.L.; Lan, T.G. Preliminary discussion on the history of cultivation and utilization of Chinese fir. *J. Nanjing For. Univ. (Nat. Sci. Ed.)* **1988**, *2*, 54–59. (In Chinese) [CrossRef]

40. Hou, B.X.; Chen, F.S.; Cheng, Z.H. Origin of distribution, utilization and cultivation history of Chinese fir in Hunan. *Hunan For. Sci. Technol.* **1995**, *3*, 1–6. (In Chinese)
41. Li, Y.X. Genetic Diversity and Genetic Divergence of *Cunninghamia lanceolata* (Lamb.) Hook. Geographical Provenances. Master's Thesis, Chinese Academy of Forestry, Beijing, China, 2015. (In Chinese)
42. Wu, Z.L. Preliminary study on the distribution of Chinese fir. *Acta Geogr. Sin.* **1955**, *3*, 273–285. (In Chinese) [CrossRef]
43. Lenoir, J.; Gégout, J.C.; Marquet, P.A.; de Ruffray, P.; Brisse, H. A significant upward shift in plant species optimum elevation during the 20th century. *Science* **2008**, *320*, 1768–1771. [CrossRef]
44. van Vuuren, D.P.; Edmonds, J.; Kainuma, M.; Riahi, K.; Thomson, A.; Hibbard, K.; Hurtt, G.C.; Kram, T.; Krey, V.; Lamarque, J.-F.; et al. The representative concentration pathways: An overview. *Clim. Chang.* **2011**, *109*, 5. [CrossRef]
45. Shugart, H.H.; Antonovsky, M.Y.; Jarvis, P.G.; Sandford, A.P. CO_2 climatic change and forest ecosystems. In *The Greenhouse Effect, Climatic Change and Ecosystems*; Bolin, B., Doeoes, B.R., Jaeger, J., Warrick, R.A., Eds.; John Wiley and Sons: New York, NY, USA, 1986.
46. Parmesan, C.; Yohe, G. A globally coherent fingerprint of climate change impacts across natural systems. *Nature* **2003**, *421*, 37–42. [CrossRef]
47. Chen, G.X.; Yu, K.W.; Liao, L.P.; Xu, G.S. Effect of human activities on forest ecosystems: N cycle and soil fertility. *Nutr. Cycl. Agroecosyst.* **2000**, *57*, 47–54. [CrossRef]
48. Liang, D.L.; Wang, B.; Jiang, L.L.; Chen, K. The population and its spatial characteristic analysis from 2000 to 2010 in Taiwan. *J. Shanxi Norm. Univ. (Nat. Sci. Ed.)* **2016**, *30*, 90–97. (In Chinese) [CrossRef]
49. Yang, X.Y.; Zhang, S.B.; Lyu, Y.Q.; Zhao, Y.; Lyu, S.H. Characteristics and future projections of summer extreme precipitation in Sichuan Province, China. *J. Mt. Sci.* **2020**, *17*, 1696–1711. [CrossRef]
50. Xu, C.C.; Wu, W.X.; Ge, Q.S. Impact assessment of climate change on rice yields using the ORYZA model in the Sichuan Basin, China. *Int. J. Climatol.* **2018**, *38*, 2922–2939. [CrossRef]
51. Li, N.; Bai, R.; Wu, L.; Li, W.; Chen, M.; Chen, X.; Fan, C.H.; Yang, G.S. Impacts of future climate change on spring phenology stages of rubber tree in Hainan, China. *Ying Yong Sheng Tai Xue Bao* **2020**, *31*, 1241–1249. (In Chinese) [CrossRef]
52. Wu, Z.Q. Preliminary study on growth and environmental factors of *Cunninghamia lanceolata* in the south of Yangtze River in Anhui province. *Lin Ye Ke Xue Yan Jiu* **1998**, *1*, 34–36. (In Chinese)
53. Tian, D.L.; Xiang, W.H.; Chen, X.Y.; Yan, W.D.; Fang, X.; Kang, W.X.; Dan, X.W.; Peng, C.H.; Peng, Y.Y. A long-term evaluation of biomass production in first and second rotations of Chinese fir plantations at the same site. *Forestry* **2011**, *84*, 411–418. [CrossRef]

forests

MDPI

Article

Assessing Forest-Change-Induced Carbon Storage Dynamics by Integrating GF-1 Image and Localized Allometric Growth Equations in Jiangning District, Nanjing, Eastern China (2017–2020)

Jiawei Liu [1], Boxiang Yang [1], Mingshi Li [1,*] and Da Xu [2,3,*]

1 Co-Innovation Center for Sustainable Forestry in Southern China, Nanjing Forestry University, Nanjing 210037, China; liuliawei@njfu.edu.cn (J.L.); ybx0329@njfu.edu.cn (B.Y.)

2 Zhejiang Forest Resources Monitoring Centre, Hangzhou 310020, China

3 Zhejiang Forestry Survey Planning and Design Company Limited, Hangzhou 310020, China

* Correspondence: nfulms@njfu.edu.cn (M.L.); xdnbblnfu518zjfr@gmail.com (D.X.); Tel.: +86-25-85427327 (M.L.); +86-13750837216 (D.X.)

Abstract: Forest and its dynamics are of great significance for accurately estimating regional carbon sequestration, emissions and carbon sink capacity. In this work, an efficient framework that integrates remote sensing, deep learning and statistical modeling was proposed to extract forest change information and then derive forest carbon storage dynamics during the period 2017 to 2020 in Jiangning District, Nanjing, Eastern China. Firstly, the panchromatic band and multi-spectral bands of GF-1 images were fused by using four different methods; Secondly, an improved Mask-RCNN integrated with Swin Transformer was devised to extract forest distribution information in 2020. Finally, by using the substitution strategy of space for time in the 2017 Forest Management and Planning Inventory (FMPI) data, local carbon density allometric growth equations were fitted by coniferous forest and broad-leaved forest types and compared, and the optimal fitting was accordingly determined, followed by the measurements of forest-change-induced carbon storage dynamics. The results indicated that the improved Mask-RCNN synergizing with the Swin Transformer gained an overall accuracy of 93.9% when mapping the local forest types. The carbon storage of forest standing woods was calculated at 1,449,400 tons in 2020, increased by 14.59% relative to that of 2017. This analysis provides a technical reference for monitoring forest change and lays a data foundation for local agencies to formulate forest management policies in the process of achieving dual-carbon goals.

Keywords: GF-1; image fusion; Swin Transformer; Mask-RCNN; carbon density growth equation

Citation: Liu, J.; Yang, B.; Li, M.; Xu, D. Assessing Forest-Change-Induced Carbon Storage Dynamics by Integrating GF-1 Image and Localized Allometric Growth Equations in Jiangning District, Nanjing, Eastern China (2017–2020). *Forests* **2024**, *15*, 506. https://doi.org/10.3390/f15030506

Academic Editors: Daniela Dalmonech, Alessio Collalti and Gina Marano

Received: 24 January 2024
Revised: 1 March 2024
Accepted: 6 March 2024
Published: 8 March 2024

1. Introduction

As the largest organic "carbon pool" in terrestrial ecosystems, forest provides about 80% of the global above-ground vegetation biomass [1], and its carbon storage approximately accounts for 46.6% of the terrestrial ecosystems' carbon stock [2]. Thus, widespread forest dynamics inevitably alters the carbon sequestration capability of forest ecosystems and then promotes global climate change and the greenhouse effect [3]. Therefore, accurate acquisition of forest dynamics information contributes to evaluating the carbon sink potential of forest ecosystems in the near future, which lays an underlying data basis for assessing the degree of achieving the dual-carbon goals in China [4].

The traditional means for capturing forest dynamics information mainly rely on massive in situ surveys. However, this manner has the drawbacks of high time and labor costs, poor timeliness and potential low accessibility, therefore making it difficult to meet the needs of dynamic monitoring over wide forested regions [5,6]. Fortunately, the emergence and advancement of remote sensing technology have overcome the defects of the traditional survey means. In particular, high-spatiotemporal-resolution remote sensing data contain more structural details and spectral and phenological variations; although such an analysis process tends to be more complex, it is more conducive to the fine identification and mapping of forest types and even tree species [7,8]. In recent decades, pixel-based classification methods have been widely used in forest information extraction based on medium- or high-resolution imagery. For example, Huang et al. combined Sentinel-2 spectral features and radar data backscattering features to classify tree species of typical plantation forests in the tropics via the random forest method [9]. Although the classical statistics-based or machine learning methods have operational speed and simplicity, the "salt and pepper phenomenon" is quite common in classification results [10]. To minimize the negative effects of pixel-based analysis methods, Object-based Image Analysis (OBIA) has been proposed and applied in forest-related remote sensing efforts because it can make full use of shallow information such as spectral, texture and geometric features and the spatial topology of features in medium- or high-resolution remote sensing imagery for classification [11]. For example, Mao et al. combined Sentinel active and passive remote sensing data to develop an object-oriented SNIC + RF algorithm to classify the land cover of Qianjiangyuan National Park, obtaining an overall accuracy of 93.98% [12]. However, OBIA only considers the shallow features within the segmented objects; it is prone to cause missegmentation and misclassification in complex situations [13]. Therefore, determining how to adequately extract and utilize effective information from medium- or high-resolution remote sensing images becomes the key to improving classification accuracy.

Along with the escalation in computer processing power, there has been a fast development of deep learning in a wide range of application areas [14]. Deep learning methods are representation learning methods with multiple levels of representation, obtained by composing simple but non-linear modules that each transform the representation at one level (starting with the raw input) into a representation at a higher, slightly more abstract level. Deep learning extends standard machine learning by discovering intermediate representations that can be used to solve more complex problems [15]. Convolutional neural networks (CNNs), first proposed by Yann LeCun for image processing, are some of the most widely used deep neural networks currently [16]. CNNs and their derived network models, such as the fully convolutional networks (FCNs), U-net and DeepLab V+, have been widely used in high-resolution image classification, and they have shown their robustness and strong generalization ability as well as the capability of extracting and utilizing high-level features from remote sensing images [17]. In addition, CNN-based model improvements continue to confirm these observed advantages. For example, He et al. combined the features of urban objects in high-resolution images and the characteristics of urban forest itself and proposed the Object-Based U-Net-Dense Net-Coupled Network (OUDN) based on a CNN for extracting urban forests, and its extraction accuracy reached 0.997 [18].

In particular, the Mask-Region Convolutional Neural Network (Mask-RCNN) model is a flexible and generalized instance segmentation framework [19] that is commonly used for target recognition and extraction [20,21]. Based on this, Xie et al. proposed a Modified Mask-RCNN model for urban forest extraction after hyperpixel segmentation of GF-2 images and transferred the model to UAV image classification, achieving a high overall accuracy of 92.48% [22]. Shi et al. also applied the Modified Mask-RCNN model to extract urban–suburban fragmented land cover types from several types of high-resolution satellite images in Nanjing, and its accuracy was higher than that of the object-oriented decision-tree-based classification model [23]. However, due to the limited receptive field of the model structure, Mask-RCNN produced a fuzzy target edge segmentation and missed small targets [24,25]. In contrast, the Transformer technology builds a global information model by capturing long-distance dependencies of the entire feature graph with an attention mechanism [26,27]. In particular, the Swin Transformer model, an improved version of Transformer, has global modeling capability, and its hierarchical network structure and sliding window information interaction mode expand the receptive field to some extent to reduce the amount of computation, making it more suitable for multi-scale target classification recognition and extraction [28]. For example, Gao et al. used an instance segmentation optimization method by incorporating Swin Transformer to effectively solve the problem of difficult segmentation for multi-larval individual image recognition in complex real-world scenes, and they achieved a satisfactory result [29]. Although these existing deep learning models have had different levels of success in extracting and recognizing objects of interest from remote sensing images, coupling Mask-RCNN with Swin Transformer to accurately identify different forest types including coniferous forest, broad-leaved forest and shrub forest from 2 m resolution satellite images has been rarely attempted, and it deserves further investments and tests.

A forest growth model is an important tool for studying tree growth and stand harvests; it helps to conceptualize and abstract the complex phenomenon and process of tree growth and to simulate and predict tree change and future development trends [30]. The whole stand model is a kind of stand growth and harvest model with nearly a hundred years of history; it describes the total amount of the whole stand and the growth process of average individual trees [31]. According to whether the density factor is introduced into the model as an independent variable, the whole stand model can be divided into two categories. One is the density-independent model, an example of a model belonging to this category is the traditional stand harvest table in Europe and America, with little practical significance. Another type of model is the density-related model, which takes the index of stand density as an independent variable to simulate the change in stand growth or harvest. Such models are widely used at present [32]. For example, the Gompertz model is often used to describe the growth of certain plants and the law of economic activities [33]. Zhao et al. applied the Gompertz growth model to predict the output value of forest products; the relative error was only 0.0082%, and the correlation coefficient reached 0.9992 [34]. The Richards model has been widely used in various fields of forest growth, and it has the advantages of an accurate description of the growth process and the strongest applicability [35]. For example, by taking stand age, site index and stand density index as independent variables, Jiang et al. studied the whole stand model of a Chinese fir plantation with variable density based on the Richards model, achieving a high accuracy of 82.48% [36]. Based on the Richards model, Feng et al. established a full-stand model of Beijing's Platyphelus orientales plantation based on the compatibility of the harvest model and showed a strong applicability [37]. Wang et al. used the logistic model to study the population quantitative characteristics of Taxus chinensis, a rare and endangered plant, and fitted the S-shaped curve of population growth, with a fitting accuracy of more than 90% [38]. Overall, the density-related whole stand models are simple to understand and easy to use, and they can directly predict the growth and harvest of the stand per unit area; thus, the total harvest of the whole stand can be easily derived. With the introduction of new statistical methods, such as the mixed-effect model, machine learning and quantile regression, their estimation accuracy and application range can be further improved [39]. For example, Zhang et al. studied a model of *Poplus* spp. and analyzed the distribution characteristics under different classes of environmental indicators based on the KNN model and RF model, and a high accuracy and good fitting effect were accordingly achieved [40]. However, the model simulation coefficients of the above-mentioned density-dependent models are species-dependent and site-specific; thus, to better predict the growth of particular stands by using these models, it is necessary to make the model parameters localized.

The estimation methods of vegetation carbon storage are mainly divided into three types [41]: (1) The first is survey-based estimation, namely estimation using regional forest survey data, and this method usually gives the most accurate results but with extremely high labor and time costs [42]. (2) The second is remote-sensing-based estimation, for which commonly used data include optical remote sensing data, LiDAR data and synthetic aperture radar (SAR) satellite data [43]. For instance, Vincent et al. used WorldView-2 and LiDAR data to perform a fine estimation of vegetation carbon storage in Auckland, New Zealand, and the accuracy reached up to 95.9% [44]. LiDAR can effectively and accurately measure tree height and three-dimensional spatial structure to derive a very high carbon estimate for individual trees or forest stands, but it is often subjected to the limitation of high operational cost [45]. Vatandaşlar et al. used SAR data to estimate the carbon stocks of Mediterranean forests, and the conclusion fully proved that the total carbon stocks of forest ecosystems could be estimated using appropriate SAR images and could be applied to forestry with good accuracy [46]. (3) The third is process-based estimation, which mainly refers to mechanism modeling. The mechanism modeling method can estimate forest carbon storage and quantitatively describe the forest carbon cycle process, but it needs a lot of parameters or variables to drive the process model, and in the reality of model application, these parameters are frequently difficult to obtain in an accurate manner [47].

The major aim of this study was to propose a forest change analysis framework that integrates deep learning and forest stand growth modeling to accurately assess the carbon storage dynamics at a district or county scale. Specifically, the deep learning Mask-RCNN model was improved by replacing ResNet101 with Swin Transformer for the accurate extraction of forest types first, and then the optimal stand carbon density growth equation was determined to accurately calculate forest growth, jointly supporting the evaluation of forest carbon storage dynamics during the period 2017–2020.

2. Materials and Methods

2.1. Study Area

The study area is located in Jiangning District, Nanjing City, Jiangsu Province (118°28′ E~119°06′ E, 31°37′ N~32°07′ N) (Figure 1). Jiangning District belongs to the northern subtropical monsoon climate zone, with an average annual temperature of 15.7 °C and an average annual precipitation of 1072.9 mm. The area contains a variety of landforms, including low hills, highlands, plains and basins. The terrain is high in the north and south and low in the center. The area is rich in water systems, with two major water systems, the Yangtze River and the Qinhuai River, and many small rivers and reservoirs are scattered throughout the area. Figure 1 shows the location of the study area and its vegetation cover. During the past few decades, Jiangning District had rapid economic development and an intense change in land use. Thus, Jiangning's forest was frequently disturbed by human activities. Choosing Jiangning District as a study area can test the reliability and effectiveness of the proposed framework when quantifying the dynamic changes of forest carbon stocks under rapid changes, to provide a reference and data support for local forest change monitoring [48].

2.2. Data

In this study, two GF-1 images including a panchromatic and 4 multi-spectral (PMS) bands, acquired on 17 August and 5 September 2020, respectively, were collected and mosaicked to cover the entire study area of Jiangning District, and the images were fused and acted as the major input for subsequent fine land cover classification. The PMS camera of the GF-1 satellite provides a 2 m resolution panchromatic band and 4 8 m spatial resolution multi-spectral bands (red, green, blue and near-infrared) simultaneously. The camera has a high radiometric resolution, with a quantification level of 10 bits, and its image swath is 60 km. Additionally, a DEM with a spatial resolution of 12.5 m covering the entire study area was compiled from the Beijing KOSMOS Image Mall (a satellite image vendor) to support subsequent land cover classifications. In addition, the 2017 Forest Management and Planning Inventory (FMPI) data were collected from the Nanjing Greening and Gardening Bureau. These data were acquired by a complete field survey of all forest stands involved. Before implementing this survey, all forest stands must be delineated from current aerial photographs or large-scale topographic maps by visual interpretation, followed by deploying well-trained experienced investigators to each forest stand to capture forest stand information including tree species composition, dominant tree species, age, average height, average DBH, crown closure and disturbance type and severity. Finally, the inventory results were subjected to a quality check by limited sample plot measurements and cross-comparison. According to the needs of forest classification in the study area, the vector data of 2017 were converted into raster data. The 2017 FMPI data were used as the benchmark of forest status for the dynamic assessment. All the spatial data mentioned here were resampled to 2 m resolution to facilitate the change analysis.

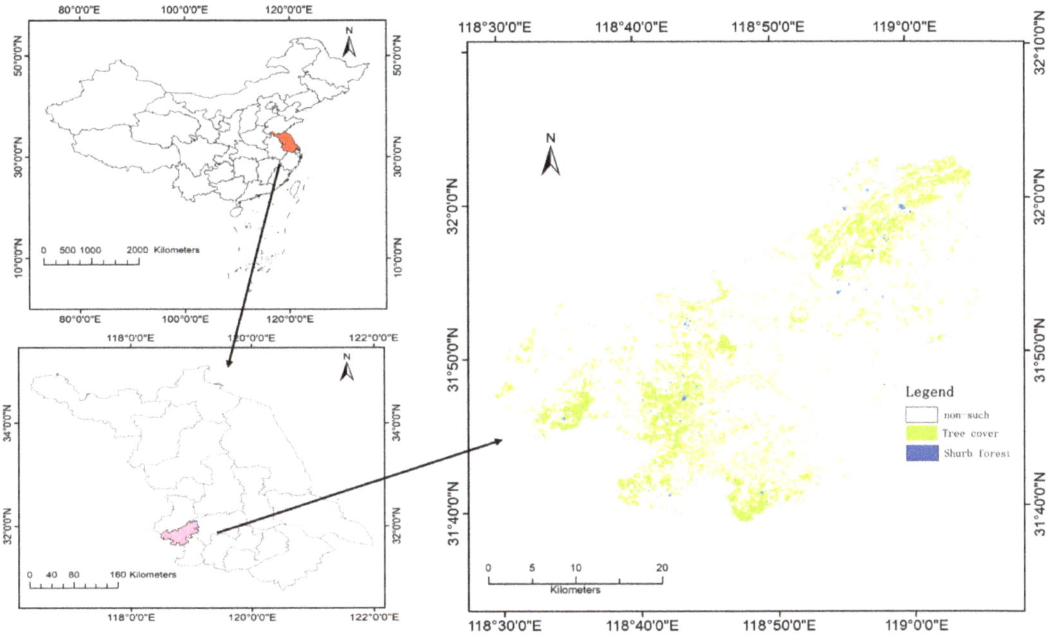

Figure 1. Geographical location and its vegetation cover of the study area (**Upper Left**: the map of China's territory; **Lower Left**: the map of the administrative boundaries of Jiangsu Province; **Right**: the 2022 vegetation cover map of the study area).

2.3. Image Fusion

The post-atmospheric correction panchromatic band and multi-spectral bands of GF-1 images were fused to generate new higher-quality images with a 2 m spatial resolution by using the Brovey transform, GS transform, NND transform and wavelet transform fusion algorithms. The Brovey transform method multiplies multi-spectral images and panchromatic high-resolution images on the basis of normalization to enhance image information [49,50]; GS transform eliminates the strong correlation between the bands and reduces the redundant information by orthogonalizing the image data [51,52]; the NNDiffuse transform method can significantly improve the image processing quality and speed, and the fused images will be better when all the multi-spectral bands are covered by the panchromatic wavelength and the wavelengths of the multi-spectral bands do not overlap [53]. Wavelet transform first decomposes the panchromatic band into four new images (approximation, horizontal details, vertical details and diagonal details) and then resamples the multi-spectral bands to make their spatial resolution consistent with that of the panchromatic band; on this basis, a principal component analysis of the these resampled multi-spectral bands is implemented, followed by a histogram-matching operation between the approximation image and the first principal component image. Once this is completed, the substitution of the approximation image with the histogram-matched first principal component image and the implementation of a wavelet reconstruction analysis are conducted to obtain the fused new images [54,55]. These fusion algorithms were implemented in ENVI and Matlab packages. The evaluation of the final fused image quality is an extremely important step in the process of image fusion [56]. Generally, the evaluation includes the intuitive subjective qualitative evaluation by human visual observation and the objective quantitative evaluation by calculating quantitative statistical indicators [57,58]. The quantitative assessment indicators used in this analysis were the mean standard deviation, the mean correlation coefficient and the mean gradient [59,60].

2.4. Forest Distribution Extraction Modeling and Validation

First, because water can absorb a lot of solar energy, the reflectance of water is much lower than that of most ground objects, and its reflectance decreases with the increase in wavelength in the visible–shortwave infrared range. The normalized difference water index (NDWI) widens the reflectance gap between the near-infrared band with the weakest reflectance and the green band with the strongest reflectance for a water body through the calculation of a ratio, making the water object information more prominent [61]. In the analysis, NDWI was therefore used to classify the image into water body and non-water body classes by specifying a threshold of 0.2. Green vegetation strongly absorbs red light and strongly reflects near-infrared waves; on this basis, the normalized vegetation index (NDVI), defined as the ratio between the difference and the sum of the reflectance of the near-infrared channel and the reflectance of the visible channel, was used to highlight the signal of vegetation. NDVI can reflect vegetation coverage and growth status and can effectively remove some radiation errors. In an area covered by vegetation, the NDVI value is positive, and with the increase in vegetation coverage, the NDVI value will be increased, but it is easily saturated in high-vegetation-coverage areas [62]. Hence, NDVI was used to classify the non-water body region of the GF-1 image into vegetation areas and non-vegetation areas. In the vegetation areas, combined with the actual situation of the study area, five major land cover types, namely coniferous forests, broad-leaved forests, shrub forests, grassland and cropland, were identified as the scheme of classification. And 600 forest samples, 200 grassland samples and 200 cropland samples were picked up from the fused GF-1 false-color composite based on our local knowledge to train the classification models. Due to the complex structures of different forest types, it is difficult to accurately characterize their differences in a single remote sensing band. Therefore, we used the original multi-spectral reflectance bands [63], NDVI, Gray-Level Co-occurrence Matrix-based textural measures [64], and topographic variables including elevation, slope and aspect [65] as the inputs for the classifications of machine learning algorithms. Specifically, support vector machine (SVM), random forest (RF), Mask-RCNN and Mask-RCNN integrated with Swin Transformer were compared in terms of the classification performance. Among them, SVM is a linear classifier with the largest interval defined on the feature space, and there are several types of kernel functions available to potentially optimize the algorithm performance. For example, the Gaussian kernel function can map finite-dimensional data to a higher-dimensional space, which may give SVM a high accuracy in image classification [66]. In this study, the Gaussian kernel function was selected to extract forest types based on the SVM algorithm, with the C parameter set to 1 and the gamma parameter set to 0.1. RF is a data-driven non-parametric classification algorithm. The major parameters including the number of variables m and the number of trees T in the algorithm need to be adjusted according to specific application scenarios. In this analysis, parameters $T = 100$ and $m = 5$ were set after multiple tries. Further, ResNet101 was adopted as the backbone network part of the image feature extraction network model. Then, the sliding window of the region of interest (ROI) corresponding to the candidate frame was generated to obtain a feature map with a high-quality candidate frame, which was used for subsequent target frame location, mask recognition and image classification. The network structure consisted of the backbone feature extraction network, Region Proposal Network (RPN) and classification structure. The main feature extraction part included the Swin Transformer and the Feature Pyramid Network (FPN), which can extract features from input images and obtain feature maps with multi-scale semantic information. The proposed ROI network was mainly composed of a CNN for realizing the segmentation so as to screen the approximate candidate box of the target location. The classification structure mainly consisted of two parts: category branch and mask branch. Figure 2 displays the integrated framework that combines Mask-RCNN with Swin Transformer, which was used to extract forest distribution information.

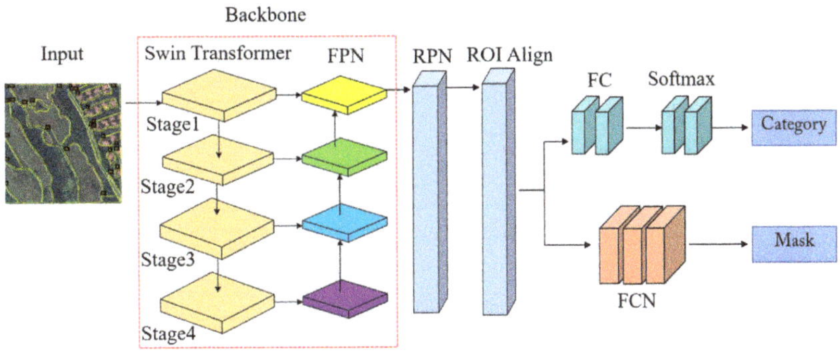

Figure 2. The integrated framework that combines Mask-RCNN with Swin Transformer for forest type extraction.

The Mask-RCNN model combined with Swin Transformer obtained internal parameters through iterative training. After multiple experimental tests, the learning rate was set to 0.005, the training batch size was 4, and 100 iterations were performed in the training phase. To implement the classifications using the deep learning algorithms, a sample library was established according to the actual situation of the study area. The fused GF-1 false-color composite was first cut into 780 × 780 image blocks in TensorFlow (https://www.tensorflow.org/). In order to ensure the consistency and full calculation of the number of each type as far as possible, sample images with uniform distribution and complete types were selected for annotation. The sample images were rotated and flipped to increase the number of available sample images. In total, this study used 864 labeled training map sheets and 432 labeled validation map sheets. The online annotation tool VIA (VGG Image Annotator, version 2) was used to annotate the training and validation datasets (sample images); VIA is an open-source image annotation software developed by the Visual Geometry Group, does not need to be downloaded and installed, can run offline in the browser, is simple to use and can annotate points, lines and polygons.

2.5. Calculation of Forest Carbon Storage Benchmark in 2017

In the current analysis, considering the availability of data and the operability, economy and accuracy of the used methods, we only measured or calculated the carbon storage of forest living woods, excluding the carbon storage of forest soil, litter and debris, dead wood and understory shrubs and herbs. For the measurement of forest living wood carbon storage based on the 2017 FMPI data, the forest biomass conversion factor method was adopted. This method has been widely used in the estimation of forest biomass and carbon storage in wide regions because there is a good regression relationship between forest stock volume, forest biomass and carbon storage by dominant tree species, and the estimation of forest carbon storage via forest biomass has a high accuracy [67]. Thus, the carbon storage of each forest stand in Jiangning District was calculated by using Equation (1), while the carbon storage of shrub forest was calculated by using Equation (2). To minimize the calculation complexity of shrub forests, we did not differentiate tree species of shrub forests, and we just focused on different average biomass density values of shrub forests at the two time points. The parameters involved in the calculation processes for tree species are summarized in Table A1 in Appendix A. Since the parameters of BEF (biomass expansion factor), wood basic density, carbon content and root-to-stem ratio of some minor tree species were unavailable in the study area, the parameters of similar tree species from the same family or genus were used for an approximate calculation. Finally, we summed the measurements of all the forest stands of living woods and shrub forest to obtain the total carbon storage benchmark in 2017.

$$C_z = \sum_{i,j} \{ A_{i,j} \cdot V_{i,j} \cdot BCEF_{i,j} \cdot (1 + R_{i,j}) \cdot CF_{i,j} \} \quad (1)$$

$$C_z = M_i \cdot A_i \cdot CFD \tag{2}$$

where C_z is the total carbon storage of forest stands (tons of carbon); A_{ij} is the area of the jth forest stand with dominant tree species i (hm^2); V_{ij} is the stock volume per unit area of the jth forest stand with dominant tree species i (m^3/hm^2); $BCEF_{ij} = BEF_{ij} \cdot D_i$; BEF_{ij} is the biomass expansion factor of tree species i in climatic zone j, which is the ratio of above-ground biomass over the biomass of tree trunks; R_{ij} is the ratio of root over stem of tree species i in climatic zone j; CF_{ij} is the carbon content of tree species i in climatic zone j (ton carbon/ton dry matter; C/t d.m); M_i is shrub layer biomass per unit area (hm^2); A_i is shrub layer area (hm^2); CFD is carbon content ratio (C/t d.m).

2.6. Derivation of the 2020 Forest Carbon Storage

Based on the outcomes of Section 2.4, we could easily obtain the spatial distributions of conventional forest stands (e.g., coniferous forest and broad-leaved forest) and the shrub forest type. For the shrub forest, we used Equation (2) to calculate the carbon storage in 2020 by specifying its newly updated biomass density per unit area, derived from limited harvesting measurements of shrubs, and its distribution area. For the conventional forest stands, their calculation of the 2020 carbon storage was divided into two parts, including the persisting forest part (pixels that were forest type in both 2017 and 2020), forest gain part (pixels that were non-forest type in 2017 but were forest in 2020). The 2020 carbon storage measurement of the persisting forest was based on the optimal local carbon density growth allometric equations. Based on the 2017 FMPI data in Nanjing, we fit the carbon density growth equations including the logistic (Equation (3)), Richards (Equation (4)) and Gompertz (Equation (5)) growth models (carbon density against forest age) of the coniferous forests and broad-leaved forests in Nanjing City by adopting the space for time substitution strategy in consideration of the availability of the inventory data of different tree species with different stand ages via RStudio package, and we selected the optimal equations for calculating the carbon storage of the persisting forests in Jiangning by specifying a 3-year growth duration in the optimal equations. For the part of forest gain, its initial forest age was set to 1.5 years (residing in the nursery with an average duration of 1.5 years for this mid-latitude region with good hydrothermal conditions); similarly, the growth duration was also set to 3 years to calculate the carbon storage by following the optimal equations for different forest types.

$$y = \frac{a}{(1 + be^{-ct})} \tag{3}$$

$$y = a\left(1 - e^{-ct}\right)^{b} \tag{4}$$

$$y = ae^{-be^{-ct}} \tag{5}$$

where y represents the carbon density of the forest stand; t represents the average age of the forest stand; and a, b and c are the model parameters to be solved.

2.7. Validation Strategies

For the validation of forest type classification, we identified another set of validation samples with the same size of 1000 pixels as the training set to construct confusion matrices. The validation sample points were selected by stratified random sampling to ensure an unbiased estimate of accuracy; 400 random points were generated in the non-forest region, and 600 random points were generated in the forest region. Then, the actual type attributes of these points were visually interpreted based on the fused GF-1 false-color composite coupled with our local knowledge. On this basis, the overall accuracy (OA) and Kappa coefficient were calculated to evaluate the classification accuracy. The formula for calculating the Kappa coefficient is written in Equation (6):

$$\text{Kappa} = \frac{P_o - P_e}{1 - P_e} \tag{6}$$

Here, P_o is the overall accuracy of classification, which refers to the probability that the classification results are consistent with the true reference data, and is calculated by dividing the amount of correctly classified pixels by the total of the pixels involved in the validation. P_e is the estimate of the chance agreement, which refers to the completely random assignment of pixels to classes (e.g., dicing to determine the class of a pixel). Thus, the Kappa coefficient indicates the difference degree between the actual classifications made by a classifier and the random classifications made by chance. When the Kappa coefficient value ranges from 0.61 to 0.80, there is a substantial difference between the results of an automated classifier and the results of a random chance classifier, suggesting an indication of good classification performance. When the Kappa value ranges from 0.81 to 1.00, there is an almost perfect difference between the results of an automated classifier and the results of a random chance classifier, showing an indication of excellent classification performance.

For the validation of the fitted carbon density growth models based on the 2017 FMPI data, we collected similar research studies (Table A7 in Appendix A) and cross-compared our simulated carbon density results of coniferous forest and broad-leaved forest with the results of those similar studies to prove the effectiveness and reliability of our simulated carbon density growth models.

Figure 3 shows the overall flow chart of this study.

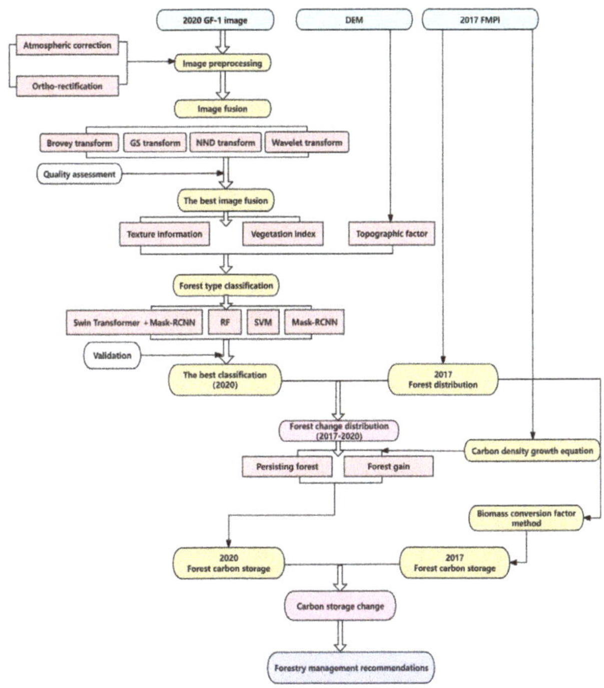

Figure 3. The flow chart of the current analysis.

3. Results

3.1. Image Fusion

Table A2 in the Appendix A lists the objective evaluation statistics of the four image fusion methods. It was found that the GS fusion method achieved the highest mean correlation coefficient, mean gradient and information entropy values, 0.9743, 63.2020 and 7.6548, respectively, and the db2 (mother wavelet)-based wavelet transform fusion method achieved the worst fusion performance, evidenced by the lowest correlation coefficient, mean gradient and information entropy values, 0.9081, 1.1443 and 3.8584, respectively. The order of mean correlation coefficients of the four image fusion methods was GS > NND > wavelet transform > Brovey, indicating that the fusion of GS and NND images was the closest to the original image, and the fidelity effect was good. The order of information entropy was GS > NND > Brovey > wavelet transform, which reflects that GS and NND have strong detail-expressive force and contain a large amount of information. The average gradient was ordered as GS > Brovey > NND > wavelet transform, indicating that GS and Brovey images are clearer.

Thus, GS-based image fusion was selected as the optimal input for subsequent forest classification analysis. Figure 4 shows the visual effects of the four fused images. Subjectively, Brovey and GS fusion methods could maintain high spectral fidelity compared to the false-color composite of the original multi-spectral bands, but wavelet transform and NNDiffuse methods had obvious spectral deviation. Based on the objective and subjective evaluations, the GS fusion method was determined to be the best fusion method among the four fusion methods, and its fused images were used to support subsequent classification applications.

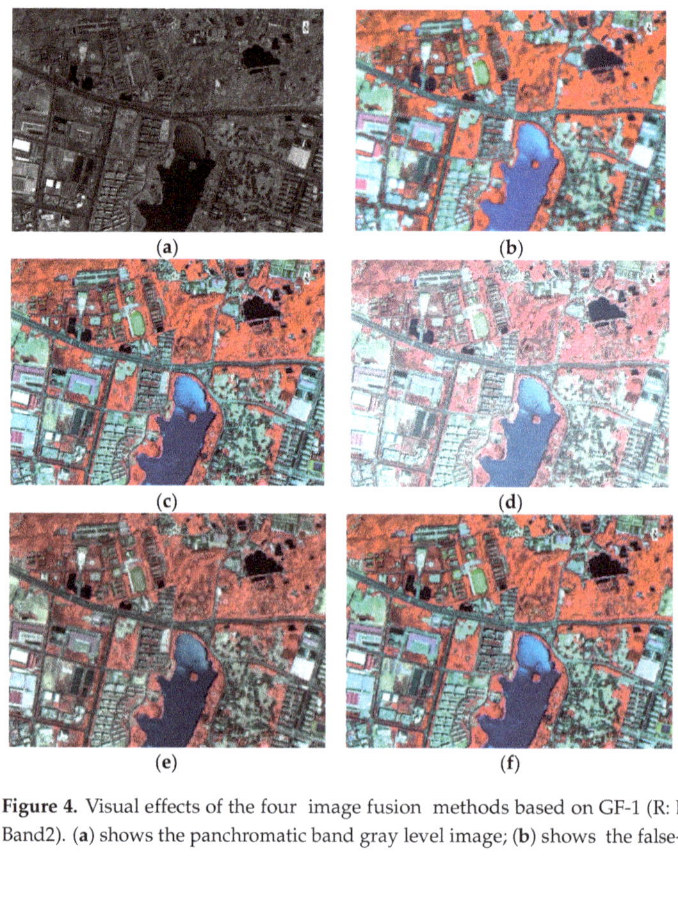

(a) (b) (c) (d) (e) (f)

Figure 4. Visual effects of the four image fusion methods based on GF-1 (R: Band4, G: Band3, B: Band2). (**a**) shows the panchromatic band gray level image; (**b**) shows the false-color composite of the original multi-spectral band combination (R: NIR band, G: red band, B green band); (**c**) shows the false-color composite of the Brovey-fused images; (**d**) shows the false-color composite of the db2-based wavelet-fused images; (**e**) shows the false-color composite of the NND-fused images; (**f**) shows the false-color composite of the GS-fused images.

3.2. Validation

3.2.1. Forest Classification Verification

Tables A3–A6 in Appendix A display the independent validation statistics of the four classification algorithms applied to extract the forest distributions. It was found that the SVM model had an overall classification accuracy of 85.8% and a Kappa coefficient of 0.787 (Table A3, in which the user accuracy of broad-leaved forest, coniferous forest, shrub forest and non-forest reached 90.31%, 76.16%, 49.28% and 91.49%, respectively), the RF model achieved an overall accuracy at 87.8% and a Kappa coefficient of 0.817 (Table A4, in which the user accuracy of broad-leaved forest, coniferous forest, shrub forest and non-forest reached 91.97%, 80.00%, 53.03% and 92.62%, respectively), the Mask-RCNN achieved an overall accuracy of 90.1% and a Kappa coefficient of 0.851 (Table A5, in which the user accuracy of broad-leaved forest, coniferous forest, shrub forest and non-forest reached 93.57%, 85.52%, 63.24% and 92.96%, respectively), and the Swin Transformer coupled with Mask-RCNN had an overall accuracy of 93.9% and a Kappa coefficient of 0.908 (Table A6, in which the user accuracy of broadleaf forest, coniferous forest, shrub forest and non-forest reached 94.94%, 90.13%, 81.13% and 96%, respectively). Obviously, the Swin Transformer coupled with Mask-RCNN outperformed the other three classification models, and its classification results were retained to support subsequent forest distribution change analysis.

3.2.2. Cross-Comparison of the Carbon Density Fitting Results

Figure 5 shows the fitting effects of forest carbon density against forest age by coniferous forest type and broad-leaved forest type, based on the logistic, Richards and Gompertz models. The fitting R^2 values of the logistic model, Richards model and Gompertz model for the coniferous forest type were estimated at 0.78, 0.75 and 0.77 respectively, and the fitting R^2 values of the three models for the broadleaf forest type were 0.71, 0.70 and 0.71, respectively. Apparently, the logistic model was the best model for simulating carbon density growth in both coniferous forest and broad-leaved forest; thus, it was used to predict the growth of carbon density of each forest stand by specifying the forest type first. Clearly, the fitting performance of the logistic model outperformed the other two models in both coniferous forests and broad-leaved forests, and all the models performed better when fitting coniferous forest data than when fitting broad-leaved forest data in the current study area (Figure 5).

Figure 6 shows the cross-validation effects of the fitted models by intercomparing our fitting results with other existing similar studies (Table A7 in Appendix A). In the comparison of coniferous forests, the carbon density results obtained by Justin (Drawing G) [68] were slightly higher than our research results in each age group, and the carbon density results of Li (Drawing C) [69] were also higher than our results in near-mature forest, mature forest and over-mature forest. At the same time, we maintained a high degree of agreement with the carbon density results obtained by Liu (Drawing E) [70], Yan (Drawing F) [71], Wise (Drawing H) [72] and Riahi (Drawing I) [73]. In the comparison of broad-leaved forests, we found that our carbon density results were lower than those of Li (K) [69], Yue (P) [74] and Liu (N) [70] and slightly higher than those of Lan (Drawing L) [75] and Hu (Drawing Q) [76]. But they all reasonably fall within the range of other available findings.

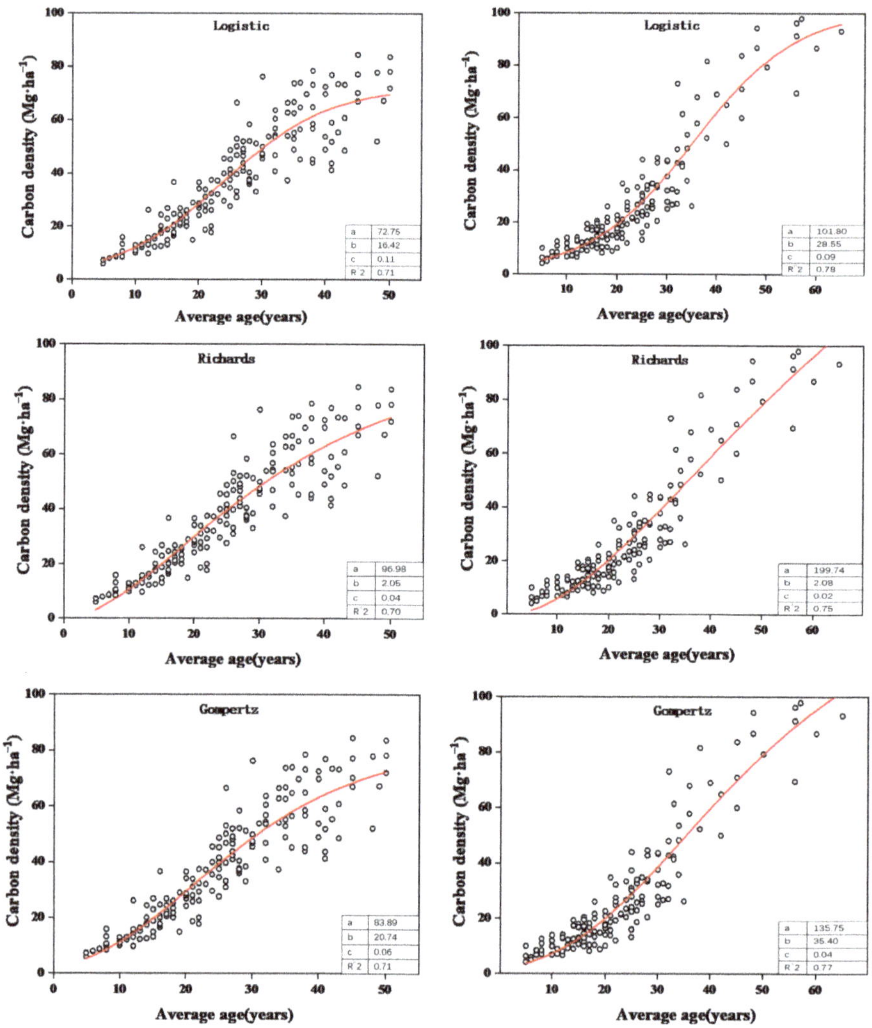

Figure 5. The fitting relationships between carbon density and forest age for broad-leaved forest (**left**) and coniferous forest (**right**) based on the 2017 FMPI data.

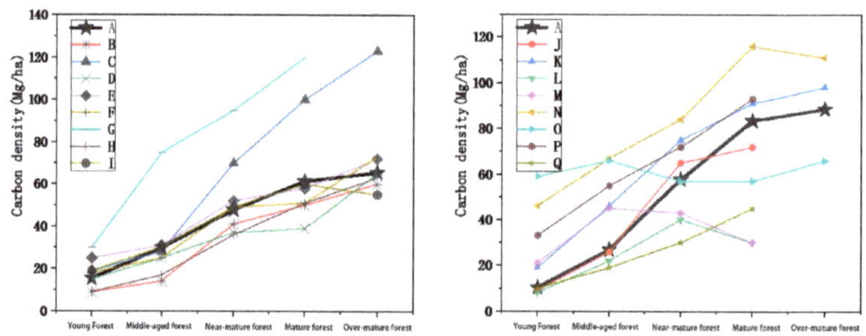

Figure 6. Comparison of the simulated coniferous forest (**left**) and broad-leaved forest carbon densities (**right**) with the results of other similar studies (A is the simulated results in the current analysis).

The cross-validation effect comparison with other existing similar literature shows that the simulated carbon densities of the coniferous forest and the broad-leaved forest at different age groups in the current work were reasonably within the ranges of other existing research results (Figure 6), indicating that the simulated models were reliable and applicable and that they could be used to calculate the carbon densities of persisting forests and newly gained forests in 2020.

3.3. Result Analysis

3.3.1. Forest Classification Extraction Results

Figure 7 displays the forest distribution pattern created from the 2017 FMPI results, Figure 8 shows the 2020 forest type distribution results extracted using the Swin Transformer coupled with the Mask-RCNN model. The forests were principally distributed in the northeastern and southeastern portions of Jiangning District in both 2017 and 2020 (Figures 7 and 8). And there was a total forest area of 41,913.805 hm^2 in 2017 and 42,261.315 hm^2 in 2020, with a net increase of about 347.51 hm^2. Figure 9 conveys the forest change information during the period 2017 to 2020, which was derived from the spatial overlay analysis of the 2017 forest distribution and the 2020 forest distribution. The statistical results of the forest change information map (Figure 9) showed the following: (1) There was a forest gain of 637.164 hm^2 during the period 2017 to 2020, and because the newly planted forest lands in 2020 did not take on the spectral signature of the forest type on the GF-1 remote sensing image, the value of 637.164 hm^2 might underestimate the actual areas of afforestation that occurred during the period 2017 to 2020. (2) there was a forest loss of 289.654 hm^2 during the period 2017 to 2020, and simultaneously, an analysis of forest reduction areas showed that it was mainly due to the construction demand of cities and farmland, and the demand of urban development caused forest reduction. For some regions, forest loss was characterized by small and scattered patches, mainly due to the conversion of forests to farmland. (3) The main reason for the increase in forest in some areas in the past three years was afforestation and greening efforts made by Jiangning District in response to the national policy.

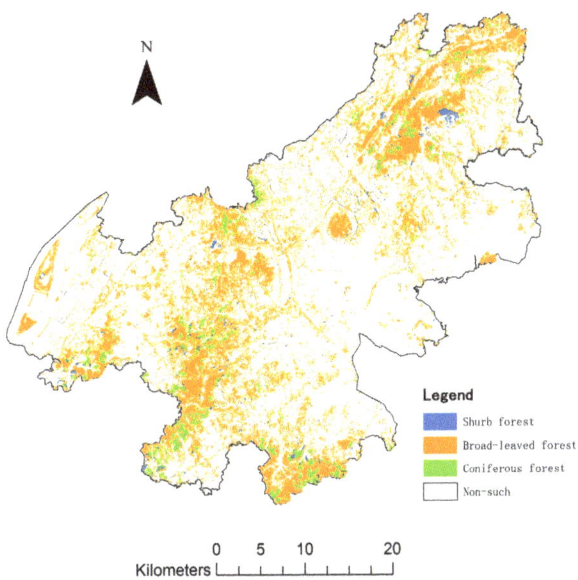

Figure 7. Forest distribution map of Jiangning in 2017.

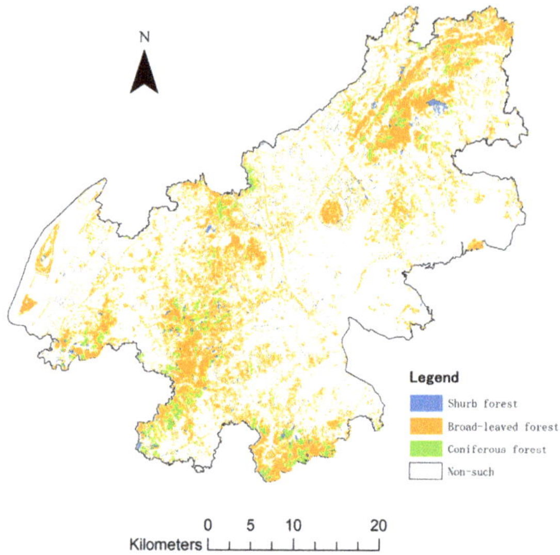

Figure 8. Forest distribution map of Jiangning in 2020.

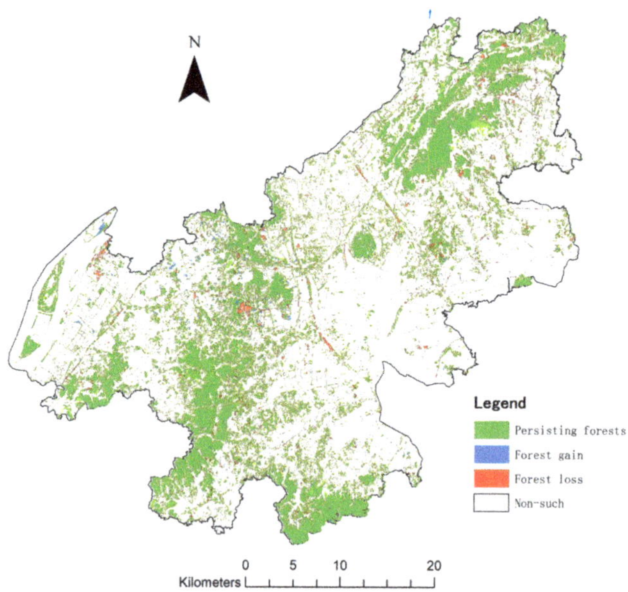

Figure 9. Forest change map of Jiangning during 2017–2020.

3.3.2. Forest Carbon Storage Measurement and Its Dynamics

Based on the methods mentioned in Sections 2.5 and 2.6, the forest carbon storage of 2017 was calculated at 125.40×10^4 t, and the shrub forests' carbon storage was derived at 1.085×10^4 t; thus, the total carbon storage of coniferous forest, broad-leaved forest and shrub forest totaled 126.49×10^4 t in 2017. Based on the distributions of the persisting forests and the newly gained forests, the 2017 FMPI data and the fitted carbon equations, the carbon storage of the persisting forests in 2020 was calculated at 142.53×10^4 t, the carbon storage of forest gains was estimated at 1.04×10^4 t and the carbon storage of the shrub forests in 2020 was estimated at 1.142×10^4 t; thus, the total carbon storage of the forests in 2020 of Jiangning District was estimated at 143.54 × 10^4 t. Comparing the 2017 total carbon storage with the 2020 carbon storage, we found that there was a net increase in carbon storage of 2.41 × 10^4 t.

4. Discussion

(1) Forest type extraction effectiveness and reliability

The initial backbone of the improved Mask-RCNN model in this study was replaced by the Swin Transformer model, and the modified framework produced an extraction accuracy that was similar to or even higher than that in other studies that also used this type of approach [28,29]. Cong et al. compared the extraction accuracy of the improved Mask-RCNN model fused with Swin Transformer with the convolutional neural network model embedded with UNet3+ in bell pepper instance detection, and they found that the average detection accuracy, average detection recall, average segmentation accuracy and F1 score were 98.1%, 99.4%, 94.8% and 98.8%, respectively, indicating that the improved Mask-RCNN model fused with Swin Transformer could effectively segment different classes of bell peppers under overexposure, bell pepper overlap and leaf occlusion conditions [77]. Jamali et al. compared multiple models within RF, SVM, VGG-16, 3D CNN, and Swin Transformer to classify coastal wetlands in Saint John, New Brunswick, Canada, and this work demonstrated that Swin Transformer has great potential in classifying complex coastal landscapes [78]. Our results in the current study are consistent with the findings of the above-mentioned two studies and confirm the effectiveness and high accuracy of the proposed framework for forest type extraction applications.

(2) Forest growth model

According to the fitting results of the three forest growth models, logistic, Gompertz and Richards, the logistic model had the best effect. Xu used five growth theory models, namely Compertz, Korf, logistic, Mischerlich and Schumacher, to fit the unit stock of the Masson pine forest and Chinese fir forest in Longquan, Zhejiang, and found that the logistic equation had the best fitting effect, and the coefficient of determination was above 0.85 [79]. Similarly, Rong et al. took Jinguuling Forest Farm in Jilin as a research object and established a stand stock growth model by using five theoretical growth equations: Richards, logistic, single molecule, Gompertz and Korf. Their results showed that the logistic model was the best in Larix artificial forest, mixed broad-leaved forest and mixed coniferous natural forest [80]. The results of the two analyses are consistent with the results of our study, which fully indicates the validity and credibility of the results of this study.

(3) Carbon storage measurement

Based on the 2017 FMPI data, this analysis focused only on the carbon pool of living trees. Forest carbon storage was measured by dominant tree species categories in each forest stand, and the local species-specific parameter values of BEF, carbon content, root-to-stem ratio and basic density of wood compiled from the Second Forestry Carbon Sink Measurement and Monitoring Work Program of Jiangsu Province were used, which made the calculation results more accurate and in line with the forests of Jiangning District of Nanjing City. Further, if the empirical parameters of soil and AGB (above-ground biomass) sample plots surveyed by Fang et al. were taken as a basis, a rough estimation of the soil carbon stock, litter carbon pool and understory shrub and grass carbon pool in the forests of Jiangning District could be performed [42]; thus, our calculation results related to different carbon pools could be expanded to support more a comprehensive assessment of forest carbon fixation capability. However, these parameters are generally regional-scale-oriented, and application of them to the local scale, e.g., Jiangning District, should be locally corrected. Due to the unavailability of field sample data on soil, litter and understory shrub and grass in Jiangning's forests, the empirical estimations may have some uncertainties and deserve further field sampling validations.

(4) Usage of the measured carbon storage dynamics

The forest change map of Jiangning District (2017–2020) (Figure 7) and the measured forest carbon dynamics provide key references for the development of carbon sink forestry, sustainable forest management and forest protection and utilization measures [81]. The information on forest change and carbon storage will enable administrative agencies to grasp the change magnitude of forest carbon sequestration capacity and future development potential in the whole region, to more accurately and efficiently identify areas suitable for forest growth, and to regularly inventory, update, and release data on carbon-stock-related indicators based on the framework proposed in the current work, helping to continuously explore new paths for forest carbon stock measurement and monitoring [82]. Additionally, the change information will facilitate the adoption of forest quarantine, biological and chemical control, prediction and forecasting, and damage assessment to prevent forest disasters from occurring and to improve forest management effectiveness [83]. Ultimately, this study can provide a technical reference for efficient monitoring of forest status change, and its findings lay an underlying data basis for local agencies to develop targeted forest management policies or measures when pursuing the double-carbon goals (achieving peak CO_2 emissions before 2030 and carbon neutrality before 2060).

5. Conclusions

In this study, we developed an efficient framework that integrates deep learning, statistical modeling and GF-1 remote sensing images for the timely evaluation of changes in forest distribution and forest living carbon storage. We particularly fit the local optimal carbon density growth equations by forest type to facilitate the dynamic assessment of standing trees' carbon storage in Jiangning's forests between 2017 and 2020. The results showed that Mask-RCNN fused with Swin Transformer had the highest capability in extracting coniferous forest, broad-leaved forest and shrub forest compared to the other three models. Additionally, the net increase in forest area and standing tree carbon storage indicated the effectiveness of the forest management practices implemented by the local agencies. The study findings provide key data and a methodological basis for the development of carbon sink forestry, sustainable forest management and targeted measures for forest conservation and utilization and contribute to pursuing the double-carbon goals (achieving peak CO_2 emissions before 2030 and carbon neutrality before 2060 in China).

Author Contributions: Conceptualization, M.L. and D.X.; methodology, J.L. and B.Y.; software, J.L. and B.Y.; validation, J.L. and B.Y.; formal analysis, B.Y.; investigation, J.L. and B.Y.; resources, M.L. and D.X.; data curation, M.L.; writing—original draft preparation, J.L., writing—review and editing, J.L. and M.L.; visualization, J.L., supervision, M.L.; project administration, M.L.; funding acquisition, M.L. and D.X. All authors have read and agreed to the published version of the manuscript.

Funding: This work was jointly funded by the National Natural Science Foundation of China, grant number 31971577, and the Priority Academic Program Development of Jiangsu Higher Education Institutions (PAPD).

Data Availability Statement: The data will be made available by the authors on request.

Conflicts of Interest: Da Xu was employed by Zhejiang Forestry Survey Planning and Design Company Limited. The remaining authors declare that the research was conducted in the absence of any commercial or financial relationships that could be construed as a potential conflict of interest. Zhejiang Forestry Survey Planning and Design Company Limited had no role in the design of the study; in the collection, analyses, or interpretation of data; in the writing of the manuscript, or in the decision to publish the results.

Appendix A

Table A1. The involved calculation parameters for dominant tree species (groups) in China.

Number	Dominant Species	BEF	Basic Density of Wood D (t/m^3)	Root-to-Stem Ratio	Carbon Content Ratio (tC/t d.m)
1	Red Pine	1.4251	0.4137	0.1920	0.5141
2	Black Pine	1.8920	0.4500	0.2180	0.5146
3	Horsetail Pine	1.2940	0.4482	0.1730	0.5271
4	Overseas Pine	1.4209	0.4894	0.2813	0.5156
5	Wetland Pine	1.3780	0.3590	0.2680	0.5311
6	Torch Pine	1.5680	0.4354	0.3380	0.5361
7	Other Pines	1.3410	0.4649	0.1810	0.4963
8	Fir	1.2990	0.3071	0.2030	0.5127
9	Willow Fir	1.2710	0.2893	0.2680	0.5331
10	Metasequoia	1.3630	0.2740	0.3510	0.5083
11	Pond fir	1.3580	0.3700	0.3133	0.5156
12	Cypress	1.4580	0.4722	0.2190	0.5088
13	Yew (Sequoia)	1.4477	0.3913	0.2197	0.5156
14	Other Fir	1.3340	0.3765	0.2420	0.5185
15	Oak	1.2880	0.6119	0.2890	0.4798
16	Birch	1.4210	0.5270	0.2530	0.4914
17	Water, Hu, Yellow	1.2930	0.4523	0.2210	0.4620
18	Ash	1.3120	0.5462	0.3190	0.4803
19	Walnut	1.3088	0.4302	0.2863	0.4803
20	Camphor	1.2490	0.4649	0.2580	0.4916
21	Nan	1.2490	0.4807	0.2580	0.5002
22	Elm	1.3683	0.4868	0.2504	0.4803
23	Mullein	1.4090	0.5161	0.1990	0.5115
24	Maple	1.2860	0.4860	0.3370	0.4803
25	Other Hardwoods	1.3850	0.6062	0.2410	0.4901
26	Lime	1.3831	0.4177	0.1997	0.4392
27	Sassafras	1.3130	0.4758	0.2610	0.4848
28	Poplar	1.3940	0.3644	0.1850	0.4502
29	Willow	1.3940	0.4409	0.1850	0.4803
30	Paulownia	1.7870	0.2367	0.2360	0.4695
31	Eucalyptus	1.1930	0.5901	0.1790	0.4748
32	Acacia	1.3860	0.5843	0.2070	0.4666
33	Mullein	1.3440	0.6768	0.1950	0.4893
34	Neem	1.3884	0.4389	0.1890	0.4803
35	Other Soft Broadleaf	1.2730	0.4222	0.2150	0.4502
36	Conifer Mix	1.3646	0.3902	0.2086	0.5168
37	Broadleaf Mix	1.2815	0.5222	0.2351	0.4796
38	Needle–Broadleaf Mix	1.3230	0.4754	0.2218	0.4893

Table A2. Objective evaluation measures of different image fusion algorithms.

Statistics / Fusion Methods	Band	Correlation Coefficient	Mean Gradient	Information Entropy
Panchromatic			18.1610	6.2657
Multi-spectra	Blue Green Red NIR		10.0761	6.5217
GS	Blue Green Red NIR	0.9743	63.2020	7.6584

Table A2. *Cont.*

Fusion Methods \ Statistics	Band	Correlation Coefficient	Mean Gradient	Information Entropy
Brovey	Green Red NIR	0.8843	14.6312	4.6582
Wavelet Transform	Blue Green Red NIR	0.9081	1.1443	3.8584
NNDiffuse	Blue Green Red NIR	0.9094	10.0635	6.9215

Table A3. Accuracy validation statistics of forest type classifications based on SVM.

		Classification Result					
		Broad-leaved forest	Coniferous forest	Shrub forest	Non-forest	Subtotal	Producer accuracy (%)
Reference samples	Broad-leaved forest	354	15	12	13	394	89.85
	Coniferous forest	17	115	10	11	153	75.16
	Shrub forest	10	8	34	9	61	55.74
	Non-forest	11	13	13	355	392	90.56
	Subtotal	392	151	69	388	1000	
	User accuracy (%)	90.31	76.16	49.28	91.49		
					OA = 85.8%		Kappa = 0.787

Table A4. Accuracy validation statistics of forest type classifications based on RF.

		Classification Result					
		Broad-leaved forest	Coniferous forest	Shrub forest	Non-forest	Subtotal	Producer accuracy (%)
Reference samples	Broad-leaved forest	355	13	12	14	394	90.10
	Coniferous forest	15	124	8	6	153	81.05
	Shrub forest	9	8	35	9	61	57.38
	Non-forest	7	10	11	364	392	92.86
	Subtotal	386	155	66	393	1000	
	User accuracy (%)	91.97	80.00	53.03	92.62		
					OA = 87.8%		Kappa = 0.817

Table A5. Accuracy validation statistics of forest type classifications based on Mask-RCNN.

		Classification Result					
		Broad-leaved forest	Coniferous forest	Shrub forest	Non-forest	Subtotal	Producer accuracy (%)
Reference samples	Broad-leaved forest	364	11	9	10	394	92.39
	Coniferous forest	13	124	7	9	153	81.05
	Shrub forest	5	4	43	9	61	70.49
	Non-forest	7	6	9	370	392	94.39
	Subtotal	389	145	68	398	1000	
	User accuracy (%)	93.57	85.52	63.24	92.96		
					OA = 90.1%		Kappa = 0.851

Table A6. Accuracy validation statistics of forest type classifications based on Mask-RCNN combined with Swin Transformer.

		Classification Result					
		Broad-leaved forest	Coniferous forest	Shrub forest	Non-forest	Subtotal	Producer accuracy (%)
Reference samples	Broad-leaved forest	375	8	5	6	394	95.18
	Coniferous forest	10	137	3	3	153	89.54
	Shrub forest	6	5	43	7	61	70.49
	Non-forest	4	2	2	384	392	97.96
	Subtotal	395	152	53	400	1000	
	User accuracy (%)	94.94	90.13	81.13	96.00		
					OA = 93.9%		Kappa = 0.908

Table A7. List of similar studies used to verify the effectiveness and reliability of the simulated carbon density equations.

Letter	Stand Type	Source	Letter	Stand Type	Source
B	Coniferous plantation forests	Ali [84]	J	Broad-leaved plantation forests	Li [69]
C	Coniferous natural forests	Li [69]	K	Broadleaf natural forests	Li [69]
D	Mixed conifer forests	Li [85]	L	Mixed broadleaf forests	Lan [75]
E	Mixed coniferous forests	Liu [70]	M	Mixed broadleaf forests	Li [85]
F	Coniferous forests	Yan [71]	N	Mixed broadleaf forests	Liu [70]
G	Coniferous plantations	Justine [68]	O	Mixed broadleaf forests	Yang [86]
H	Coniferous forests	Wise [72]	P	Broadleaf forests	Yue [74]
I	Coniferous forests	Riahi [73]	Q	Broad-leaved natural forests	Hu [76]

References

1. Qureshi, A.; Badola, R.; Hussain, S. A review of protocols used for assessment of carbon stock in forested landscapes. *Environ. Sci. Policy* **2012**, *16*, 81–89. [CrossRef]
2. Lal, R.; Smith, P.; Jungkunst, H.F.; Mitsch, W.J.; Lehmann, J. The carbon sequestration potential of terrestrial ecosystems. *J. Soil Water Conserv.* **2018**, *73*, 145A–152A. [CrossRef]
3. Cui, J.; Glatzel, S.; Wang, B. Long-term effects of biochar application on greenhouse gas production and microbial community in temperate forest soils under increasing temperature. *Sci. Total Environ.* **2021**, *767*, 145021. [CrossRef]
4. Zhang, J.; Lin, H.; Li, S. Accurate gas extraction (AGE) under the dual-carbon background: Green low-carbon development pathway and prospect. *J. Clean. Prod.* **2022**, *377*, 134372. [CrossRef]
5. Leite, R.V.; Mohan, M.; Cardil, A. Individual Tree Attribute Estimation and Uniformity Assessment in Fast-Growing *Eucalyptus* spp. Forest Plantations Using Lidar and Linear Mixed-Effects Models. *Remote Sens.* **2020**, *12*, 3599. [CrossRef]
6. Forrester, D.I.; Benneter, A.; Bouriaud, O. Diversity and competition influence tree allometric relationships developing functions for mixed-species forests. *J. Ecol.* **2017**, *105*, 761–774. [CrossRef]
7. Atkins, J.W.; Costanza, J.; Dahlin, K.M. Scale dependency of lidar-derived forest structural diversity. *Methods Ecol. Evol.* **2023**, *14*, 708–723. [CrossRef]
8. Madonsela, S.; Cho, M.A.; Ramoelo, A.; Mutanga, O. Investigating the relationship between tree species diversity and landsat-8 spectral heterogeneity across multiple phenological stages. *Remote Sens.* **2021**, *13*, 2467. [CrossRef]
9. Huang, C.; Zhang, C.; Liu, Q. Multi-Feature Classification of Optical and SAR Remote Sensing Images for Typical Tropical Plantation Species. *Sci. Silvae Sin.* **2021**, *57*, 80–91. (In Chinese)
10. Nartišs, M.; Melniks, R. Improving pixel-based classification of GRASS GIS with support vector machine. *Trans. GIS* **2023**, *27*, 1865–1880. [CrossRef]
11. Liu, X.; Bo, Y. Object-Based Crop Species Classification Based on the Combination of Airborne Hyperspectral Images and LiDAR Data. *Remote Sens.* **2015**, *7*, 922–950. [CrossRef]
12. Mao, L.; Li, M. Integrating Sentinel Active and Passive Remote Sensing Data to Land Cover Classification in a National Park from GEE Platform. *Geomat. Inf. Sci. Wuhan Univ.* **2023**, *48*, 756–764. (In Chinese)
13. James, G.C.B.; Katerina, P. Using deep convolutional neural networks to forecast spatial patterns of Amazonian deforestation. *Methods Ecol. Evol.* **2023**, *13*, 2622–2634. [CrossRef]
14. Zhao, R.; Yan, R.; Chen, Z. Deep learning and its applications to machine health monitoring. *Mech. Syst. Signal Process.* **2019**, *115*, 213–237. [CrossRef]
15. LeCun, Y.; Bengio, Y.; Hinton, G. Deep learning. *Nature* **2015**, *521*, 436–444. [CrossRef]

16. LeCun, Y.; Boser, B.; Denker, J.S. Backpropagation applied to handwritten zip code recognition. *Neural Comput.* **1989**, *1*, 541–551. [CrossRef]
17. Jeon, E.; Kim, S.; Park, S.; Kwak, J.; Choi, I. Semantic segmentation of seagrass habitat from drone imagery based on deep learning: A comparative study. *Ecol. Inform.* **2021**, *66*, 101430. [CrossRef]
18. He, S.; Du, H.; Zhou, G. Intelligent map of urban forests from high-resolution remotely sensed imagery using object-based u-net-densenet-coupled network. *Remote Sens.* **2020**, *12*, 3928. [CrossRef]
19. Hameed, K.; Chai, D.; Rassau, A. Score-based mask edge improvement of Mask-RCNN for segmentation of fruit and vegetables. *Expert Syst. Appl.* **2022**, *190*, 116205. [CrossRef]
20. Sievänen, R.; Salminen, O.; Lehtonen, A. Carbon stock changes of forest land in Finland under different levels of wood use and climate change. *Ann. For. Sci.* **2014**, *71*, 255–265. [CrossRef]
21. Yuan, L.; Qiu, Z. Mask-RCNN with spatial attention for pedestrian segmentation in cyber–physical systems. *Comput. Commun.* **2021**, *180*, 109–114. [CrossRef]
22. Xie, Y.; Xu, Y.; Hu, C. Urban forestry detection by deep learning method with GaoFen-2 remote sensing images. *J. Appl. Remote Sens.* **2022**, *16*, 022206. [CrossRef]
23. Shi, F.; Yang, B.; Li, M. An improved framework for assessing the impact of different urban development strategies on land cover and ecological quality changes-A case study from Nanjing Jiangbei New Area, China. *Ecol. Indic.* **2023**, *147*, 109998. [CrossRef]
24. Ren, K.; Chen, Z.; Gu, G. Research on infrared small target segmentation algorithm based on improved mask R-CNN. *Optik* **2023**, *272*, 170334. [CrossRef]
25. Carvalho, O.L.F.; de Carvalho Junior, O.A.; Albuquerque, A.O. Instance segmentation for large, multi-channel remote sensing imagery using mask-RCNN and a mosaicking approach. *Remote Sens.* **2020**, *13*, 39. [CrossRef]
26. Vaswani, A.; Shazeer, N.; Parmar, N. Attention is all you need. In *Advances in Neural Information Processing Systems*; Curran Associates, Inc.: New York, NY, USA, 2017; Volume 30.
27. Tian, Y.; Bu, X. Lumbar spine image segmentation method based on attention mechanism and Swin Transformer model. *Meas. Test. Technol.* **2021**, *48*, 57–61.
28. Liu, Z.; Lin, Y.; Cao, Y. Swin transformer: Hierarchical vision transformer using shifted windows. In Proceedings of the IEEE/CVF International Conference on Computer Vision, Virtual, 11–17 October 2021; pp. 10012–10022. [CrossRef]
29. Gao, J.; Zhang, X.; Guo, Y. Research on the optimized pest image instance segmentation method based on the Swin Transformer model. *J. Nanjing For. Univ. (Nat. Sci. Ed.)* **2023**, *47*, 1. (Translated from Chinese) [CrossRef]
30. Gupta, R.; Sharma, L.K. The process-based forest growth model 3-PG for use in forest management: A review. *Ecol. Model.* **2019**, *397*, 55–73. [CrossRef]
31. Pretzsch, H.; Biber, P. Fertilization modifies forest stand growth but not stand density: Consequences for modelling stand dynamics in a changing climate. *For. Int. J. For. Res.* **2022**, *95*, 187–200. [CrossRef]
32. Li, C.; Barclay, H.; Huang, S. Modelling the stand dynamics after a thinning induced partial mortality: A compensatory growth perspective. *Front. Plant Sci.* **2022**, *13*, 1044637. [CrossRef] [PubMed]
33. Satoh, D. Discrete Gompertz equation and model selection between Gompertz and logistic models. *Int. J. Forecast.* **2021**, *37*, 1192–1211. [CrossRef]
34. Zhao, J.; Hu, H.; Wang, J. Forest Carbon Reserve Calculation and Comprehensive Economic Value Evaluation: A Forest Management Model Based on Both Biomass Expansion Factor Method and Total Forest Value. *Int. J. Environ. Res. Public Health* **2022**, *19*, 15925. [CrossRef] [PubMed]
35. Clément, J.B.; Sous, D.; Bouchette, F. A Richards' equation-based model for wave-resolving simulation of variably-saturated beach groundwater flow dynamics. *J. Hydrol.* **2023**, *619*, 129344. [CrossRef]
36. Jiang, X.; Wen, S.; Yu, X. Studies on the Variable—Density Whole Stand Model of Cunninghamia lanceolata Plantations and Its Application. *J. Fujian For. Sci. Technol.* **2000**, *27*, 22–25. (Translated from Chinese)
37. Feng, Z.; Xiong, N.; Wang, J.; Li, X. Establishment and Study of Full Stand Models for Sidebar Artificial Forests in Beijing City. *J. Beijing For. Univ.* **2008**, *S1*, 214–227. (Translated from Chinese)
38. Wang, M.; Wei, X.; Wei, Y.; Wang, M. Population structure and dynamic characteristics of rare and endangered plant Ormosia hosiei in Sichuan and Guizhou Province. *Chin. Guihaia* **2023**, *45*, 179. (Translated from Chinese)
39. Luo, M.; Wang, Y.; Xie, Y. Combination of feature selection and catboost for prediction: The first application to the estimation of aboveground biomass. *Forests* **2021**, *12*, 216. [CrossRef]
40. Zhang, B.; Liu, G.; Feng, Z. Constructing a Model of Poplus spp. Growth Rate Based on the Model Fusion and Analysis of Its Growth Rate Differences and Distribution Characteristics under Different Classes of Environmental Indicators. *Forests* **2023**, *14*, 2073. [CrossRef]
41. Piao, S.; Fang, J.; Ciais, P. The carbon balance of terrestrial ecosystems in China. *Nature* **2009**, *458*, 1009–1013. [CrossRef]
42. Fang, J.; Wang, G.; Liu, G. Forest biomass of China: An estimate based on the biomass-volume relationship. *Ecol. Appl.* **1998**, *8*, 1084.
43. Wulder, M.; White, J.; Nelson, R. Lidar sampling for large-area forest characterization: A review. *Remote Sens. Environ.* **2012**, *121*, 196–209. [CrossRef]
44. Vincent, W.; Gao, J. Towards refined estimation of vegetation carbon stock in Auckland, New Zealand using WorldView-2 and LiDAR data: The impact of scaling. *Int. J. Remote Sens.* **2019**, *40*, 8727–8747.

45. Nelson, R.; Hyde, P. Investigating RaDAR–LiDAR synergy in a North Carolina pine forest. *Remote Sens. Environ.* **2007**, *110*, 98–108. [CrossRef]
46. Vatandaşlar, C.; Abdikan, S. Carbon stock estimation by dual-polarized synthetic aperture radar (SAR) and forest inventory data in a Mediterranean forest landscape. *J. For. Res.* **2022**, *33*, 827–838. [CrossRef]
47. Wang, S.; Kobayashi, K.; Takanashi, S. Estimating divergent forest carbon stocks and sinks via a knife set approach. *J. Environ. Manag.* **2023**, *330*, 117114. [CrossRef] [PubMed]
48. Cao, X.; Ke, C.; Ran, J. Research on urban land use expansion based on GIS technology—A case study of Jiangning District, Nanjing. *Resour. Sci.* **2008**, *30*, 385–391. (Translated from Chinese)
49. Gillespie, A.R.; Kahle, A.B.; Walker, R.E. Color enhancement of highly correlated images-Channel Ratio and 'Chromaticity' Transformation Techniques. *Remote Sens. Environ.* **1987**, *22*, 343–365. [CrossRef]
50. Hamid, R.S. Improved Adaptive Brovey as a New Method for Image Fusion. *arXiv* **2018**, arXiv:1807.09610.
51. Ehlers, M.; Klonus, S.; Johan, A. Multi-sensor image fusion for pansharpening in remote sensing. *Int. J. Image Data Fusion* **2010**, *1*, 25–45. [CrossRef]
52. Witharana, C.; Larue, M.A.; Lynch, H.J. Benchmarking of data fusion algorithms in support of earth observation based Antarctic wildlife monitoring. *J. Photogramm. Remote Sens.* **2016**, *113*, 124–143. [CrossRef]
53. Chen, Y.; Sun, K. Quality evaluation of Gaofen-2 image fusion method. *Surv. Mapp. Sci.* **2017**, *42*, 35–40.
54. Vijayaraj, V.; Younan, N.H.; O'Hara, C.G. Quantitative analysis of pansharpened images. *Opt. Eng.* **2006**, *45*, 828. [CrossRef]
55. Ducay, R.; Messinger, D.W. Image fusion of hyperspectral and multispectral imagery using nearest-neighbor diffusion. *J. Appl. Remote Sens.* **2023**, *17*, 024504. [CrossRef]
56. Oliveira, R.J.; Caldeira, B.; Teixidó, T. Geophysical data fusion of ground-penetrating radar and magnetic datasets using 2D wavelet transform and singular value decomposition. *Front. Earth Sci.* **2022**, *10*, 1011999. [CrossRef]
57. Yuan, Y.; Zhang, J.; Chang, B. Objective quality evaluation of visible and infrared color fusion image. *Opt. Eng.* **2011**, *50*, 033202. [CrossRef]
58. Guo, M.; Li, B.; Shao, Z. Objective image fusion evaluation method for target recognition based on target quality factor. *Multimed. Syst.* **2022**, *28*, 495–510. [CrossRef]
59. Sarkar, J.; Rashid, M. Visualizing mean, median, mean deviation, and standard deviation of a set of numbers. *Am. Stat.* **2016**, *70*, 304–312. [CrossRef]
60. Edjabou, M.E.; Martín-Fernández, J.A.; Scheutz, C. Statistical analysis of solid waste composition data: Arithmetic mean, standard deviation and correlation coefficients. *Waste Manag.* **2017**, *69*, 13–23. [CrossRef]
61. Zhu, H.; Jia, G.; Zhang, Q. Detecting Offshore Drilling Rigs with Multitemporal NDWI: A Case Study in the Caspian Sea. *Remote Sens.* **2021**, *13*, 1576. [CrossRef]
62. Liao, L.; Song, J.; Wang, J. Bayesian method for building frequent Landsat-like NDVI datasets by integrating MODIS and Landsat NDVI. *Remote Sens.* **2016**, *8*, 452. [CrossRef]
63. Roy, D.P.; Li, J.; Zhang, H.K. Examination of Sentinel-2A multi-spectral instrument (MSI) reflectance anisotropy and the suitability of a general method to normalize MSI reflectance to nadir BRDF adjusted reflectance. *Remote Sens. Environ.* **2017**, *199*, 25–38. [CrossRef]
64. Zheng, G.; Li, X.; Zhou, L.; Yang, J.; Ren, L. Development of a gray-level co-occurrence matrix-based texture orientation estimation method and its application in sea surface wind direction retrieval from SAR imagery. *IEEE Trans. Geosci. Remote Sens.* **2018**, *56*, 5244–5260. [CrossRef]
65. Wang, F.; Bhattacharya, A.; Gelfand, A.E. Process modeling for slope and aspect with application to elevation data maps. *Test* **2018**, *27*, 749–772. [CrossRef]
66. Ring, M.; Eskofier, B.M. An approximation of the Gaussian RBF kernel for efficient classification with SVMs. *Pattern Recognit. Lett.* **2016**, *84*, 107–113. [CrossRef]
67. Nonini, L.; Fiala, M. Estimation of carbon storage of forest biomass for voluntary carbon markets: Preliminary results. *J. For. Res.* **2021**, *32*, 329–338. [CrossRef]
68. Justin, M.F. Dynamics of biomass and carbon sequestration across a chronosequence of masson pine plantations. *J. Geophys. Res. Biogeosciences* **2017**, *122*, 578–591. [CrossRef]
69. Li, Q. *Carbon Stock and Carbon Sequestration Potential of Arboreal Forests in China from 2010 to 2050*; China Academy of Forestry Sciences: Beijing, China, 2016; (Translated from Chinese)
70. Liu, W. Carbon stock and allocation characteristics of public welfare forests in Baisanzuo National Park. *J. Ecol.* **2021**, *40*, 1–10. (Translated from Chinese)
71. Yan, R. Dynamic analysis and potential prediction of carbon stock in arboreal forests in Chongqing. *For. Econ.* **2019**, *41*, 70–77, (Translated from Chinese)
72. Wise, M. Implications of limiting CO_2 concentrations for land use and energy. *Science* **2009**, *324*, 1183–1186. [CrossRef]
73. Riahi, K. Scenarios of long-term socio-economic and environmental development under climate stabilization. *Technol. Forecast. Soc. Change* **2007**, *74*, 887–935. [CrossRef]
74. Yue, J. *Carbon Sequestration Dynamics, Potential and Impact mechanisms of Major Forest Types in Gansu*; University of Chinese Academy of Sciences (Research Centre for Soil and Water Conservation and Ecological Environment, Ministry of Education, Chinese Academy of Sciences): Yangling, China, 2018; (Translated from Chinese)

75. Lan, T.; Gu, J.; Wen, Z. Spatial distribution characteristics of carbon storage density in typical mixed fir and broadleaf forests. *Energy Rep.* **2021**, *7*, 7315–7322. [CrossRef]
76. Hu, W.; Wang, X. Comparison of ecosystem carbon density between sharp-toothed Quercus serrata and cork oak forests. *J. For. Environ.* **2017**, *37*, 8–15, (Translated from Chinese)
77. Cong, P.; Li, S.; Zhou, J. Research on instance segmentation algorithm of greenhouse sweet pepper detection based on improved mask RCNN. *Agronomy* **2023**, *13*, 196. [CrossRef]
78. Jamali, A.; Mahdianpari, M. Swin transformer and deep convolutional neural networks for coastal wetland classification using sentinel-1, sentinel-2, and LiDAR data. *Remote Sens.* **2022**, *14*, 359. [CrossRef]
79. Xu, Q. *Construction and Simulation of Stand Growth Model*; Zhejiang Agricultural Forest University: Hangzhou, China, 2015.
80. Rong, J.; Liu, D. Growth model of stand stock of main forest types in overcut forest area of northeast China. *Appl. Res.* **2011**, *25*, 30–34. [CrossRef]
81. Kaya, A.; Bettinger, P.; Boston, K.; Akbulut, R.; Ucar, Z. Optimisation in forest management. *Curr. For. Rep.* **2016**, *2*, 1–17. [CrossRef]
82. Paul, T.; Kimberley, M.O.; Beets, P.N. Natural forests in New Zealand–a large terrestrial carbon pool in a national state of equilibrium. *For. Ecosyst.* **2021**, *8*, 34. [CrossRef]
83. Runting, R.K.; Ruslandi; Griscom, B.W. Larger gains from improved management over sparing–sharing for tropical forests. *Nat. Sustain.* **2019**, *2*, 53–61. [CrossRef]
84. Ali, A.; Ahmad, A.; Akhtar, K. Patterns of biomass, carbon, and soil properties in masson pine (*Pinus massoniana* Lamb) plantations with different stand ages and management practices. *Forests* **2019**, *10*, 645. [CrossRef]
85. Li, Y. Carbon stocks in forest ecosystems and their distribution characteristics in Zhejiang Province. *J. Plant Ecol.* **2016**, *40*, 354–363, (Translated from Chinese)
86. Yang, J.; He, Y.; Caspersen, J.P. Delineating individual tree crowns in an uneven-aged, mixed broadleaf forest using multispectral watershed segmentation and multiscale fitting. *IEEE J. Sel. Top. Appl. Earth Obs. Remote Sens.* **2016**, *10*, 1390–1401. [CrossRef]

forests

MDPI

Article

Habitat Distribution Pattern of Rare and Endangered Plant *Magnolia wufengensis* in China under Climate Change

Xiaodeng Shi [1,2,3,†], Qun Yin [2,3,†], Ziyang Sang [4], Zhonglong Zhu [2,3], Zhongkui Jia [2,3,*] and Luyi Ma [2,3,*]

[1] Zhejiang Academy of Forestry, Hangzhou 310023, China
[2] Magnolia Wufengensis Research Center, Beijing Forestry University, Beijing 100083, China
[3] Key Laboratory for Silviculture and Conservation of the Ministry of Education, College of Forestry, Beijing Forestry University, Beijing 100083, China
[4] Wufeng Magnolia Technology Development Co., Ltd., Yichang 443400, China
* Correspondence: jiazk@163.com (Z.J.); maluyi@bjfu.edu.cn (L.M.)
† These authors contributed equally to this work.

Abstract: *Magnolia wufengensis* is a newly discovered rare and endangered species endemic to China. The primary objective of this study is to find the most suitable species distribution models (SDMs) by comparing the different SDMs to predict their habitat distribution for protection and introduction in China under climate change. SDMs are important tools for studying species distribution patterns under climate change, and different SDMs have different simulation effects. Thus, to identify the potential habitat for *M. wufengensis* currently and in the 2050s (2041–2060) and 2070s (2061–2080) under different climate change scenarios (representative concentration pathways RCP2.6, RCP4.5, RCP6.0, and RCP8.5) in China, four SDMs, Maxent, GARP, Bioclim, and Domain, were first used to compare the predicted habitat and explore the dominant environmental factors. The four SDMs predicted that the potential habitats were mainly south of 40° N and east of 97° E in China, with a high distribution potential under current climate conditions. The area under the receiver operating characteristic (ROC) curve (AUC) (0.9479 ± 0.0080) was the highest, and the Kappa value (0.8113 ± 0.0228) of the consistency test and its performance in predicting the potential suitable habitat were the best in the Maxent model. The minimum temperature of the coldest month (−13.36–9.84 °C), mean temperature of the coldest quarter (−6.06–12.66 °C), annual mean temperature (≥4.49 °C), and elevation (0–2803.93 m), were the dominant factors. In the current climate scenario, areas of 46.60×10^4 km^2 (4.85%), 122.82×10^4 km^2 (12.79%), and 96.36×10^4 km^2 (10.03%), which were mainly in central and southeastern China, were predicted to be potential suitable habitats of high, moderate, and low suitability, respectively. The predicted suitable habitats will significantly change by the 2050s (2040–2060) and 2070s (2060–2080), suggesting that *M. wufengensis* will increase in high-elevation areas and shift northeast with future climate change. The comparison of current and future suitable habitats revealed declines of approximately 4.53%–29.98% in highly suitable habitats and increases of approximately 6.45%–27.09% and 0.77%–21.86% in moderately and lowly suitable habitats, respectively. In summary, these results provide a theoretical basis for the response to climate change, protection, precise introduction, cultivation, and rational site selection of *M. wufengensis* in the future.

Keywords: climate change; environmental factors; introduction; *Magnolia wufengensis*; species distribution models; suitable habitats

Citation: Shi, X.; Yin, Q.; Sang, Z.; Zhu, Z.; Jia, Z.; Ma, L. Habitat Distribution Pattern of Rare and Endangered Plant *Magnolia wufengensis* in China under Climate Change. *Forests* **2023**, *14*, 1767. https://doi.org/10.3390/f14091767

Academic Editors: Daniela Dalmonech, Alessio Collalti and Gina Marano

Received: 4 August 2023
Revised: 24 August 2023
Accepted: 29 August 2023
Published: 31 August 2023

1. Introduction

According to the Sixth Assessment Report (AR6) of the Intergovernmental Panel on Climate Change (IPCC), the global surface temperature from 2011 to 2020 is 1.1 °C higher than that from 1850 to 1900, and the global temperature rise may reach 1.5 °C, or face the risk of temporarily breaching 1.5 °C in the near term [1]. The escalating global surface temperature presents a profound and imminent threat to the survival of plants [2], which will lead to the reduction of the suitable habitat area for some plants, habitat fragmentation, and even the acceleration of the loss of global biodiversity [3–5]. Rare and endangered plants (REPs) constitute a vital component of biodiversity, playing a pivotal role in maintaining ecosystem health [6,7]. REPs are highly susceptible to environmental changes due to their special physiological characteristics [8,9], such as narrow distribution ranges, high environmental requirements, weak natural regeneration, and few genetic resources [10,11]. According to recent research, approximately one-third of the world's tree species face the risk of extinction primarily due to human activity and climate change-induced habitat loss [12]. Over a hundred tree species have already vanished from the wild, and unless immediate and decisive action is taken, many more are on the brink of extinction [13]. The protection of rare and endangered plants is the premise and basis for the protection of biodiversity and should be one of the most urgent tasks at present.

Magnolia wufengensis is a rare and endangered species of the Magnoliaceae family that was published in 2006 [14,15]; it is a group-building tree in forest ecosystems. *M. wufengensis* is a lofty, deciduous arbor species with abundant variations in flower shape, flower color, and petal number. The shapes of its tepals are ovoid, obovate, narrowly obovate, and long lanceolate. The tepal color changes from white and pale red to red and purple–red; the number of tepals varies from 9 to 46 [16]. Thus, *M. wufengensis* has great aesthetic value, can be used for urban greening and mountain afforestation, and has great potential for promotion and utilization.

However, the natural distribution range of *M. wufengensis* is very narrow, and it is found only in the central and western parts of Wufeng County, Hubei Province, and the Three Gorges area, where it is endemic [17]. Moreover, due to excessive deforestation by humans and the fragility of its own genetic resources, the *M. wufengensis* population is small, its habitat is severely fragmented, and it is very scarce in the wild, with fewer than 2000 trees [18]. According to the standards of the IUCN (International Union for the Conservation of Nature), *M. wufengensis* was endangered at a level of EN A2c, which indicates that it is endemic to China [17]. Thus, it is in a critically endangered state and is in urgent need of protection. Introduction and cultivation are effective ways to protect and preserve germplasm resources of endangered plants [19]. Introduction and cultivation require an understanding of the environmental conditions that affect its growth and the selection of suitable habitat areas to be successful [20]. However, there are still some problems that need to be considered in the introduction and cultivation planning. One is the lack of detailed national growth data on *M. wufengensis,* and it is difficult to determine the range of suitable habitats across the country. Due to the huge ornamental value and economic value of *M. wufengensis*, many nurseries in China blindly follow trends and introduce species in some areas that are not suitable, resulting in poor and slow growth of *M. wufengensis*, which has caused large setbacks in the early stage [21]. The second is how to adapt to global climate change. Global climate change has become one of the main threats to rare and endangered plants in the future [22]. Future climate change will cause changes in temperature and water conditions in plant growth areas, which will significantly affect the distribution pattern of plants [23].

A viable solution to address the aforementioned problems is the utilization of species distribution models (SDMs), which involve integrating species existence data with environmental information to simulate and predict habitats suitable for species growth while also mapping the distribution of potential suitable habitats for species across space and time [24,25]. Because SDMs can reveal the relationship between the suitability of the habitat and the environment for species, they are widely used in ecological research [26], especially in predicting the distribution of species and their hotspots [27], managing invasive species [28], protecting of endangered species [11], and verifying the relationship between climate change and species distribution [29]. The commonly used SDMs mainly include the CLIMEX model, Genetic Algorithm for Ruleset Production (GARP), MaxEnt model, Bioclim model, and Domain model [30–32]. Each mode is based on different principles and algorithms, and the applicable conditions are also quite different [33,34].

While some studies have focused on the potential suitable habitat areas of *M. wufengensis* in the early stages [21,35], there was no model screening and a lack of potential habitat simulation at the Chinese scale. The objective of this study is to employ an SDM to predict the potential habitat distribution of *M. wufengensis* in China, considering the implications of future climate change. We believe that the SDMs should be screened first, and the most suitable model can be selected to effectively predict its potential suitable habitat distribution. Meanwhile, as China is the global distribution center of Magnolia plants, there are related species distributed from south to north [36]. We hypothesize that *M. wufengensis*, a newly discovered species of the genus Magnolia, can grow in most parts of China. With global warming in the future, the north-south boundary of China's climate may move northward [25]. According to the prediction results of most sympatrically distributed species [37,38], this may also cause the distribution range of *M. wufengensis* to move northward or even expand as a whole.

Therefore, this study aims to utilize four ecological niche models, Maxent, GARP, Bioclim, and Domain, all of which can be constructed solely with species occurrence point data, to predict the potential habitat of *M. wufengensis* in China. These four models have demonstrated the ability to yield reliable predictions even when distribution sample data and environmental variables are limited [33]. Specifically, we aimed to answer the following questions: (1) Which SDMs best simulate the potential habitat of *M. wufengensis* in China? (2) What are the main environmental factors affecting the habitat distribution of *M. wufengensis* in China? What is the threshold? (3) Where is the potential habitat distribution of *M. wufengensis* under future climate change? How has it changed from the current distribution? Our study will provide scientific guidance for the response to climate change, ex situ conservation, resource protection, precise introduction, cultivation, and rational site selection of *M. wufengensis* in China.

2. Materials and Methods

The method of this study consists of five major steps: (1) collecting and processing occurrence data and environmental factors from different sources; (2) modeling and comparing suitable habitat, the area under the receiver operating characteristic (ROC) curve (AUC) and the Kappa value with Maxent, GARP, Bioclim, and Domain; (3) simulating suitable habitat based on the best model under current and future climate change; (4) evaluating and exploring dominant environmental factors and thresholds; and (5) calculating the centroid of the suitable area under different climate scenarios and exploring the shift of the centroid. The step-by-step flow chart is shown in Figure 1.

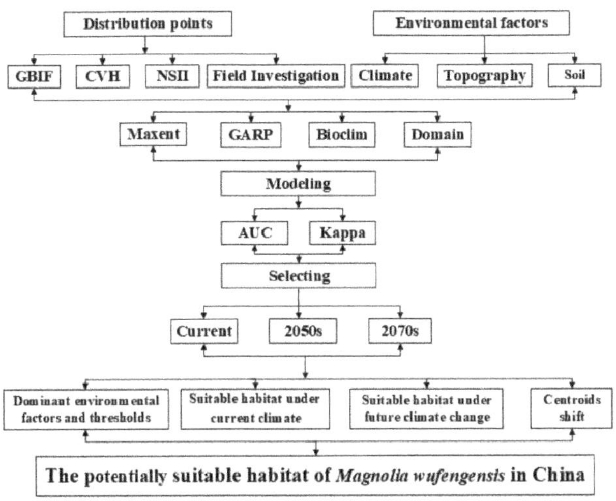

Figure 1. Flow chart for predicting potential suitable habitat.

2.1. Species Distribution Points

The distribution point data of *M. wufengensis* were obtained from data collection by the (1) Global Biodiversity Information Facility (GBIF, https://www.gbif.org/, accessed on 15 July 2019), (2) Chinese Virtual Herbarium (CVH, http://v5.cvh.org.cn/, accessed on 15 July 2019), (3) National Specimen Information Infrastructure (NSII, http://mnh.scu.edu.cn/, accessed on 15 July 2019), and (4) a field investigation (direct observation combined with GPS (Global Positioning System) technology) [21] of the main *M. wufengensis* introduction areas in China from June 2016 to October 2018 (Figure 2). After removing incorrect and duplicate distribution points and retaining distribution points spaced at least 5 km apart from adjacent points, a total of 49 *M. wufengensis* distribution record points were obtained.

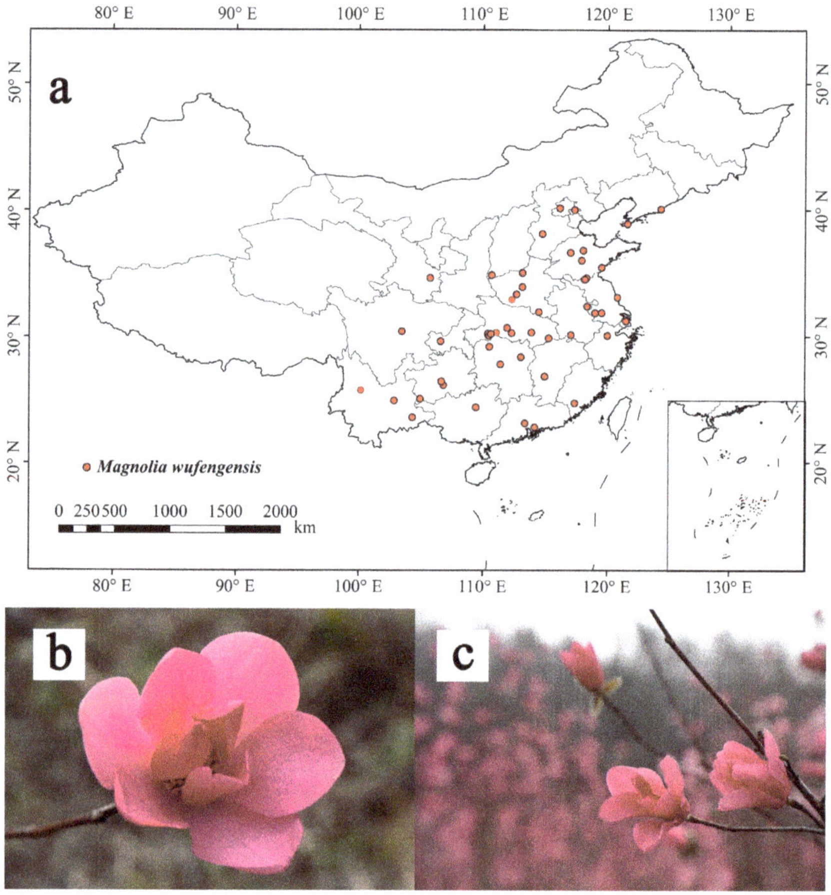

Figure 2. (**a**) Distribution points of *M. wufengensis* in China; (**b**,**c**) *M. wufengensis*.

2.2. Environmental Factors

We downloaded 19 climatic factors and 3 topographical variables (elevation, aspect, and slope) for 3 periods (a current period and the future periods of the 2050s (2040–2060) and 2070s (2060–2080)) from the World Climate Database (http://www.worldclim.org/, accessed on 15 July 2019) (Table S1). Here, we chose the CCSM4 climate system model, which has great advantages in climate simulation [39]. For the future climate data in the 2050s and 2070s, we selected four RCPs: RCP2.6, RCP4.5, RCP6.0, and RCP8.5.

We downloaded 11 soil factors from the Harmonized World Soil Database (HWSD, http://www.fao.org/soils-portal/soil-survey/soil-maps-anddatabases/harmonized-world-soil-database-v12/en/ accessed on 15 July 2019) (Table S1).

To reduce the complexity of the model and prevent overfitting of the model due to the correlation among various environmental factors, which affect the accuracy of the model [40,41], we used R software (Version 3.6.1) to perform Pearson's analysis and selected environmental factors with a correlation lower than 0.8.

Finally, 9 climatic factors, 3 topographic factors, and 4 soil factors were chosen to simulate the suitable habitat of *M. wufengensis* [21] (Table 1).

Table 1. Environmental factors were ultimately selected to simulate suitable habitat.

Category	Variable	Description	Unit
Climate	bio1	Annual mean temperature	°C
	bio2	Mean diurnal range (mean of monthly (max temp–min temp))	°C
	bio6	Min temperature of the coldest month	°C
	bio10	Mean temperature of the warmest quarter	°C
	bio11	Mean temperature of the coldest quarter	°C
	bio12	Annual precipitation	mm
	bio13	Precipitation of the wettest month	mm
	bio14	Precipitation of the driest month	mm
	bio15	Precipitation seasonality (coefficient of variation)	/
Topographic	Elevation		m
	Slope		°
	Aspect		rad
Soil	t-bulk	Topsoil bulk density	kg/dm^3
	t-ph	Topsoil pH (H_2O)	/
	t-clay	Topsoil clay fraction	%
	t-oc	Topsoil organic carbon	%

Environmental factors from different sources were resampled in ArcGIS to ensure a consistent spatial resolution (2.5 arcminute resolution) of the data and convert the data into a format that can be recognized by all the SDMs [42]. The base map for data processing was obtained from the 1:4,000,000 China administrative division map of the standard map service of the Ministry of Natural Resources of the People's Republic of China (http://bzdt.ch.mnr.gov.cn/, accessed on 15 July 2019).

2.3. Model Simulation

We used MaxEnt (MaxEnt version 3.4.1) [34], GARP models (Desktop-GARP version 1.1.6) [43], Bioclim, and Domain (based on DIVA-GIS 7.5) [44,45].

Using the "Sample Points" tool in DIVA-GIS software, 75% of the known distribution points were designated as the training data, while the remaining 25% were combined with randomly selected background points, totaling ten times the size of the distribution points, to form the testing data. To compare the distinctiveness of predictions generated by four different models, ten sets of training data and corresponding testing data were randomly generated. The training datasets were utilized for model predictions, while the testing datasets were used for model validation [46].

Upon converting 16 environmental factor datasets into the required format for each model, they were imported into the respective models, and the previously generated training data were also imported, along with the configuration of relevant parameters. The Maxent model's setup followed the guidelines outlined by Phillips et al. (2006) [34], the GARP model was configured based on Anderson et al. (2003) [43], the Bioclim settings were derived from Booth et al. (2013) [44], and the Domain model's configuration referred to Carpenter et al. [45].

Maxent model procedure: First, 16 environmental data files (.asc) were loaded into the Maxent software through the "Browse" function. Second, the training data of *M. wufengensis* were imported into the Maxent model. Finally, we selected the "Response Curves" and "Jackknife Test", which are used to analyze the environmental factors affecting the distribution of *M. wufengensis*. For the rest, we adopted the default settings.

Desktop-GARP procedure: First, 16 environmental data files (.asc) were processed by the "Dataset Manager" in Desktop-GARP, converted into a format recognized by Desktop-GARP (.raw), and loaded into the software in the form of a dataset. Second, the training data of *M. wufengensis* (Upload Data Points) were loaded into the Desktop-GARP software. The default parameter settings were selected, all 4 genetic rules were selected, the model was run 100 times, the maximum number of iterations was 1000, and the convergence limit was 0.01. Due to the instability of the model operation, according to the method of Anderson et al. (2003) [43], the "best subsets" were enabled, and the internal testing features were activated to select the 10 best models, which were added and superimposed in ArcGIS. A final grid map with a range value of 0–10 was obtained, i.e., the predicted potential distribution map of *M. wufengensis*, and the grading calculation was then performed.

Both the Bioclim and Domain models were simulated based on DIVA-GIS. First, the 16 environmental data files (.asc) were converted into a format recognized by DIVA-GIS (.raw), and the stack dataset was then generated. Second, the training data of *M. wufengensis* were inserted into "Data". Finally, the environmental dataset was added in stack format to the Modeling-Bioclim/Domain module, and predictions of the Bioclim and Domain models were then generated.

2.4. Model Evaluation

In this study, the accuracy of the model was evaluated using the area under the curve (AUC) and Cohen's kappa. The AUC value is widely employed due to its threshold-independent nature, making it a robust measure for model performance assessment [47,48]. The prediction results of each model were converted into "grd" format in DIVA-GIS. Then, the testing datasets were imported, the evaluation file was created, and the AUC and Kappa values were output [49].

The range of the AUC is between 0 and 1. An AUC < 0.7 suggests that the prediction performance is extremely poor, values between 0.7 and 0.8 indicate moderate performance, values between 0.8 and 0.9 suggest good performance, and values between 0.9 and 1.0 indicate excellent performance [50]. In other words, the closer the value is to 1, the better the model fit. The evaluation criteria for Kappa were excellent, 1.0–0.81; very good, 0.80–0.61; good, 0.60–0.41; fair, 0.40–0.21; and fail, <0.20 [51,52].

2.5. Suitable Habitat Partitions under Current and Future Conditions

According to this study, which was repeated 10 times, each model produced 10 sets of prediction results. The prediction result with the largest AUC for each model was selected as the base map, and ArcGIS was used for raster format conversion and reclassification. The species existence probabilities derived from each model were used to classify the predicted potential suitable area for *M. wufengensis* of the different models, as shown in Table 2, and the suitable habitat distribution maps of the different models were then obtained [43,52].

Table 2. Classification standards of suitable habitats for different SDMs.

Suitability	GARP	Maxent	Bioclim	Domain
Unsuitable	[0, 2]	[0, 0.2]	0	[0, 90]
Lowly suitable	(2, 4]	(0.2, 0.4]	(0, 5%]	[91, 93]
Moderately suitable	(4, 6]	(0.4, 0.6]	(5%, 10%]	[94, 96]
Highly suitable	(6, 10]	(0.6, 1.0]	(10%, 27%]	[97, 100]

3. Results

3.1. Prediction Results of Four Models

In this study, four SDMs were utilized to predict the distribution map of the current potential suitable habitat for *M. wufengensis* in China (Figure 3). The representation of highly suitable, moderately suitable, lowly suitable, and unsuitable habitats is denoted by the red, yellow, green, and white areas, respectively (the same applies below).

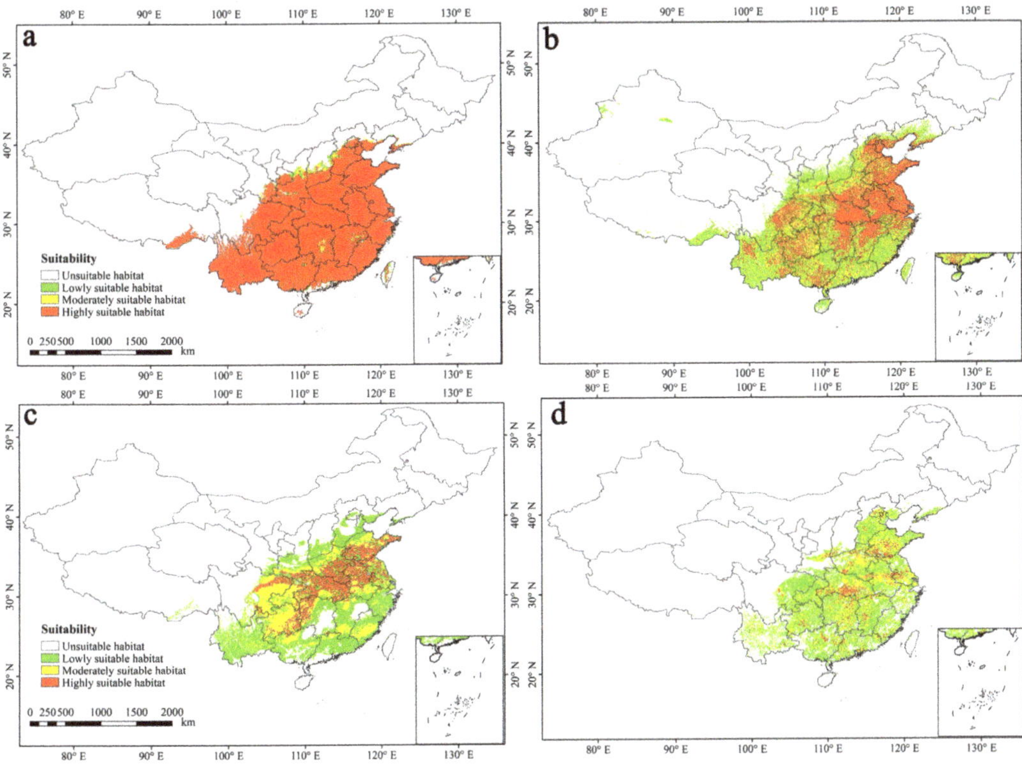

Figure 3. Distribution map of suitable habitat for *M. wufengensis* in China simulated by four SDMs ((**a**) GARP; (**b**) Maxent; (**c**) Bioclim; and (**d**) Domain).

The suitable habitats of *M. wufengensis* predicted by the four SDMs were all distributed south of 40° N and east of 90° E in China. In terms of climate zone, the potential suitable habitats were mainly in the subtropical temperate and humid monsoon climate zones, and some areas were in the temperate monsoon climate zone, mainly including Hebei, Shandong, Henan, Anhui, Jiangsu, Hubei, Hunan, Zhejiang, Jiangxi, Fujian, Guangdong, Guangxi, Yunnan, Chongqing, Guizhou, and parts of eastern Sichuan, Shaanxi, Shanxi, Gansu, and Liaoning.

Due to the different principles and algorithms of the different models, there were some differences in their results. The distribution map of the suitable habitats for *M. wufengensis* obtained by the GARP simulation (Figure 3a) showed that most of the suitable habitat was highly suitable, while there were very limited moderately and lowly suitable habitats. The range of suitable habitats of different grades for *M. wufengensis* simulated by the Maxent model was relatively distinct (Figure 3b) and was similar to Bioclim's prediction (Figure 3c). The range of highly suitable habitats for *M. wufengensis* simulated by the Domain model was very small and scattered (Figure 3d), and most of the suitable habitats were lowly suitable habitats.

The areas of different grades of suitable habitats for *M. wufengensis* simulated by the four models in Figure 4 show that the GARP model simulated the largest area of highly suitable habitat at 273.05×10^4 km^2, which was 2.86 times that simulated by the Maxent model (95.26×10^4 km^2), 6.90 times that simulated by the Bioclim model (39.53×10^4 km^2), and 31.00 times that simulated by the Domain model (8.81×10^4 km^2). The GARP model similarly simulated the moderate and low suitability habitats for *M. wufengensis* as the smallest areas, at 5.96×10^4 km^2 and 4.95×10^4 km^2, respectively, which was 5.63% and 5.58% of the moderately and lowly suitable habitats simulated by the Maxent model, 9.00% and 4.45% of those simulated by the Bioclim model, and 9.18% and 4.72% of those simulated by the Domain model, respectively. In general, the GARP model predicted the largest range of suitable habitats for *M. wufengensis*. The suitable habitat of *M. wufengensis* obtained by the Domain simulation was the smallest and most scattered. The prediction range of the Maxent model was closer to that of the Bioclim model, the suitable habitats of different grades were also clearer, and the distribution was more reasonable.

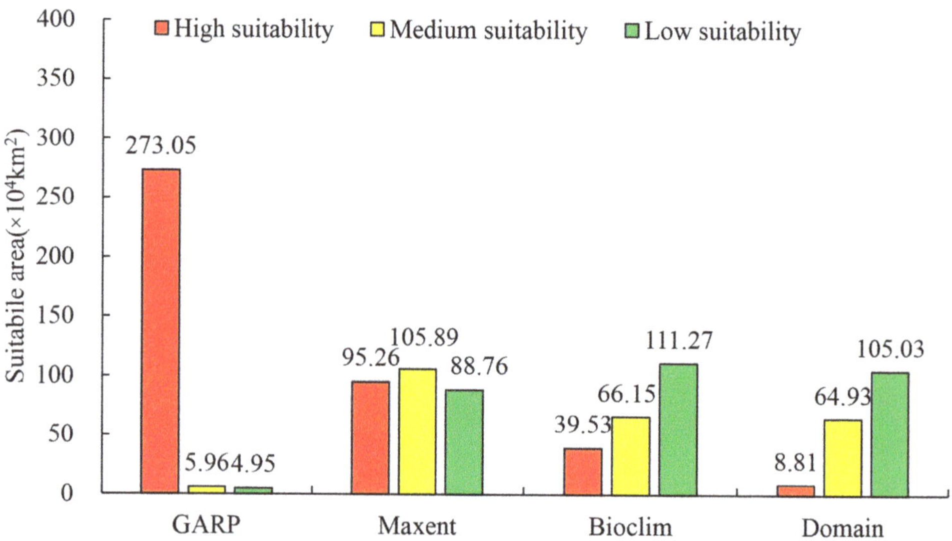

Figure 4. The suitable habitats for *M. wufengensis* of different grades in China simulated by four schemes.

3.2. Model Accuracy Evaluation

The four SDMs simulated the mean values of the AUC and Kappa for the current potential suitable habitat for *M. wufengensis* in China, as shown in Figure 5. Figure 5 shows that the average AUC of the prediction results of the four models is above 0.85, which far exceeds that of a random model (AUC = 0.5), indicating that the four models have a relatively good predictive effect for the suitable habitat of *M. wufengensis*. The standard deviation of the AUC values of the four models was ordered as follows: Domain > Bioclim > Maxent > GARP. The results of this study showed that for the suitable habitat of *M. wufengensis*, the Maxent model (0.9479 ± 0.0080) had the best predictive results, followed by the Domain (0.9367 ± 0.0287) and GARP (0.8719 ± 0.0039) models, and the Bioclim (0.8513 ± 0.0177) model had the lowest level of prediction. Figure 5 shows that the average Kappa values of the four SDMs are above 0.7, and the consistency of the models is significant. The four SDMs predicted the potential suitable habitats for *M. wufengensis* with high accuracy, and they can be used to predict the distribution of potential suitable habitats. The Kappa values were in the following order: Maxent (0.8113 ± 0.0228) > Domain (0.7629 ± 0.0531) > Bioclim (0.7166 ± 0.0372) > GARP (0.6969 ± 0.0200). In summary, combined with the AUC and Kappa values of the model, the Maxent model best predicted the potential suitable habitat for *M. wufengensis*, with the most significant consistency Table 3.

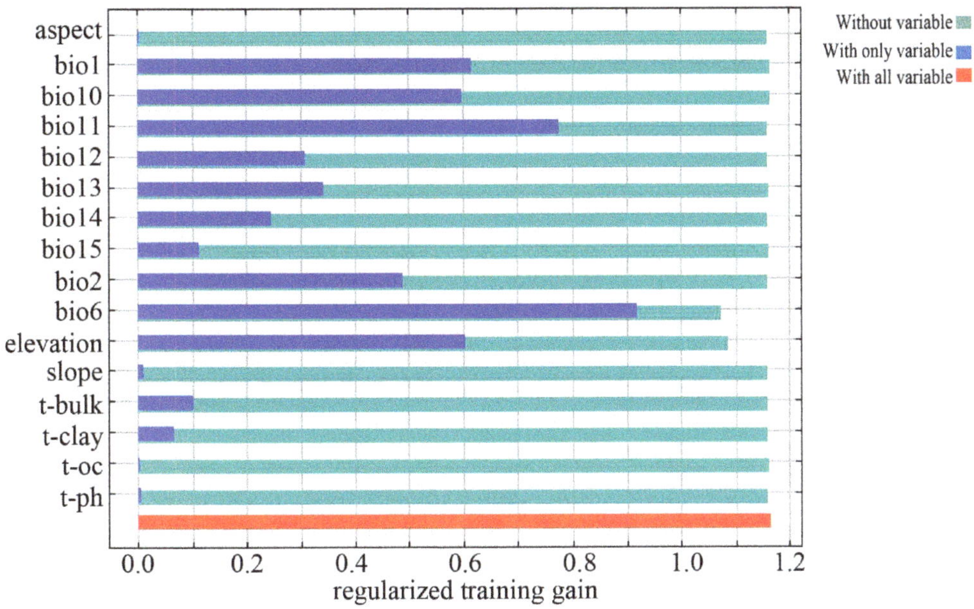

Figure 5. Jackknife test for the importance of the variables.

Table 3. Comparison of AUC and Kappa values of the results of the four SDMs. Different small letters indicate significant differences among treatments as assessed by Duncan's test ($p < 0.05$).

Model	Maxent	GARP	Bioclim	Domain
AUC	0.9479 ± 0.0080 a	0.8719 ± 0.0039 b	0.8512 ± 0.0177 c	0.9367 ± 0.0287 a
Kappa	0.8113 ± 0.0228 a	0.6969 ± 0.0200 c	0.7166 ± 0.0372 c	0.7629 ± 0.0531 b

3.3. Evaluation of Environmental Factors

Through the above analysis, it can be concluded that the Maxent model is the best model for predicting the potential suitable habitats of *M. wufengensis*. When simulating the potential suitable habitat for a species, adopting default parameters is customary, but it will cause overfitting, reduce the accuracy of the research results, and directly affect the transferability of the model [53–55]. Therefore, in this study, we used the Maxent model with parameters optimized by the ENMeval package in R software to simulate the potential suitable habitats of *M. wufengensis* in China. The relevant parameters were the regularization multiplier value (RM) of 3.5, and the feature combination (FC) was LQ (Table S2). At this time, the model fits the species distribution points well and has significantly decreased the complexity and reduced the degree of overfitting.

After predicting the potential suitable habitat of *M. wufengensis* through the Maxent model with optimized parameters, the percent contributions and the permutation importance of each environmental factor were determined (Table S3). Among the 16 environmental factors, the minimum temperature of the coldest month (bio6), elevation (28.73%), and mean temperature of the coldest quarter (bio11) ranked in the top 3, with cumulative values as high as 96.07% and 92.32%, respectively (Table S3).

According to the results of the jackknife test (Figure 5), using factors other than the ones examined, the regularized training gain increased the most in bio6, followed by bio11 and bio1.

The dominant environmental factors for the potential suitable habitats of *M. wufengensis* in this study were bio6, bio11, bio1, and elevation.

The response curve (Figure 6) shows that when the distribution probability of suitable habitats for *M. wufengensis* was greater than the threshold (0.2), the intervals of the dominant environmental factors restricting the distribution of *M. wufengensis* were bio6 (−13.36–9.84 °C), bio11 (−6.06~−12.66 °C), bio1 (≥4.49 °C), and elevation (0–2803.93 m).

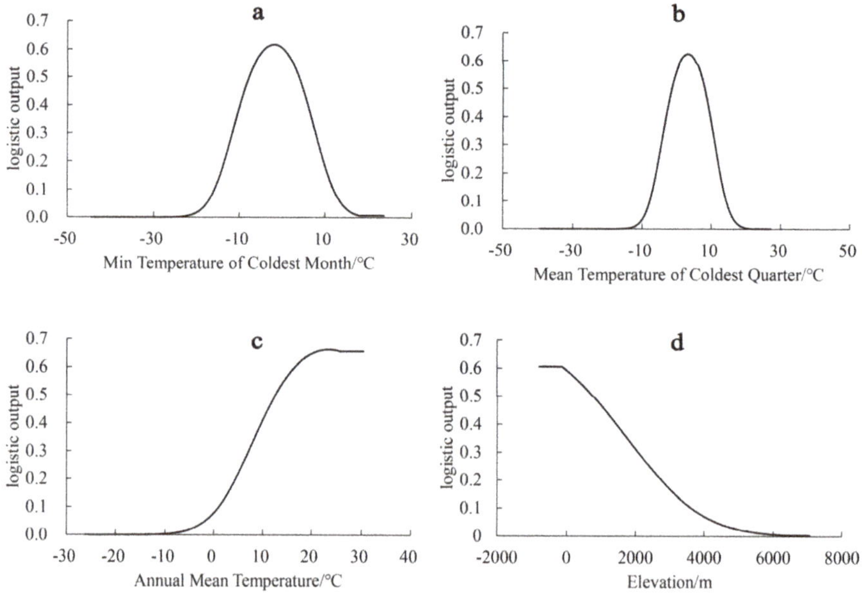

Figure 6. Responses of *M. wufengensis* to the four main environmental factors.

3.4. *Current Potential Suitable Habitats in China*

The results for *M. wufengensis* in the current potentially suitable habitats in China under the optimized parameters are shown in Figure 7.

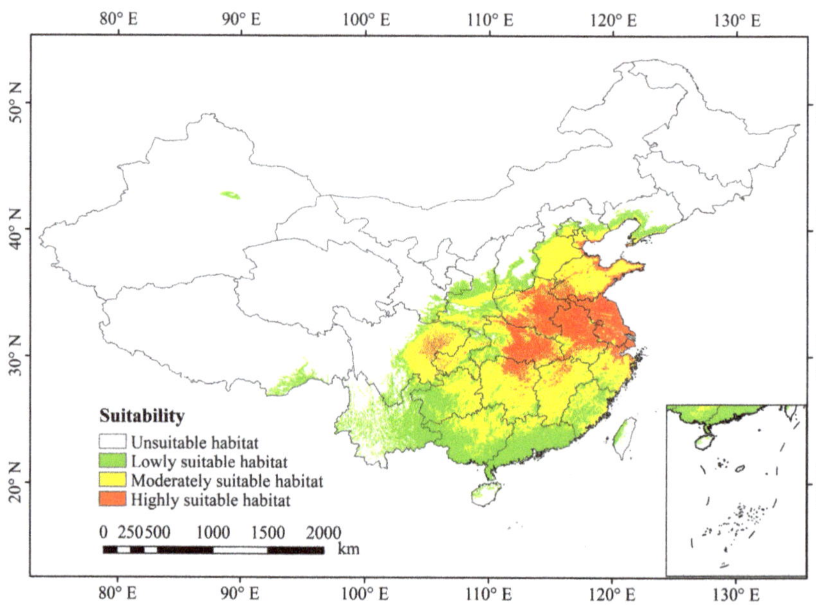

Figure 7. Suitable habitats for *M. wufengensis* in China under the current period.

The highly suitable habitats for *M. wufengensis* were highly concentrated and were mainly distributed in East China and Central China, covering Hubei, Henan, Anhui, Jiangsu, Shanghai, the northern part of Hunan, Jiangxi, Zhejiang, southern Shandong, and eastern Sichuan. The moderately suitable habitats mainly surrounded the highly suitable habitats, covering eastern Sichuan, Hunan, Jiangxi, Zhejiang, Fujian, Shandong, southeastern Hebei, southern Shanxi, eastern Guizhou, the northern parts of Guangxi and Guangdong, and Chongqing, Beijing, and Tianjin. The lowly suitable habitats were mainly in Yunnan, southwestern Guizhou, the southern parts of Guangxi and Guangdong, southern Fujian, central Sichuan, southeastern Gansu, the southern parts of Shaanxi and Shanxi, and small areas of Hebei and Liaoning. Currently, although Xinjiang and Tibet do not have *M. wufengensis* populations, according to the MaxEnt prediction, Turpan in Xinjiang and Linzhi in Tibet nevertheless have small areas of lowly suitable habitats, indicating that the model has high transferability. Under the current climate scenario, the areas of highly, moderately, and lowly suitable habitat for *M. wufengensis* were 46.60×10^4 km^2, 122.82×10^4 km^2, and 96.36×10^4 km^2, accounting for 4.85%, 12.79%, and 10.03% of China's total land area, respectively.

3.5. Potential Suitable Habitat and Dynamic Changes in the Future

Four climate change scenarios (RCP2.6, RCP4.5, RCP6.0, and RCP8.5) in the 2050s and 2070s were selected, and the optimized MaxEnt model was used to simulate the distribution of potential suitable habitat for *M. wufengensis* (Figure 8), the dynamic changes relative to the current suitable habitat (Figure 9), and the changes in area (Table 4) under future climate change conditions.

Table 4. Suitable habitat area change for *M. wufengensis* under different climate change scenarios.

Climate Scenario	High Suitability		Medium Suitability		Low Suitability		No Suitability	
	Area (10^4 km^2)	Change Rate (%)	Area (10^4 km^2)	Change Rate (%)	Area (10^4 km^2)	Change Rate (%)	Area (10^4 km^2)	Change Rate (%)
Current	46.59	/	122.82	/	96.36	/	694.83	/
RCP2.6 (2050)	32.63	−29.98	130.74	6.45	117.43	21.86	679.81	−2.16
RCP4.5 (2050)	33.48	−28.15	154.86	26.09	97.10	0.77	675.16	−2.83
RCP6.0 (2050)	36.34	−22.00	156.09	27.09	92.26	−4.25	675.91	−2.72
RCP8.5 (2050)	37.85	−18.77	144.75	17.86	101.48	5.31	676.53	−2.63
RCP2.6 (2070)	33.58	−27.94	147.68	20.24	103.54	7.45	675.81	−2.74
RCP4.5 (2070)	34.01	−27.02	149.15	21.44	102.21	6.07	675.23	−2.82
RCP6.0 (2070)	36.65	−21.33	143.50	16.84	100.99	4.80	679.46	−2.21
RCP8.5 (2070)	44.48	−4.54	144.27	17.46	99.31	3.06	672.55	−3.21

The results showed that the potential suitable habitat of *M. wufengensis* will undergo various changes under future climate change scenarios. The highly suitable habitats for *M. wufengensis* first decreased and then increased with increasing RCP value and time, but the area was smaller than the current area and exhibited serious fragmentation in the 2050s and 2070s (Figures 8 and 9). Among the various predictions, the highly suitable habitats were the smallest at only 32.62×10^4 km^2 under RCP2.6 in the 2050s, which was 29.98% lower than the current area (Table 4), and the areas that exhibited the main reductions were in southwestern Shandong, central Henan, the northern part of Anhui and Jiangsu, which are characterized by plains and would become moderately suitable habitats (Figure 9a). Under the RCP8.5 scenario in the 2070s, the highly suitable habitats were the closest to those of the current period (Figures 8h and 9h; Table 4). However, under future climate change, the highly suitable habitat for *M. wufengensis* increased significantly and was concentrated in the Sichuan Basin (Figure 8).

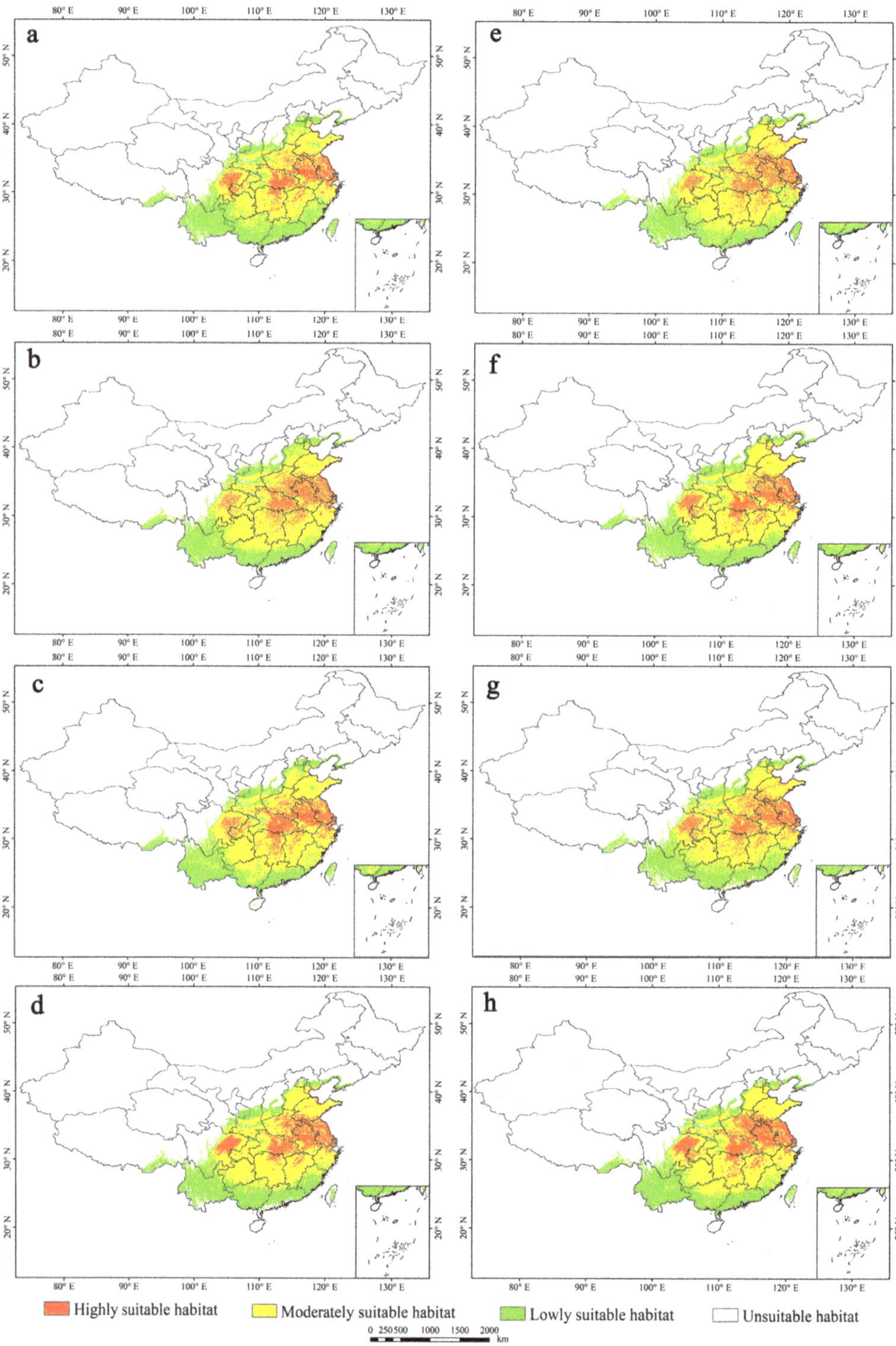

Figure 8. Distribution of suitable habitats for *M. wufengensis* under future climate change scenarios ((**a–d**) 2050s: RCP2.6, RCP4.5, RCP6.0, and RCP8.5; (**e–h**) 2070s: RCP2.6, RCP4.5, RCP6.0, and RCP8.5).

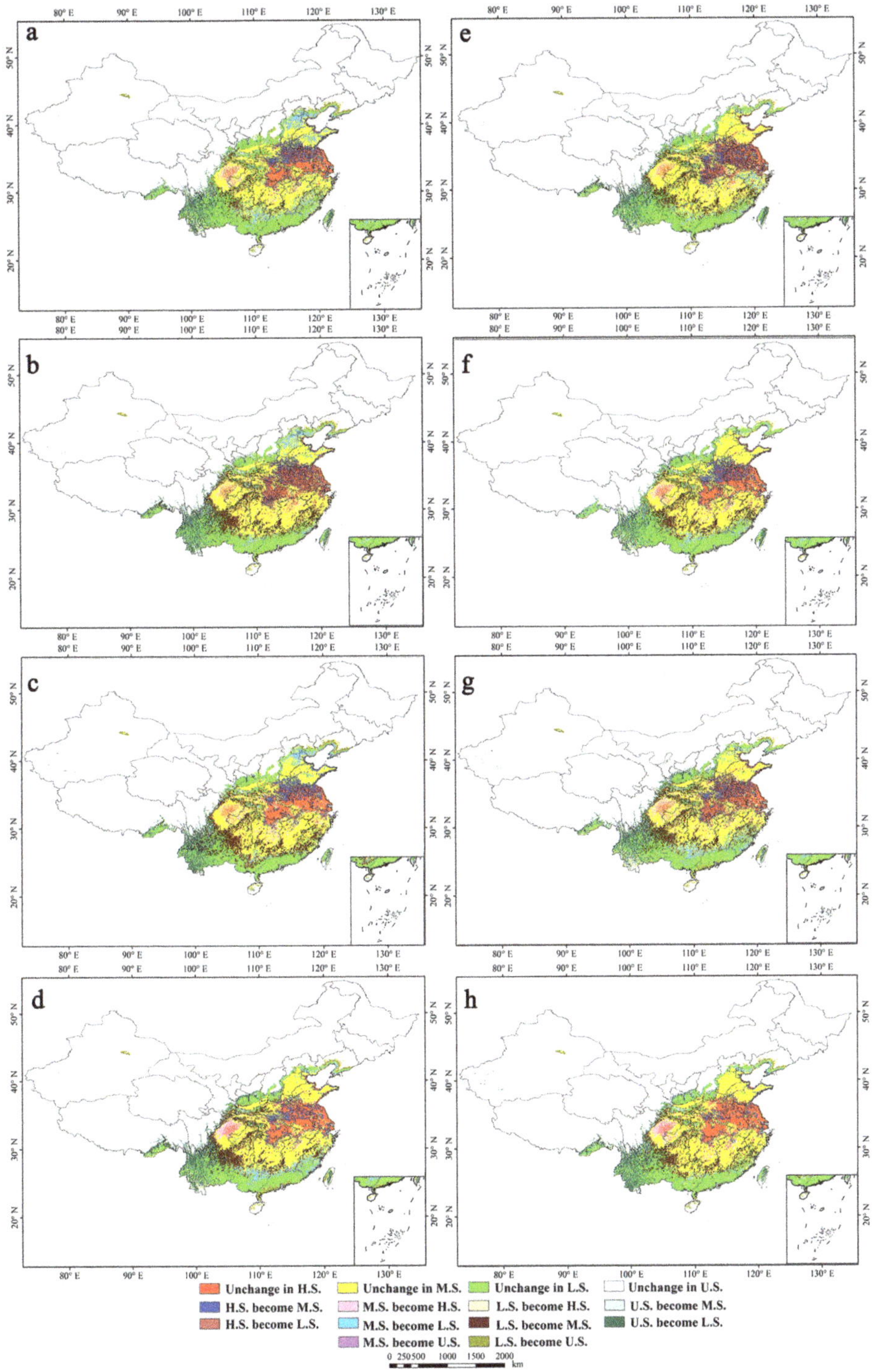

Figure 9. Dynamic changes in potential suitable habitats for *M. wufengensis* under future climate change scenarios ((**a–d**) 2050s: RCP2.6, RCP4.5, RCP6.0, and RCP8.5; (**e–h**) 2070s: RCP2.6, RCP4.5, RCP6.0, and RCP8.5).

Moderately suitable habitats for *M. wufengensis* under future climate change will increase compared to those of the current period (Table 4). Under the RCP2.6 scenario in the 2050s, the increased area was the smallest at only 6.45% (Table 4). The area of moderately suitable habitats reached a maximum of 156.09×10^4 km^2 under RCP6.0 in the 2050s, which was 27.09% larger than the current area, and the areas of increase were mainly in Shandong, southern Henan, Jiangsu, northern Anhui, the southwestern part of Guizhou, and the central part of Guangxi. There was a trend of expansion toward highly and lowly suitable habitats (Figures 8c and 9c).

The lowly suitable habitat for *M. wufengensis* showed a trend of first increasing and then decreasing under the RCP2.6 and RCP8.5 scenarios in both the 2050s and 2070s and gradually increasing under the RCP4.5 scenario, while the suitable habitats first decreased and then increased under the RCP6.0 scenario (Table 4). Under the RCP2.6 scenario in the 2050s, the lowly suitable habitat reached a maximum area (117.43×10^4 km^2), which was 21.86% higher than the current area (Table 4). The areas of increase were mainly in southeastern Tibet and most of Yunnan and Taiwan Island, which tended to extend toward moderately suitable and unsuitable areas (Figures 8a and 9a), while under the RCP6.0 scenario in the 2050s, the area of lowly suitable habitat was 4.25% less than the current area, with little change (Figure 9c). In the future, under different climate scenarios, the increased areas will mainly be high-elevation mountain areas. For example, the area of the lowly suitable habitat for *M. wufengensis* on Taiwan Island will increase and move to the central high-elevation area, and on Hainan Island, this area will gradually decrease to only central high-elevation areas (Figures 8 and 9).

Overall, the area of suitable habitats for *M. wufengensis* will decrease under future climate change scenarios, with an overall decrease of 4.53%–29.98%; the area of moderately suitable habitats will continue to increase in the future, with an increase of 6.45%–27.09%; the area of lowly suitable habitats will increase by 0.77%–21.86%; and the area of unsuitable habitats will gradually decrease by 2.16%–3.21% in the future.

3.6. Centroid Shifts in Different Suitable Habitats

Figure 10 indicates that the centroids of the highly suitable habitats (Figure 10a) and the moderately suitable habitats for *M. wufengensis* (Figure 10b) migrate mainly to the northeast under future climate change (Figure 10(a1,b1)). The centroids of the lowly suitable habitats (Figure 10c) migrate mainly northward (Figure 10(c1)). In general, the centroid of the suitable habitats for *M. wufengensis* moves more to the northeast.

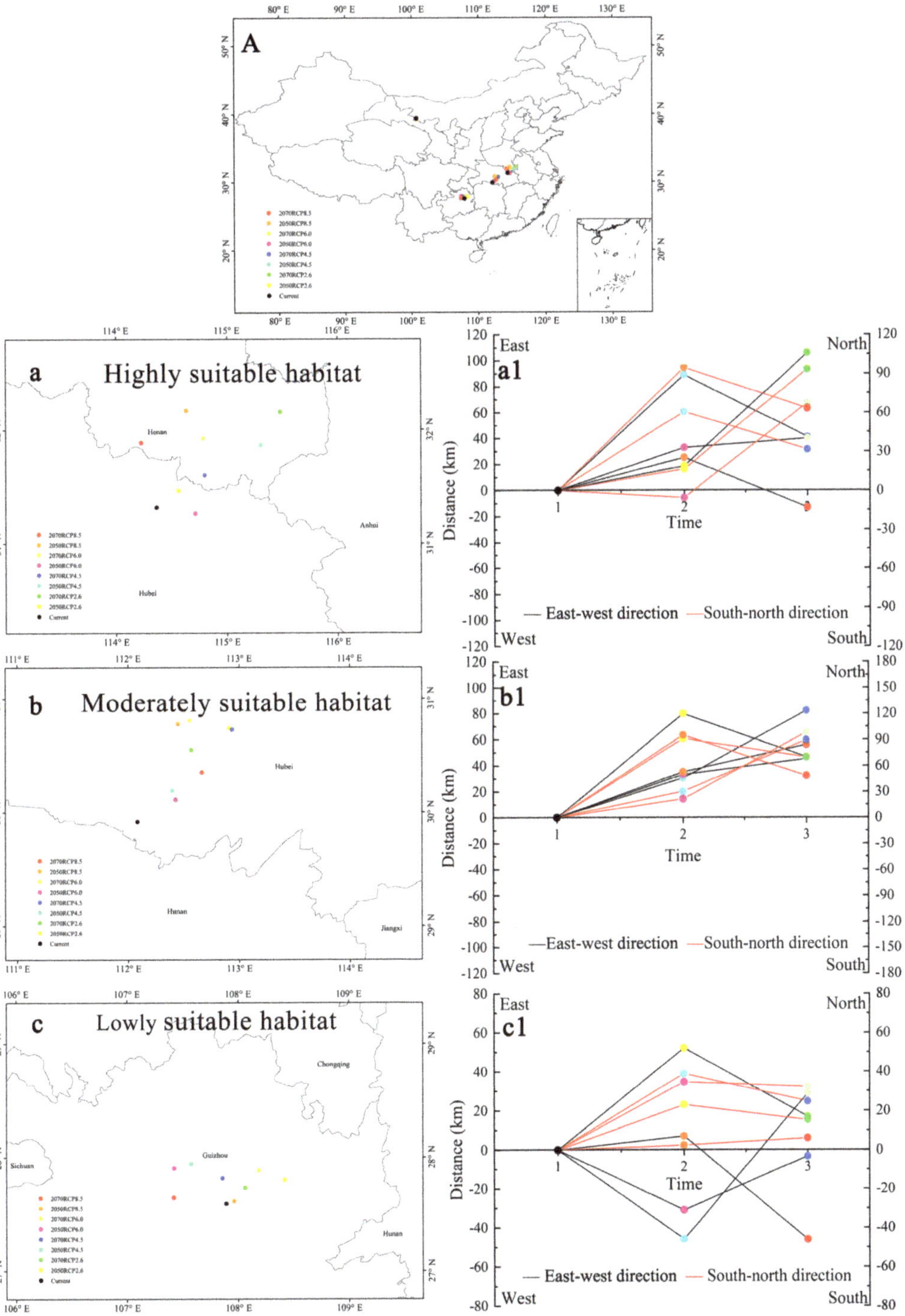

Figure 10. Centroid migration in suitable habitats for *M. wufengensis* under future climate scenarios in China. (**a1–c1**) indicates the distances that centroids of different grades of suitable habitat migrate in two directions (north–south and east–west) under future climate change.

4. Discussion

4.1. Model Performance

M. wufengensis is an endangered and rare tree species in China. To provide better protection for it, we used four SDMs to study its suitable habitat area. While the four SDMs employed in this study demonstrated promising outcomes in simulating the distribution of potential suitable habitats for *M. wufengensis*, some variations were observed in the predicted suitable habitats among the different models. This is mainly because the different SDMs are based on related mathematical algorithms and fit data in different ways [56]. In this study, the GARP model exhibited the broadest range and the largest area of suitable habitats predicted for *M. wufengensis*. This can be attributed to the GARP model's foundation on the principle of the genetic rule algorithm [57], which undergoes continuous iteration to screen and evaluate the dataset. By simulating the ecological requirements of the species, the model effectively determines the potential suitable areas within the study region. Related studies have further confirmed that the GARP model commonly faces the issue of overprediction, displaying a tendency to predict species fitness beyond their known niches [58,59], presumably due to GARP's failure to model lesser relationships in the data [60]. In addition, the GARP model's rulemaking process did not include factors that caused species to spread to these incorrect areas [61]. Compared to the GARP model, the Bioclim model exhibits a smaller overall suitable habitat area, yet with distinct levels of specificity. This distinction primarily arises from the environmental envelope principle that underpins the Bioclim model [44]. This model assumes that the environmental climate within the envelope is similar to the actual distribution area's climate, allowing a certain species to grow and reproduce normally within this environment [52]. As a result, the predictions are relatively conservative. In contrast to the GARP model, the Domain model predicts the majority of the suitable habitat as low suitability areas. This discrepancy arises from the fact that the Domain model classifies based on a similarity matrix between points [45]. This is largely influenced by sample points, leading to the appearance of certain suitable habitat areas in nearly all locations where distribution points exist [46]. With fewer sample points, the level of suitability naturally tends to be lower. In comparison, while the potential suitable habitats of *M. wufengensis* predicted by the Maxent model closely resembled those of the GARP model, the distinctions between the different habitat grades were more pronounced, resulting in clearer and more defined predicted outcomes. This is because the Maxent model does not make any assumptions about that which is unknown under the constraints of the maximum entropy theory algorithm and runs completely according to the maximum entropy of that which is closest to their actual states, so its predictions of suitable habitats for species are more accurate. Zhang et al. (2016) [62] also found that the area of suitable habitat predicted by the Maxent model was smaller than that of the GARP model for the distribution of *Arceuthobium sichuanense* in China; however, the level of local detail in their study was more distinct. Moreover, the Maxent model predicted that the suitable habitats of *M. wufengensis* can reach Yining, Bole, and Turpan in Xinjiang, which is an area that was not predicted by the other three models, reflecting the good transferability of the Maxent model.

The most important assumption of the Bioclim model is that species can grow, develop, and reproduce in places where the climatic conditions are similar or comparable to those in their current living environment [45], so the results of its predictions of species distribution are relatively conservative. The Domain model is greatly affected by the sampling of the original distribution point of the species when predicting their distribution area and will spread around according to the sampling distribution point, resulting in the expansion or contraction of the ecological niche of the species distribution area [52,60]. Thus, the results of this study about these four models are consistent with previous studies. Wang et al. (2020) [49] compared and analyzed the predictive effects of these four models when researching and predicting the potential suitable areas of *Pseudolarix amabilis* and found that the prediction range of GARP was wider, the prediction range of Bioclim was smaller, and the Maxent model demonstrated higher accuracy in its prediction results compared to the other models. Duan et al. (2020) [63] studied the potential distribution of *Ammopiptanthus* species in China and found that the Maxent model yielded the highest AUC value and exhibited the best overall performance in predicting the species' distribution. Elith et al. (2006) [60] demonstrated that the Maxent model outperformed the Bioclim, Domain, and GARP models when utilizing 16 different SDMs to predict the distribution of potential suitable areas for 226 species.

4.2. Dominant Environmental Factors

The dominant environmental factors limiting the distribution of *M. wufengensis* in China were the minimum temperature of the coldest month (bio6), mean temperature of the coldest quarter (bio11), annual mean temperature (bio1), and elevation. Zhu (2012) found that extremely low temperatures and their duration were important limiting factors during the overwintering of *M. wufengensis*. Yang (2015) [64] found that during an actual introduction process in North China, *M. wufengensis* under 3–4 years old could not naturally overwinter, and they recommended that appropriate cold protection measures be taken to protect *M. wufengensis* from low-temperature freezing damage. Currently, the most advanced landscaping methods are to build windproof barriers and use insulation cotton to wrap and cover trees [65]. This observation aligns with the finding that the Northeast China region is unsuitable for the survival of *M. wufengensis*, while the North China region comprises habitats with moderate and low suitability for the species. Liang (2010) [16] reported that the semilethal temperature of *M. wufengensis* is −15 °C, which closely aligns with the lower limit of −13.36 °C derived from the response curve of bio6 in this study. Numerous researchers studying other plants with sympatric distributions have also reached similar conclusions. For instance, Yan et al. (2019) [66] found that bio11 significantly influenced the distribution of *Pinus massoniana*. Zhang et al. (2018) [38] observed that temperature exerted a substantial impact on the potential geographical distribution of *Sorbus amabilis*. Similarly, Li et al. (2016) [67] analyzed the importance of eight climate variables for *Quercus chenii* using the MaxEnt model, and their findings highlighted that the caloric index played a pivotal role in limiting the species' geographical distribution, followed by the water index.

Elevation affects the growth of *M. wufengensis* by affecting soil physicochemical properties, soil enzyme activity, light conditions, and air temperature, thereby limiting the vertical geographic distribution of the species. The upper limit of the distribution of *M. wufengensis* under natural conditions is 1400–2000 m [14,15], which is within the upper limit of 2803.93 m obtained from the elevation response curve in this study.

The results of this study showed that precipitation had little effect on the distribution of *M. wufengensis* in China. However, precipitation plays a vital role in plant growth. *M. wufengensis* is a succulent root that is sensitive to water, and it easily dies if it receives too much water [68]. The current highly suitable habitats for *M. wufengensis* are mainly in East China. These areas are also traditional subtropical monsoon climate areas in China. One of the most notable features of these areas is that rain and heat occur in the same period, and it is easy to form a high-temperature and high-humidity environment. In the actual investigation, pests and diseases were detected. The main diseases of *M. wufengensis* were sooty blotch on the surface of leaves and root rot, which causes the plants to rot from the root stem, and the main pests were sucking pests such as *Tetranychus cinnabarinus*, *Liriomyza*, and *Pseudococcus* [21]. Therefore, suitable site conditions should be considered when introducing *M. wufengensis*. For example, in low-lying areas, high beds should be made, and in the summer, cleaning and drainage should be performed as appropriate. If sooty blotch occurs, it is necessary to spray the plant with a 0.3%–0.5% mass fraction of Bordeaux mixture for prevention and control, and root rot should be controlled by continuously spraying 2–3 times with 500 times carbendazim solution, which can mostly control the development of the disease. In addition, sucking pests should be sprayed 500–1000 times with omethoate for prevention [21].

4.3. Current Suitable Habitats

Our study revealed that the potential suitable habitats for *M. wufengensis* are currently widespread across southeast China, with areas of high suitability observed in Hubei, Henan, Anhui, Sichuan, Zhejiang, Jiangsu, and Hunan. These results are consistent with a previous study [21], which used the fuzzy similarity priority ratio method based on the principle of climatic similarity and revealed that the suitable habitats of *M. wufengensis* were mainly distributed in most areas from the eastern Huanghuai Plain to the northern Guangdong and Guangxi Hills. According to other studies [69,70], there are 17 species of *Magnolia* in East China, 20 species in Central China, and 29 species in South China, demonstrating a high richness of *Magnolia* species in this study area and relatively reliable research results, which can provide a scientific basis for the effective protection and precise introduction and cultivation of *M. wufengensis.* Interestingly, unlike previous studies, our study showed that Turpan in central Xinjiang and Linzhi in southern Tibet are also suitable for *M. wufengensis,* providing evidence that the Maxent model has good transferability.

4.4. Changes to the Suitable Habitats in the Future

In the context of future global climate change, rare and endangered plants will face higher extinction risks than other plants due to their narrow habitats, sparse populations, and weak natural regeneration [71]. Understanding the potential contraction or expansion of their habitats is of great significance for the conservation of these rare and endangered species. According to our research results, although the highly suitable habitats of *M. wufengensis* show a decreasing trend under different climate change models in the future, the moderately and lowly suitable habitats show an increasing trend. This indicates that although some areas may not be suitable for the growth of *M. wufengensis*, there will be some new areas suitable for its growth, which is very beneficial for its conservation and introduction. Gao et al. (2022) [72] found that the highly suitable habitat of the rare and endangered plant *Firmiana kwangsiensis* in Guangxi is expected to decrease over time, and it will become extinct in some areas, but it can also adapt to some new areas. Li et al. (2019) [73] showed that in the case of future global warming, the potential suitable habitats of *Osmanthus yunnanensis* expand to the east and north, and the areas of suitable distribution increase; the potential suitable habitats of *Osmanthus delavayi* expand to the west and north, and the highly suitable habitats decrease. Zhang et al. (2018) [38] found that the overall geographical distribution area of *Sorbus amabilis* contracted, and the degree of fragmentation increased and migrated to high-altitude areas in the future. The results of this study are basically consistent with the above conclusions. The suitable habitats of *M. wufengensis* in the future showed characteristics of habitat fragmentation, and the centroids of the suitable habitats generally moved to high elevations in the northeast direction. The main limitation of the potential habitat distribution of *M. wufengensis* is the low-temperature factor. In the future, the temperature in northern China will rise, and *M. wufengensis* will be able to survive winter in some areas, so the potential suitable habitats will expand to high-latitude areas. In Guizhou, Yunnan, Guangxi, Sichuan, Hubei, Hunan, Anhui, Taiwan, and other places with high mountain ranges, the areas of suitable habitat significantly increased. The mountain ranges in these areas have a high average elevation, and the temperature will increase due to global warming in the future, which can meet the growth conditions of *M. wufengensis.* Therefore, the potential suitable habitats for *M. wufengensis* will migrate to higher altitudes at the same time.

In summary, judicious utilization of SDMs can effectively and swiftly discern the potential distribution areas of rare and endangered species, facilitating the formulation of subsequent conservation policies. Additionally, these models can anticipate the species' future distribution based on environmental factors in forthcoming time frames, enabling the proactive development of corresponding conservation plans and thereby maximizing species richness preservation. However, in practical applications, certain models rely solely on algorithmic logic, lacking robust ecological interpretations [74]. Some models even depend solely on expert experience, exhibiting considerable subjectivity [33]. Moreover, the influence of sample quality and quantity is significant for some models [33]. Therefore, during the actual modeling process, the selection of the most suitable model should be based on the algorithms and theoretical underpinnings specific to different species distribution models.

5. Conclusions

In this study, four SDMs (Maxent, GARP, Bioclim, and Domain) were used for the first time to comprehensively predict and analyze potential suitable habitats for the introduction of the rare and endangered plant *M. wufengensis* in China. The GARP model simulated the widest range of suitable habitats. The Domain model simulated the smallest range, and its layers were not clear. The Bioclim model results were similar to those of the Maxent model, and the Maxent model had the best performance and the best simulation effect based on the AUC and Kappa statistics. The low-temperature factor was the dominant environmental factor affecting the distribution of *M. wufengensis* in China. The potential suitable habitat for *M. wufengensis* was mainly distributed in the areas south of 40° N and east of 97° E in China, with a high distribution potential under current climate conditions. Under future climate scenarios, the highly suitable habitats for *M. wufengensis* in China generally showed a decreasing trend, and moderately and lowly suitable habitats showed an increasing trend. The centroid of the future potential suitable habitats of *M. wufengensis* migrated to the northeast at a high latitude.

Supplementary Materials: The following supporting information can be downloaded at https://www.mdpi.com/article/10.3390/f14091767/s1. Table S1: Environmental variables used in this study; Table S2: The performance of Maxent model using default settings and optimized settings by ENMeval. Table S3: The contribution and permutation importance of 16 environmental factors.

Author Contributions: Investigation, X.S. and Z.Z.; methodology, X.S. and Q.Y.; formal analysis, X.S. and Z.Z.; data curation, Z.S.; writing—original draft preparation, X.S. and Q.Y.; writing—review and editing, Z.S., Z.Z., Z.J. and L.M.; supervision, Z.S. and L.M.; project administration, Z.J. and L.M.; funding acquisition, X.S., Z.J. and L.M. All authors have read and agreed to the published version of the manuscript.

Funding: This research is supported by the Zhejiang Provincial Scientific Research Institute Special Project (2023F1068-4), the Special Fund for Forest Scientific Research in the Public Welfare under grant no. 201504704, and the Transformation and application of forestry intellectual property project under grant no. Intellectual property transformation 2017-11.

Acknowledgments: We are thankful to Wufeng Bo Ling *Magnolia wufengensis* Technology Development Co., Ltd. for providing help in the field survey.

Conflicts of Interest: The authors declare no conflict of interest.

References

1. IPCC. Summary for Policymakers. In *Climate Change 2021: The Physical Science Basis*; Masson-Delmotte, V., Zhai, P., Pirani, A., Connors, S.L., Péan, C., Berger, S., Caud, N., Chen, Y., Goldfarb, L., Gomis, M.I., et al., Eds.; Contribution of Working Group I to the Sixth Assessment Report of the Intergovernmental Panel on Climate Change In Press; IPCC: Geneva, Switzerland, 2021.
2. Yang, X.; Liu, B.; Bussmann, R.W.; Guan, X.; Xu, W.; Xue, T.; Yu, S. Integrated plant diversity hotspots and long-term stable conservation strategies in the unique karst area of southern China under global climate change. *Forest Ecol. Manag.* **2021**, *498*, 119540. [CrossRef]
3. Shi, M.; Wu, H.; Jiang, P.; Zheng, K.; Liu, Z.; Dong, T.; He, P.; Fan, X. Food-water-land-ecosystem nexus in typical Chinese dryland under different future scenarios. *Sci. Total Environ.* **2023**, *880*, 163183. [CrossRef] [PubMed]
4. Jiang, R.; Zou, M.; Qin, Y.; Tan, G.; Huang, S.; Quan, H.; Zhou, J.; Liao, H. Modeling of the Potential Geographical Distribution of Three Fritillaria Species Under Climate Change. *Front. Plant Sci.* **2022**, *12*, 749838. [CrossRef]
5. Wang, W.; Li, Z.J.; Zhang, Y.L.; Xu, X.Q. Current Situation, Global Potential Distribution and Evolution of Six Almond Species in China. *Front. Plant Sci.* **2021**, *12*, 619883. [CrossRef]
6. Wang, G.Z.; Gen, Q.F.; Xiao, M.Y.; Zhang, M.Y.; Zhang, Y.Y.; Wang, Z.S. Predicting *Pseudolarix amabilis* potential habitat based on four Niche models. *Acta. Ecol. Sin.* **2020**, *40*, 6096–6104.
7. Dee, L.E.; Cowles, J.; Isbell, F.; Pau, S.; Gaines, S.D.; Reich, P.B. When do ecosystem services depend on rare species? *Trends Ecol. Evol.* **2019**, *34*, 746–758. [CrossRef] [PubMed]

8. Liao, Y.; Song, X.; Ye, Y.; Gu, J.; Wang, R.; Zhao, D.; Shao, X. Climate Change May Pose Additional Threats to the Endangered Endemic Species *Encalypta buxbaumioidea* in China. *Diversity* **2023**, *15*, 269. [CrossRef]
9. Jeong, H.M.; Kim, H.R.; Hong, S.; You, Y.H. Effects of elevated CO2 concentration and increased temperature on leaf quality responses of rare and endangered plants. *J. Ecol. Environ.* **2018**, *42*, 1. [CrossRef]
10. Markham, J. Rare species occupy uncommon niches. *Sci. Rep.* **2014**, *4*, 6012. [CrossRef]
11. Yang, Z.; Bai, Y.; Alatalo, J.M.; Huang, Z.; Yang, F.; Pu, X.; Wang, R.; Yang, W.; Guo, X. Spatio-temporal variation in potential habitats for rare and endangered plants and habitat conservation based on the maximum entropy model. *Sci. Total Environ.* **2021**, *784*, 147080. [CrossRef]
12. Rivers, M.; Newton, A.C.; Oldfield, S. Global Tree Assessment Contributors. Scientists' warning to humanity on tree extinctions. *Plants People Planet* **2023**, *5*, 466–482. [CrossRef]
13. Albert, J.S.; Destouni, G.; Duke-Sylvester, S.M.; Magurran, A.E.; Oberdorff, T.; Reis, R.E.; Winemiller, K.O.; Ripple, W.J. Scientists' warning to humanity on the freshwater biodiversity crisis. *Ambio* **2021**, *50*, 85–94. [CrossRef] [PubMed]
14. Ma, L.Y.; Wang, L.R.; He, S.C.; Liu, X.; Wang, X.Q. A new species of *Magnolia* (Magnoliaceae) from Hubei, China. *Bull Bot. Res.* **2006**, *26*, 4–7.
15. Ma, L.Y.; Wang, L.R.; He, S.C.; Liu, X.; Wang, X.Q. A new variety of *Magnolia* (Magnoliaceae) from Hubei, China. *Bull Bot. Res.* **2006**, *26*, 516–519.
16. Liang, D.W. Select on superior tree *Magnolia wufengensis* and cultivar classification. *Beijing For. Univ.* **2010**, 3–8.
17. Ma, L.Y. Breeding of *Magnolia wufengensis* and its application in Land Greening. *Land Green* **2019**, *3*, 54–56.
18. Sang, Z.Y.; Ma, L.Y.; Chen, F.J.; Zhang, P.; Zhu, Y.C. Protection status and utilization countermeasure of germplasm resources of the *Magnolia wufengensis* in Wufeng County. *Hubei Agric. Sci.* **2011**, *50*, 1564–1567.
19. Liu, H.; Ren, H.; Liu, Q.; Wen, X.; Maunder, M.; Gao, J. Translocation of threatened plants as a conservation measure in China. *Conserv. Biol.* **2015**, *29*, 1537–1551. [CrossRef]
20. Zhou, J.; Yang, M.; Wen, X.; Li, N.; Ren, H. Strengthen ex situ conservation of plants and promote protection and utilization of plant resources. *Bul. Chin. Acad. Sci. (China Ver.)* **2021**, *36*, 417–424.
21. Shi, X.D.; Ma, L.Y.; Duan, J.; Sang, Z.Y.; Zhu, Z.L.; Yao, N.; Zhou, M.M.; Jia, Z.K. Potential climatically adaptive regions for the introduction of *Magnolia wufengensis*. *J. Zhejiang A&F Univ.* **2018**, *35*, 705–715.
22. Vincent, H.; Bornand, C.N.; Kempel, A.; Fischer, M. Rare species perform worse than widespread species under changed climate. *Biol. Conserv.* **2020**, *246*, 108586. [CrossRef]
23. Ye, X.Z.; Zhao, G.H.; Zhang, M.Z.; Cui, X.Y.; Fan, H.H.; Liu, B. Distribution pattern of endangered plant *Semiliquidambar cathayensis* (Hamamelidaceae) in response to climate change after the last interglacial period. *Forests* **2020**, *11*, 434. [CrossRef]
24. Booth, T.H. Estimating potential range and hence climatic adaptability in selected tree species. *Forest Ecol. Manag.* **2016**, *366*, 175–183. [CrossRef]
25. Hu, W.; Wang, Y.; Zhang, D.; Yu, W.; Chen, G.; Xie, T.; Liu, Z.; Ma, Z.; Du, J.; Chao, B.; et al. Mapping the potential of mangrove forest restoration based on species distribution models: A case study in China. *Sci. Total Environ.* **2020**, *748*, 142321. [CrossRef]
26. Chang, Q.X.; Zhong, Y.F.; Zhang, Z.; Zhao, Y.; Song, X.Q. Visual Analysis of Research Progress on Species Distribution Prediction Based on Citespace. *Forest Inv. Plan.* **2022**, *47*, 20–33.
27. Susset, E.C.; Magro, A.; Hemptinnej, L. Using species distribution models to locate animal aggregations: A case study with *Hippodamia undecimnotata* (Schneider) overwintering aggregation sites. *Ecol. Entomol.* **2017**, *42*, 345–354. [CrossRef]
28. Anibaba, Q.A.; Dyderski, M.K.; Jagodziński, A.M. Predicted range shifts of invasive giant hogweed (*Heracleum mantegazzianum*) in Europe. *Sci. Total Environ.* **2022**, *825*, 154053. [CrossRef]
29. Li, G.; Huang, J.; Guo, H.; Du, S. Projecting species loss and turnover under climate change for 111 Chinese tree species. *Forest Ecol. Manag.* **2020**, *477*, 118488. [CrossRef]
30. Wang, Y.Q.; Ma, J.F.; Li, X.Q.; Wang, Y.F.; Cao, S.; Xie, A.T.; Ye, S.F.; Dong, B.X.; Zhao, W.X.; Qin, Y.X.; et al. The distribution of *Athetis lepigone* and prediction of its potential distribution based on GARP and MaxEnt. *J. Appl. Entomol.* **2017**, *141*, 431–440. [CrossRef]
31. Abedi-Tizaki, M.; Zafari, D. Geographic distribution of phylogenetic species of the *Fusarium graminearum* species complex and their 8-ketotrichothecene chemotypes on wheat spikes in Iran. *Mycotoxin Res.* **2017**, *33*, 245–259. [CrossRef]
32. Deb, J.C.; Phinn, S.; Butt, N.; Mcalpine, C.A. Climatic-induced shifts in the distribution of Teak (*Tectona grandis*) in tropical Asia: Implications for forest management and planning. *Environ. Manag.* **2017**, *60*, 422–435. [CrossRef]
33. Bai, J.J.; Hou, P.; Zhao, Y.H.; Xu, H.T.; Zhang, B. Research progress of species habitat suitability models and their verification. *Chin. J. Ecol.* **2022**, *41*, 1423–1432.
34. Phillips, S.J.; Anderson, R.P.; Schapire, R.E. Maximum entropy modeling of species geographic distributions. *Ecol. Model.* **2006**, *190*, 231–259. [CrossRef]
35. Shi, X.D.; Yin, Q.; Sang, Z.Y.; Zhu, Z.L.; Ma, L.Y.; Jia, Z.K. Prediction of potentially suitable areas for the introduction of *magnolia wufengensis* under climate change. *Ecol. Indic.* **2021**, *127*, 107762. [CrossRef]
36. Song, C.; Liu, H. Habitat differentiation and conservation gap of *Magnolia biondii, M. denudata*, and *M. sprengeri* in China. *Peer J.* **2019**, *6*, e6126. [CrossRef]
37. Chen, Y.H.; Lu, Y.W.; Yin, X.J. Predicting habitat suitability due to climate change of southwest coniferous forest tree species. *J. Nanjing Univ.* **2019**, *43*, 1–10.

38. Zhang, X.W.; Li, Y.; Xie, Y.P.; Bao, X.M.; Fang, Y.M. Effect of climate change on potential geographical distribution of *Sorbus amabilis*. *J. Plant Resour. Environ.* **2018**, *27*, 31–41.
39. IPCC. *Climate Change and Land 2019: An IPCC Special Report on Climate Change, Desertification, Land Degradation, Sustainable Land Management, Food Security, and Greenhouse Gas Fluxes in Terrestrial Ecosystems*; IPCC: Geneva, Switzerland, 2019; pp. 99–100.
40. Aguirre-Gutiérrez, J.; Carvalheiro, L.G.; Polce, C.; van Loon, E.E.; Raes, N.; Reemer, M.; Biesmeijer, J.C. Fit-for-purpose: Species distribution model performance depends on evaluation criteria-Dutch hoverflies as a case study. *PLoS ONE* **2013**, *8*, e63708. [CrossRef]
41. Gao, M.; Ni, S.P.; Shen, L. Analysis of global potential ecological suitable producing area for *Salvia miltiorrhiza* based on MaxEnt model. *China Pharm.* **2018**, *29*, 2243–2247.
42. Dai, G.H.; Yan, J.; Huang, C.H.; Sun, C.W.; Jia, L.M.; Ma, L.Y. The effects of climate change on the development of tree plantations for biodiesel production in China. *Forests* **2017**, *8*, 207. [CrossRef]
43. Anderson, R.P.; Lew, D.; Peterson, A.T. Evaluating predictive models of species' distributions: Criteria for selecting optimal models. *Ecol. Model.* **2003**, *162*, 211–232. [CrossRef]
44. Booth, T.H.; Nix, H.A.; Busby, J.R.; Hutchinson, M.F. Bioclim: The first species distribution modelling package, its early applications and relevance to most current MaxEnt studies. *Diver. Distrib.* **2013**, *20*, 1–9. [CrossRef]
45. Carpenter, G.; Gillison, A.N.; Winter, J. DOMAIN: A flexible modelling procedure for mapping potential distributions of plants and animals. *Biodivers Conserv.* **1993**, *2*, 667–680. [CrossRef]
46. Zhang, Q.; Zhang, D.F.; Wu, M.L.; Guo, J.; Sun, C.Z.; Xie, C.X. Predicting the global areas for potential distribution of *Gastrodia elata* based on ecological niche models. *Chin. J. Plant Ecol.* **2017**, *41*, 770–778.
47. Jayasinghe, S.L.; Kumar, L. Modeling the climate suitability of tea [*Camellia sinensis* (L.) o. Kuntze] in Sri Lanka in response to current and future climate change scenarios. *Agric. Meteorol.* **2019**, *272–273*, 102–117. [CrossRef]
48. Adhikari, P.; Shin, M.S.; Jeon, J.Y.; Kim, H.W.; Hong, S.; Seo, C. Potential impact of climate change on the species richness of subalpine plant species in the mountain national parks of South Korea. *J. Ecol. Environ.* **2018**, *42*, 36. [CrossRef]
49. Wang, S.Q. Genetic diversity and population structure of the endangered species *Paeonia decomposita* endemic to China and implications for its conservation. *BMC Plant Biol.* **2020**, *20*, 510. [CrossRef]
50. Wang, Y.S.; Xie, B.Y.; Wang, F.H.; Xiao, Q.M.; Dai, L.Y. Application of ROC curve analysis in evaluating the performance of alien species' potential distribution models. *Biodiversity Sci.* **2007**, *15*, 365–372.
51. Wang, J. The Application of Kappa in Assessing Agreement. Master's Thesis, Sichuan University, Chengdu, China, 2006.
52. Yang, F.R.; Zhang, Q.; Sun, C.Z.; Xie, C.X.; Song, J.Y. Comparative evaluation of multiple models for predicting the potential distribution areas of *Astragalus membranaceus* var. *mongholicu*. *Plant Sci. J.* **2019**, *37*, 136–143.
53. Zhu, G.P.; Qiao, H.J. Effect of the Maxent model's complexity on the prediction of species potential distributions. *Biodiversity Sci.* **2016**, *24*, 1189–1196. [CrossRef]
54. Jia, X.; Wang, C.; Jin, H.; Zhao, Y.; Liu, L.J.; Chen, Q.H.; Li, B.Y.; Xiao, Y.; Yin, H. Suitable distribution areas assessments of *Pinus koraiensis* based on an optimized MaxEnt model. *Chin. J Ecol.* **2019**, *38*, 2570–2576.
55. Anderson, R.P.; Gonzalez, I. Species-specific tuning increases robustness to sampling bias in models of species distributions: An implementation with maxent. *Ecol. Model.* **2011**, *222*, 2796–2811. [CrossRef]
56. Alberto, J.A.; Decae, A.E.; Arnedo, M.A. Environmental suitability of new reported localities of the funnelweb spider *Macrothele calpeiana*: An assessment using potential distribution modelling with presence-only techniques. *J. Biogeog.* **2011**, *38*, 1213–1223.
57. Stockwell, D.; Peters, D.P. The GARP modelling system: Problems and solutions to automated spatial prediction. *Int. J. Geog. Inf. Sci.* **1999**, *13*, 143–158. [CrossRef]
58. Srivastava, V.; Griess, V.C.; Padalia, H. Mapping invasion potential using ensemble modelling. A case study on *Yushania maling* in the Darjeeling Himalayas. *Ecol. Model.* **2018**, *385*, 35–44. [CrossRef]
59. Zhang, H.; Zhao, H.X.; Wang, H. Potential geographical distribution of *Populus euphratica* in China under future climate change scenarios based on Maxent model. *Acta. Ecol. Sin.* **2020**, *40*, 6552–6563.
60. Elith, J.; Graham, C.H.; Anderson, R.P.; Dudík, M.; Ferrier, S.; Guisan, A.; Hijmans, R.J.; Huettmann, F.; Leathwick, J.R.; Lehmann, A.; et al. Novel methods improve prediction of species' distributions from occurence data. *Ecography* **2006**, *29*, 129–151. [CrossRef]
61. Ray, D.; Behera, M.D.; Jacob, J. Evaluating Ecological Niche Models: A Comparison Between Maxent and GARP for Predicting Distribution of *Hevea brasiliensis* in India. *Proc. Natl. Acad. Sci. India Sect. B Biol. Sci.* **2018**, *88*, 1337–1343. [CrossRef]
62. Zhang, C.; Chen, L.; Tian, C.M.; Li, T.; Wang, R.; Yang, Q.Q. Predicting the distribution of dwarf mistletoe (*Arceuthobium sichuanense*) with GARP and MaxEnt models. *J. Beijing For. Univ.* **2016**, *38*, 23–32.
63. Duan, Y.Z.; Wang, H.T.; Wang, C.; Du, Z.Y. Potential distribution of endangered plant *Helianthemum songaricum* in China under climate change. *J. Plant Resour. Environ.* **2020**, *29*, 55–68.
64. Yang, Y. Studies on the ecological adaptability and overwintering technique of *Magnolia wufengensis* seedlings in Beijing. *Beijing For. Univ.* **2015**, 28–97.
65. Zhu, Z.L. Study on the limiting factors and winter protection methods of *Magnolia wufengensis* after introduction to Beijing. *Beijing For. Univ.* **2012**, 41–48.
66. Yan, Y.H.; Cen, Y.F.; Zhang, P.Y.; Zhang, Y.; Liu, X.; Li, C.Y.; Xu, S. Predicting distribution pattern and future change of *Pinus massoniana* in China based on MaxEnt model. *Chin. J. Ecol.* **2019**, *38*, 2896–2901.

67. Li, Y.; Zhang, X.W.; Fang, Y.M. Responses of the distribution pattern of *Quercus chenii* to climate change following the last glacial maximum. *Chin. J. Plant. Ecol.* **2016**, *40*, 1164–1178.
68. Hao, Y.; Peng, Z.D.; Ma, L.Y. Compare Studies on the Seedlings of *Magnolia wufengensis* and *Magnolia denudate*. *Nor. Horticul.* **2010**, *4*, 101–104.
69. Liu, Y.H. A preliminary study on the taxonomy of the family Magnoliaceae. *J. Syst. Evol.* **1984**, *22*, 89–109.
70. Liu, Y.H.; Xia, N.H.; Yang, H.Q. The origin, evolution and phytogeography of Magnoliaceae. *J. Trop. Subtrop. Bot.* **1995**, *3*, 1–12.
71. Urban, M.C. Accelerating extinction risk from climate change. *Science* **2015**, *348*, 571–573. [CrossRef]
72. Gao, X.; Liu, J.; Huang, Z. The impact of climate change on the distribution of rare and endangered tree *Firmiana kwangsiensis* using the Maxent modeling. *Ecol. Evol.* **2022**, *12*, e9165. [CrossRef]
73. Li, Y.F.; Zhang, C.G.; Zhu, H.G.; Li, X.H.; Duan, Y.F. Analyses on suitable distribution areas and main climatic variables of *Osmanthus yunnanensis* and *O. delavayi*. *J. Plant Res. Environ.* **2019**, *28*, 71–78.
74. Wu, Y.; Wang, H.F.; Mu, L.Q. Research progress and prospect of species distribution models. *J. Sci. Teachers. Coll. Univ.* **2022**, *42*, 66–70.

forests MDPI

Article

3PG-MT-LSTM: A Hybrid Model under Biomass Compatibility Constraints for the Prediction of Long-Term Forest Growth to Support Sustainable Management

Jushuang Qin [1,2], Menglu Ma [1,2], Yutong Zhu [1,2], Baoguo Wu [1,3] and Xiaohui Su [1,2,*]

1 School of Information Science and Technology, Beijing Forestry University, Beijing 100083, China; qinjs123@bjfu.edu.cn (J.Q.); maml1116@bjfu.edu.cn (M.M.); zhuyutong@bjfu.edu.cn (Y.Z.); wubg@bjfu.edu.cn (B.W.)
2 Engineering Research Center for Forestry-Oriented Intelligent Information Processing, National Forestry and Grassland Administration, Beijing 100083, China
3 Research Institute of Forestry Informatization, Beijing Forestry University, Beijing 100083, China
* Correspondence: suxhui@bjfu.edu.cn

Abstract: Climate change is posing new challenges to forestry management practices. Thinning reduces competitive pressure in the forest by repeatedly reducing the tree density of forest stands, thereby increasing the productivity of plantations. Considering the impact of thinning on vegetation and physiological and ecological traits, for this study, we used Norway spruce (*Picea abies*) data from three sites in the PROFOUND dataset to parameterize the 3-PG model in stages. The calibrated 3-PG model was used to simulate the stand diameter at breast height and the stem, root, and leaf biomass data on a monthly scale. The 3PG-MT-LSTM model uses 3-PG simulation data as the input variable. The model uses a long short-term memory neural network (LSTM) as a shared layer and introduces multi-task learning (MTL). Based on the compatibility rules, the interpretability of the model was further improved. The models were trained using single-site and multi-site data, respectively, and multiple indicators were used to evaluate the model accuracy and generalization ability. Our preliminary results show that, compared with the process model and LSTM algorithm without MTL and compatibility rules, the hybrid model has higher biomass simulation accuracy and shows a more realistic biomass response to environmental driving factors. To illustrate the potential applicability of the model, we applied light (10%), moderate (20%), and heavy thinning (30%) at intervals of 10, 15, 20, 25, 30 years. Then, we used three climate scenarios—SSP1-2.6, SSP2-4.5, and SSP5-8.5—to simulate the growth of Norway spruce. The hybrid model can effectively capture the impact of climate change and artificial management on stand growth. In terms of climate, temperature and solar radiation are the most important factors affecting forest growth, and under warm conditions, the positive significance of forest management is more obvious. In terms of forest management practices, less frequent light-to-moderate thinning can contribute more to the increase in forest carbon sink potential; high-intensity thinning can support large-diameter timber production. In summary, moderate thinning should be carried out every 10 years in the young-aged forest stage. It is also advisable to perform light thinning procedures after the forest has progressed into a middle-aged forest stage. This allows for a better trade-off of the growth relationship between stand yield and diameter at breast height (DBH). The physical constraint-based hybrid modeling approach is a practical and effective tool. It can be used to measure long-term dynamic changes in forest production and then guide management activities such as thinning to achieve sustainable forest management.

Keywords: forest biomass modeling; 3-PG model; LSTM; biomass compatibility; forest thinning

Citation: Qin, J.; Ma, M.; Zhu, Y.; Wu, B.; Su, X. 3PG-MT-LSTM: A Hybrid Model under Biomass Compatibility Constraints for the Prediction of Long-Term Forest Growth to Support Sustainable Management. *Forests* **2023**, *14*, 1482. https://doi.org/10.3390/f14071482

Academic Editors: Alessio Collalti, Daniela Dalmonech and Gina Marano

Received: 27 June 2023
Revised: 14 July 2023
Accepted: 18 July 2023
Published: 19 July 2023

1. Introduction

Forests are the main part of terrestrial ecosystems and also the largest carbon storage pool on land [1–3]. Human-induced climate change could fundamentally alter forests in the

21st century, with profound implications for the world [4]. Forest biomass is a key indicator in global carbon cycle research [5]. Thinning is a common forestry management practice. It has a significant and direct impact on forest structure. The understory microclimate will change with the change of structure and then affect the growth and development of trees [6,7]. Through thinning, stand density is reduced, the interception of precipitation is decreased while the understory light level is increased, and the decomposition of forest litter is accelerated [8]. Retained wood can make full use of light, water, and soil, and forest productivity can finally be improved. Accurately estimating the synergistic effect of thinning and climate on forest growth is of great significance for optimizing the utilization of forest resources and improving the ecological quality of forests [9–11]. In recent years, researchers have extensively explored how to estimate the response of forest biomass to thinning operations and climate change [12,13].

With the in-depth study of forestry management, mechanistic models have been developed and have received more and more attention [14,15]. A process model can simulate the key mechanistic processes of photosynthesis, respiration, and carbon water balance in tree growth. A process model takes into account the impact of climate factors and human disturbance on forest growth [16]. Individual-tree process models such as MAESTRA or TRIPLEX are established based on the three-dimensional spatial information of a single tree. These models take into account the extent to which the canopy utilizes solar radiation [17,18]. Individual-tree process models exhibit a high level of precision in their simulations. However, individual-tree process models rely on field surveys, which can incur higher costs. A stand process model, such as 3-PG or FORECAST, considers the effects of stand structure, climate change, and management practices. This model effectively elucidates and simulates the influence of environmental variables on the growth of forests [19,20]. Ecosystem process models, such as BIOME-BGC or LPG-DGVM, have comprehensively incorporated the carbon cycle within the atmosphere-vegetation-soil system. Its complexity is higher, and it is more suitable for large-scale carbon-water cycle simulation [21,22]. The essence of the process model is to complete the description of the material exchange and energy flow processes between the atmosphere, vegetation, and soil with the help of complex empirical equations. However, due to the fact that the environment is everchanging and will continue to change in the future, a model established based on past knowledge may no longer hold true. Therefore, it is difficult to predict dynamic changes in forests on a long-term scale (e.g., 50a).

With the development of artificial intelligence technology, Deep Learning (DL) provides a new means for forecasting forest growth and harvest [23,24]. DL does not restrict the selection of data sources. It is able to automatically extract knowledge from data streams to provide more flexible predictions. Kraft et al. [25] used LSTM to build a global model for fitting the Normalized Difference Vegetation Index (NDVI). Their model shows that LSTM can identify the memory effect of vegetation state on climate with satisfactory fitting accuracy. In addition to LSTM, other DL methods are also widely used. Neto et al. [26] used principal component analysis (PCA) to reduce the dimensionality of input data and used an Artificial Neural Network (ANN) to estimate productivity dynamics during Eucalyptus rotations. With a limited sample size, it shows better results than other forest productivity estimation methods. Xu et al. [27] proposed a neural network model with a multi-task loss function. This model solves the compatibility problem of tree biomass estimation and improves the generalization ability of the model. Despite the numerous successes of the DL model, its limitations are also evident [28–30]. First of all, the accuracy and generalization ability of the model depend on the 'feeding' of large amounts of data. However, the collection of forestry data requires a lot of effort and material resources. The process is also cumbersome and expensive. Secondly, the quality of the collected data cannot be guaranteed during the data collection process. There may be noise present in the data. Additionally, the training of the model can be disrupted, producing completely incorrect results. Thirdly, the training process of DL models does not involve any physical mechanisms. This can lead to unreasonable simulations in certain scientific problems.

Building hybrid interpretable models is currently a very active research frontier [31,32]. The hybrid model combines the physical consistency of process models and the high data-driven performance of DL methods [30]. Reichstein et al. [32] summarized the feasible avenues for coupling physical models with DL in geoscience problems. Among them, coupling strategies have been developed in the processes of runoff prediction and biological regulation [33,34]. Coupling physical processes into the loss function of neural networks enables more seamless hybrid model building [35]. However, the increase in interpretable components may generate more errors, resulting in a decrease in system performance. Therefore, the key to hybrid modelling is to explore the optimal solution to the "accuracy versus transparency trade-off" problem [36,37].

In the context of climate change, precisely assessing the impact of thinning on forest growth presents a viable approach to enhance forest structure, ecological quality, and service functions [12,38]. At present, we still lack the technical means to effectively and accurately understand how management behaviors affect forest production. The objectives of this manuscript were to: (1) Take the 3-PG simulation value as the model input variable, construct a 3PG-MT-LSTM hybrid model to increase model interpretability, sensitivity to thinning events, and the accuracy of long-term biomass predictions. (2) Introduce biomass compatibility rules to ensure that the model converges to a solution that follows the basic physical laws, further improving the generalization ability of the model, and evaluate the multi-index accuracy of the optimized model. (3) Under the conditions of multi-climate scenarios, estimate the impact of different thinning practice intervals and intensities on the growth of Norway spruce.

2. Materials and Methods

In this study, environmental elements, stand biomass observations, and geographic location information were first obtained from the PROFOUND database [39] (https://doi.org/10.5880/PIK.2020.006, accessed on 12 December 2022). The study area biomass data information is shown in Section 2.1. Then, we used the 3-PG model to amplify the original biomass observation data at different time intervals into monthly scale data. The 3PG-MT-LSTM model was established based on the biomass compatibility rule, using 3-PG simulated values as model inputs. Based on the characteristics of monthly average temperature, carbon dioxide concentration, monthly frost days, monthly average solar radiation, monthly precipitation, and monthly average water vapor pressure difference, the model realizes the prediction n of the forest stand stem, root, foliage, aboveground, and total biomass, and DBH. Using the single-site model as a basis, the factors of elevation, slope, and aspect were added to construct a multi-site model. The specific experimental method is presented in Section 2.2. In Section 2.3, we evaluate the accuracy of single-site and multi-site models using multiple evaluation metrics. According to the evaluation results, the model is continuously adjusted and optimized. Subsequently, in conjuction with CMIP6 multi-model and multi-scenario climate data (https://esgf-node.llnl.gov/projects/cmip6/, accessed on 14 February 2023), we simulate the growth of Norway spruce at the Bily Kriz site under different thinning regimes in Section 2.4. Finally, also in this paper, according to the simulation results, we analyze the impact of thinning on Norway spruce growth and make recommendations for forest management. The research framework is shown in Figure 1.

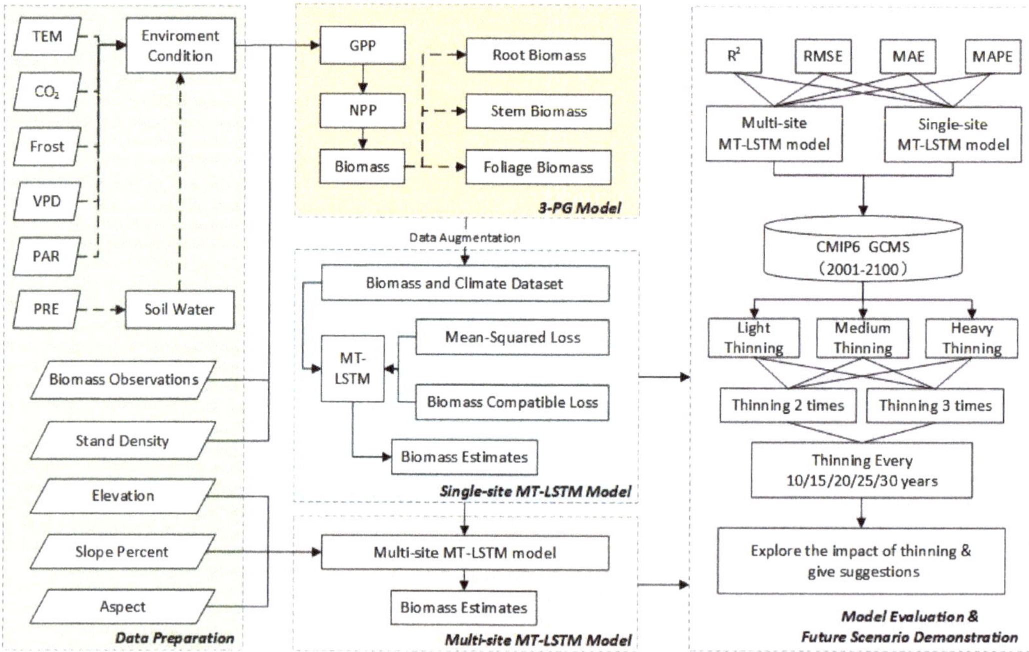

Figure 1. The research framework of this study. In the Data Preparation section, TEM, VPD, PAR, PRE represent temperature, saturated water vapor pressure difference, photosynthetically active radiation, and precipitation, respectively. In the 3-PG Model section, GPP and NPP represent gross primary productivity and net primary productivity. In the Model Evaluation & Future Scenario Demonstration section, R^2, RMSE, MAE, MAPE, and CMIP6 GCMS represent R-square, root mean squared error, mean absolute error, mean absolute percentage error, and global climate models in the Coupled Model Intercomparison Project (Phase 6), respectively.

2.1. Norway Spruce Biomass Data

Norway spruce is a large, fast-growing, and highly adaptable evergreen coniferous tree species commonly used as a timber tree throughout Europe. The growth rate of Norway spruce is at its peak between 20 and 90 years. After 90 years, the rate gradually slows down. In this study, the biomass data of Norway spruce and historical meteorological data were obtained from the PROFOUND database (Table 1).

Table 1. Basic information of biomass data.

Site	Longitude and Latitude	Elevation (m)	[a] Forest Age Range (Years)	[b] Forest Biomass Range (t/hm^2)	[c] Number of Thinning
Bily Kriz	18°19′ E, 49°18′ N	875	16–34	34.49~147.67	3
Hyytiala	24°17′ E, 61°50′ N	185	34–50	128.86~201.78	1
Solling	9°34′ E, 51°45′ N	508	85–133	250.67~372.96	4

[a] Forest Age Range refers to the age of the forest stand from the first to the last field survey conducted in the experimental plot. [b] Forest Biomass Range refers to the maximum and minimum values of forest biomass obtained during the survey and does not represent the biomass of the plot at the minimum (maximum) forest age. [c] Number of Thinning refers to the number of times the plot was artificially thinned during the entire survey process and does not include density changes caused by natural thinning.

2.2. The 3PG-MT-LSTM Hybrid Modelling Approach

2.2.1. 3-PG Model

The 3-PG (Physiological Principles in Predicting Growth) model is a process-based model that simulates forest growth and yield with a monthly time scale [19]. It is widely used for predicting biomass production and carbon sequestration in forest ecosystems. The model uses environmental variables such as temperature, precipitation, and solar radiation to predict forest growth. The model illustrates key physiological and ecological processes such as photosynthesis, respiration, and nutrient cycles in forest production. The key process formulation of the model is shown in Equation (1).

$$GPP = \alpha_{Cx} \times FPAR \times PAR \times f_T \times f_N \times f_F \times f_{age} \times min\{f_{VPD}, f_{SW}\} \quad (1)$$

Among them, GPP refers to the gross primary productivity of forest stands; α_{Cx} refers to the quantum efficiency of the vegetation canopy, which is the number of CO_2 molecules assimilated by vegetation per absorbed light quantum; $FPAR$ (Fraction of photosynthetically active radiation absorption) refers to the proportion of photosynthetically active radiation absorbed by vegetation; PAR (Photosynthetically active radiation) refers to the radiation that drives photosynthesis; and f_T, f_N, f_F, f_{age}, and f_{VPD} represent the correction factors for temperature, soil fertility, frost, stand age, vapor pressure deficit, and soil moisture, respectively.

Compared with other models, the 3-PG model is relatively simple to use and requires fewer input parameters. In addition, the model is available in multiple open-source versions such as EXCEL [40], Python [41], and R. We followed the method of 3-PG and used the 'r3pg' package of R program [42] to estimate biomass [43]. In this study, we used the Morris method to analyze the sensitivity of physiological and ecological parameters involved in the 3-PG model. The Morris sensitivity analysis method facilitates a global sensitivity analysis of parameters by calculating model outputs by changing only one input value between successive simulation runs [44]. Considering that forest growth is a dynamic process, we re-fitted and optimized the model parameters (i.e., phased localization model) after the thinning events and changes in the structure of the stand age groups. Finally, we used the built-in biomass allocation module in the 3-PG model to simulate the allocation of stand biomass among leaves, stems, and roots. The estimated values were used as inputs for the multi-task LSTM.

2.2.2. Multi-Task Learning and LSTM

MTL (Multi-task learning) is a machine learning technique that enables the joint training of multiple related tasks [45,46]. In this study, MLT is used to jointly predict the individual organ and total biomass changes in forest stands. LSTM is a type of recurrent neural network that can process sequential data by selectively remembering or forgetting past information [47,48]. The key process in LSTM is mathematically shown below.

$$f_t = \sigma_g\left(W_f x_t + W_f m_{t-1} + b_f\right) \quad (2)$$

$$i_t = \sigma_g(W_i x_t + W_i m_{t-1} + b_i) \quad (3)$$

$$o_t = \sigma_g(W_o x_t + W_o m_{t-1} + b_o) \quad (4)$$

$$c_t = f_t \odot c_{t-1} + i_t \odot \sigma_c(W_c x_t + W_c h_{t-1} + b_c) \quad (5)$$

$$m_t = o_t \odot \sigma_h(c_t) \quad (6)$$

From Equation (2) to Equation (6), σ_g, σ_c, and σ_h represent logistic sigmoid function, the input activation function, and the output activation function, respectively. f, i, o, and c represent the forget gate, input gate, output gate and cell activation vector, respectively. m depicts the hidden state vector, also known as output vector, of the LSTM units. W denotes the weight matrix (for example, W_i represents the weight matrix of input gate). The $\odot$ stands for element-wise multiplication, and b denotes the bias term.

For this study, LSTM was used to model the temporal dynamics of biomass change. The combination of multi-task learning and LSTM allows the model to capture the complex relationships between thinning rules and climate change. By sharing the LSTM layer and jointly training the model on multiple related tasks, the MT-LSTM model can learn to generalize better and make more accurate predictions [49,50].

2.2.3. Loss Function

In this study, the loss function used in the MT-LSTM model consists of mean squared error (MSE) loss function and biomass compatibility rules. MSE loss function is defined as the average of squared differences between the actual and the predicted value (Equation (7)). In the MT-LSTM model, stem (y_{stem}), root (y_{root}), leaf biomass (y_{leaf}), aboveground biomass ($y_{aboveground}$), and total biomass (y_{total}) were all estimated independently. The addition of biomass correlation ensures the compatibility of the model, which is important for accurate biomass estimation. The rule is based on the principle of mass balance, which states that the total biomass in the system must always be consistent with the sum of the components. The following equations (Equations (8) and (9)) are the biomass correlation rules added to the loss function:

$$MSE = \frac{1}{n}\sum_{i=1}^{n}\left(y_{i\_true} - y_{i\_prediction}\right)^2 \quad (7)$$

$$y_{total} = y_{aboveground} + y_{root} \quad (8)$$

$$y_{aboveground} = y_{stem} + y_{leaf} \quad (9)$$

In Equation (7), n is the number of samples, while y_{i_true} and $y_{i\_prediction}$ are the observed value of the ith sample and its corresponding predicted value, respectively. y_{total}, $y_{aboveground}$, y_{root}, y_{stem}, and y_{leaf} are the predicted values of the total, aboveground, root, stem (with bark and branch), and leaf biomass of the forest stand, respectively.

2.2.4. Constrained Hybrid Models

We used the 3-PG model to expand the observational data on stand biomass at unequal time intervals into monthly time series data. LSTM is trained using amplified data. MTL aims to enhance the generalization ability of LSTM by simultaneously learning multiple tasks. We added the biomass compatibility constraint to the LSTM loss function based on MTL. Minimize the difference between the stand's biomass and the sum of the biomass of each organ in the stand. Additionally, the superposition of unjustifiable errors in the prediction results is reduced, and the hybrid model's performance is enhanced. The 3PG-MT-LSTM model consists of two LSTM layers and one fully connected layer. The activation function of the hidden layer is ReLU (rectified linear unit). The activation function of the output layer is linear. The model optimizer is Adam (Adaptive Moment Estimation). The model learning rate is set to 10^{-4}.

2.3. *Model Evaluation and Validation*

The paper uses four evaluation indicators to assess the accuracy of the 3PG-MT-LSTM model: R^2, $RMSE$, MAE, and $MAPE$ (Equations (10)–(13)). Among them, y_i represents the observed biomass value at time i, $\hat{y}_i$ represents the estimated biomass value at time i, $\overline{y}$ represents the average value of the observed biomass.

$$R^2 = \frac{\sum_{i=1}^{n}\left(\hat{y}_i - \overline{y}\right)^2}{\sum_{i=1}^{n}\left(y_i - \overline{y}\right)^2} \quad (10)$$

$$RMSE = \sqrt{\frac{1}{n}\sum_{i=1}^{n}\left(\hat{y}_i - y_i\right)^2} \quad (11)$$

$$MAE = \frac{1}{n}\sum_{i=1}^{n}|\hat{y}_i - y_i| \quad (12)$$

$$MAPE = \frac{1}{n}\sum_{i=1}^{n}\left|\frac{\hat{y}_i - y_i}{y_i}\right| \times 100\% \quad (13)$$

R^2 measures the proportion of the variance in the dependent variable that is predictable from the independent variable. The closer R^2 is to 1, the better the model performance. *RMSE*, *MAE*, and *MAPE* measure the error between the predicted and observed values. When *MAE* and *MAPE* are used together, they provide a comprehensive evaluation of the model's fit to data of different scales. In the context of this paper's data, using both metrics can help to assess the model's performance in simulating the stand biomass of different ages. Combining the above metrics can provide a more comprehensive evaluation of a regression model's accuracy.

2.4. Future Climate Scenarios and Thinning Treatments

The Coupled Model Intercomparison Project Phase 6 (CMIP6) is a collaborative effort among climate scientists to assess and improve the performance of climate models' [51]. SSP1-2.6, SSP2-4.5, and SSP5-5.8 are three future projection scenarios in CMIP6, and their primary differences lie in greenhouse gas emissions and climate change predictions. The climate scenarios of SSP1-2.6, SSP2-4.5, and SSP5-5.8 estimate that global warming will be 1.7 °C, 2.0 °C, and 4.4 °C by 2081–2100, respectively. We used the average simulated value of three models—IPSL-CM6A-LR, MPI-ESM1-2-HR, and BCC-CSM2-MR—in the three climate scenarios (Figure A1). To study the synergic effects of climate and thinning on forest growth, we designed 18 thinning regimes for spruce from the Bily Kriz site, including an unthinning control group (Figure 2). Considering that spruce is a coniferous forest and our study plot is an even-aged stand, we only applied the low thinning method to our simulation. We designed the thinning intensity according to the number of trees. The thinning intensity is the proportion of trees cut to the total number of trees in the forest. Considering the ecological characteristics of spruce and its age groups, the stands were thinned at an intensity of 10% (light thinning), 20% (moderate thinning), or 30% (heavy thinning) in the stand stage where they are considered young (age less than 61). In the middle-aged stage, the stand was only lightly thinned. When the stand is near mature (over 90 years of age), thinning operations are no longer carried out. To distinguish the effects of thinning intervals on stand growth, we applied 10-, 15-, 20-, 25-, and 30-year intervals. We added a third thinning operation for shorter intervals (10 or 15 years) based on two thinning operations.

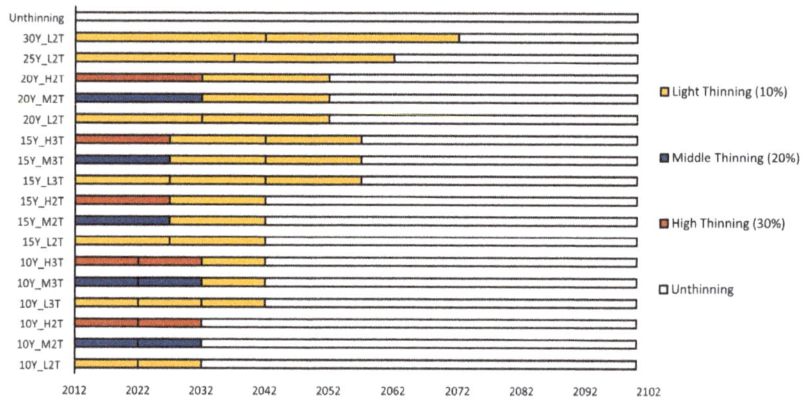

Figure 2. Thinning regimes and time. The horizontal axis represents the year. The vertical axis represents different thinning regimes. 10Y, 15Y, 20Y, 25Y, and 30Y represent the intervals of 10, 15, 20, 25, and 30 years with respect to forest stand thinning, respectively. L, M, H represent light (10%), moderate (20%), and heavy (30%) thinning intensity, respectively, which are distinguished by different colors in the figure; 2T/3T represent simulated thinning two times and three times within the set thinning period.

3. Results

3.1. 3-PG Model Calibration and *Picea asperata* Biomass Simulation

We performed Morris sensitivity analysis using the r3pg package and identified the three most sensitive parameters of the 3-PG model. They are the ratio of leaf biomass to stem productivity when DBH is 20 cm (PFS20), the minimum fraction of NPP to roots (PRn), and the canopy quantum efficiency (AlphaCx), respectively. Parameters related to biomass allocation and photosynthesis are always sensitive to model performance. The model parameterization process and results are similar. Therefore, this section uses the Solling site with the most measured data and the longest temporal range as an example to show the 3-PG calibration results. First, we calibrated the model using all the measured values at the Solling site from 1967 to 2014, as introduced in Section 2.1. There is a deviation between the estimated value (3-PG simulation) and the measured value (Observation) (Figure 3a). From 1967 to 1970, the stand was not thinned, and the estimated value of 3-PG was close to the measured value. But after the first thinning, the model did not properly capture the immediate impact of thinning. This led to a persistent underestimation of stand biomass. After the second thinning in 1981, the model continued to use the original parameters. This can lead to overly optimistic estimates of the stand's growth potential, resulting in consistently higher estimates. In order to make the estimated value reflect the true situation of the stand as much as possible, we re-optimized the sensitive parameters of the model every time a thinning event occurred. After correcting the parameters in stages, the model simulation results (3-PG staged simulation) were consistent with the observation data. They can meet the data quality requirements of deep learning. Through model simulation, we obtained a total of 3904 augmented data points to support LSTM training. With the decrease in stand density, the optimal values of sensitive parameters in the 3-PG model also change (Figure 3b). As the age of the forest increases and the density of the stand decreases, there is a corresponding increase in the PRn value. In comparison to the initial value, there is a 5.23% increase; PFS20 decreased by 7.09% compared with the initial value, and AlphaCx changed by 2.6% overall from the initial value. The performance of AlphaCx is subject to real-time thinning and exhibits notable fluctuations. A reduction of 100 trees per hectare in stand density resulted in a fluctuation of 13.67% in the parameter value. It is noteworthy that the parameters exhibit a relatively narrow range of variation with respect to their values. Nonetheless, in the 3-PG model, even minor alterations to the aforementioned three critical parameters can significantly impact the model's trajectory.

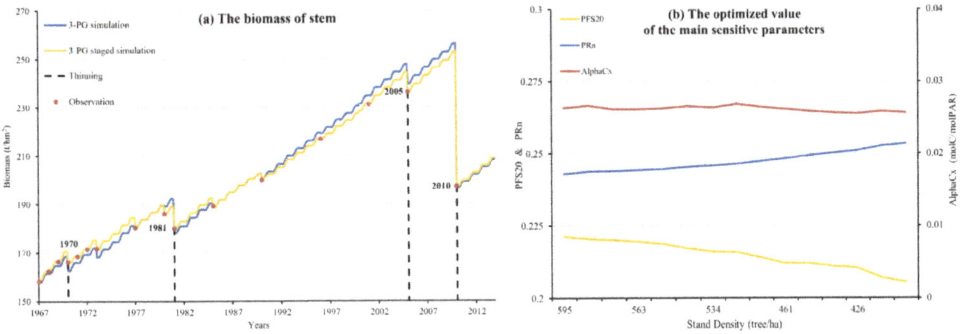

Figure 3. Taking the Solling site as an example, (**a**) comparison of the prediction results of the traditional 3-PG model and the staged 3-PG model. Among them, the 3-PG simulation and 3-PG staged simulation are estimated values after optimizing the physiological and ecological parameters of the model using all observation data and stages, respectively. Observation is the value of the observation data. The black dashed line indicates that the stand was thinned in its corresponding year; (**b**) the changes in the optimal value of the main sensitive parameters—PF20, PRn, and alphaCx—of the 3-PG model at different stages (stand density).

3.2. Calibrate and Evaluate the 3PG-MT-LSTM Model

For the present section, we initially employed the residual index to assess the efficacy of 3PG-LSTM, 3PG-MT-LSTM without compatibility rules, and 3PG-MT-LSTM with compatibility rules. Next, we analyzed whether there was a significant difference between the model residuals (Figure 4). Our findings indicate that incoporating biomass compatibility rule into the model can substantially enhance its precision. Simultaneously, the difference between the different models obtained by the Solling site with a large dataset and the joint training of multi-site data is more significant. When the datasets are small, like at the Hyytiala or Bily Kriz sites, the difference in performance between the single-task LSTM models and the multi-task LSTM models is not statistically significant. Following the incorporation of compatibility regulations, in contrast to the solitary-task LSTM model, the performance of the four models was substantially enhanced to differing extents.

Our assessment of the model was conducted based on the metrics outlined in Section 2.3 (Figure 5). The compatibility rule makes the model more capable of explaining the growth process and variation trend of distinct Norway spruce components ($R^2 > 0.9$). We observed that there is essentially no change in the prediction ability of the model for samples of varying forest ages since MAPE is comparable to the product of MAE and the average value of the data. This demonstrates that the coupling model described in this study performs better and is less susceptible to influence from outside sources. From the ratio of RMSE and MAE ($RMSE/MAE_{solling\_stem} = 6.96$, $RMSE/MAE_{hyytiala\_stem} = 6.45$, $RMSE/MAE_{bily\_stem} = 4.95$, $RMSE/MAE_{multi\text{-}model\_stem} = 7.97$), it is clear that adding the biomass compatibility rule makes it harder for the model's forecast process to avoid residual extremes. The model prioritizes minimizing the penalty values of the biomass compatibility loss function and the mean squared error (MSE) loss function. This results in models that sometimes sacrifice the accuracy of individual variables to improve overall model accuracy. Because of the volume and range of the data, the Solling site and multi-site estimation models have larger RMSE and MAE values. The inaccuracy is fairly substantial as a result of the wide variance in the stem biomass.

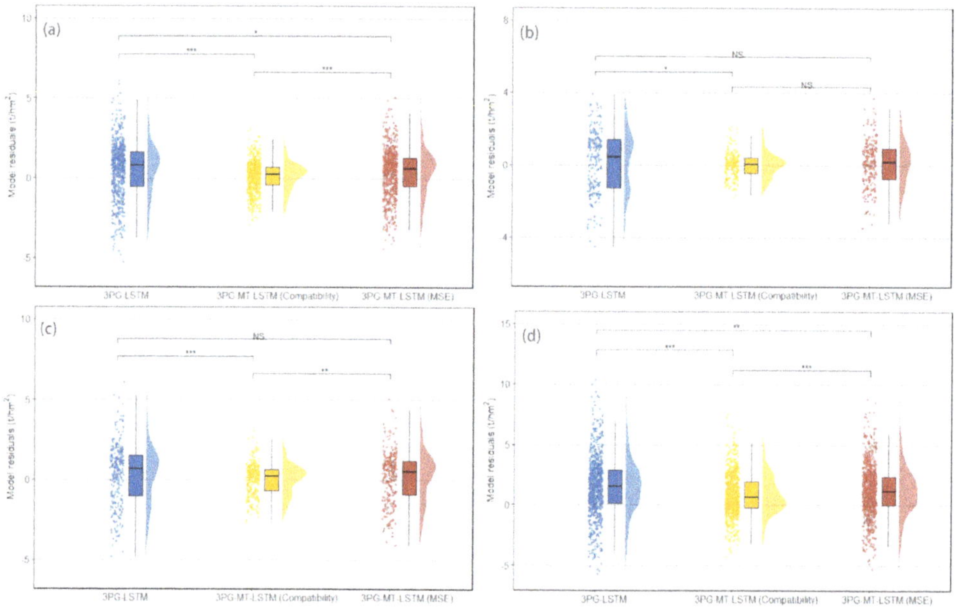

Figure 4. Residual distribution and significance analysis of three models in different sites. (**a**–**c**) represent the results obtained from training at the Solling site, Hyytiala site, and Bily Kriz site, while (**d**) represents the results obtained from training on data from all three sites combined. Asterisks refer to the significant differences in the models' residuals (NS, no significant difference; *, $p < 0.05$; **, $p < 0.01$; ***, $p < 0.001$).

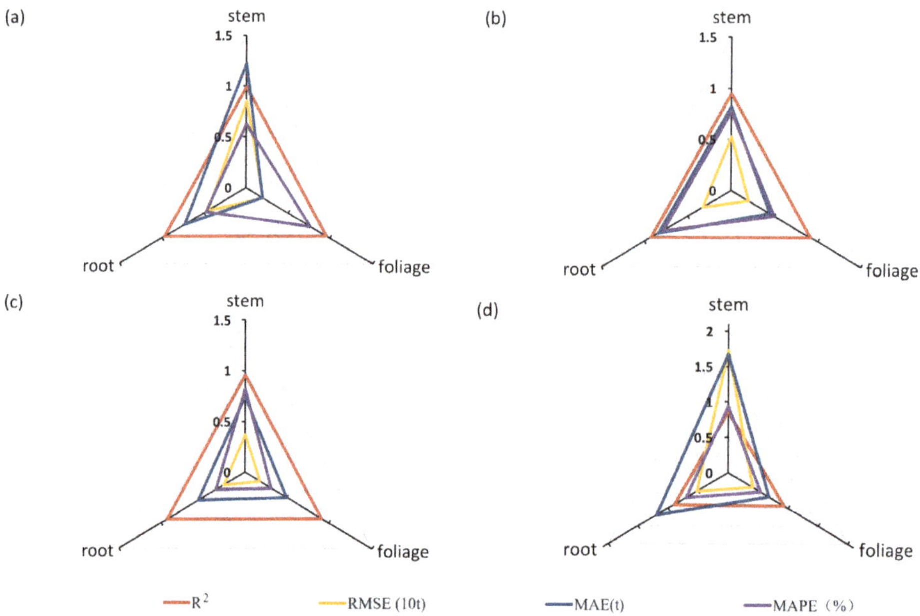

Figure 5. 3PG-MT-LSTM model evaluation results. (**a**–**d**) represent the results obtained from training at the Solling site, Hyytiala site, Bily Kriz site, and all three sites combined. Different colors represent different evaluation indicators. It should be noted that the unit of RMSE is 10 t.

3.3. Changes in Norway Spruce Biomass under Different Future Climate Scenarios

Young, middle-aged, and near-mature forests are the primary types of forests that benefit from thinning as a management strategy [52]. Considering this, in the later stages of the long-term simulation, the spruces in the Hyytiala and Solling sites will reach the mature forest stage. We chose the Bily Kriz site for simulation prediction. We estimated the changing trend of stand biomass and average diameter at breast height using the thinning regimes described and the gathered future climate data in Section 2.4. The ensemble means of three CMIP6 models were utilized to analyze the trend of each meteorological factor (Figure A1). The results indicate that there is a rising trend in the future daily minimum and maximum temperatures, as well as the daily average solar radiation, across all climate scenarios observed at the Bily Kriz site. The results also indicate that, under SSP1-2.6 and SSP2-4.5 scenarios, the precipitation in 2090s increased by 9.79% and 7.02% compared with that in 2010s. On the contrary, under the SSP5-8.5 scenario, the precipitation in 2090s decreased by 12.84% compared with that in 2010s. However, the precipitation in the SSP5-8.5 scenario decreased by 12.84%. The number of frost days had a greater impact on forest productivity. The reduction in the number of frost days persisted because of the steady escalation of the daily minimum temperature within the contexts of the SSP2-4.5 and SSP5-8.5 scenarios. In contrast to the frost days recorded in the 2010s, those observed in the 2090s exhibited a reduction of 40.26% and 69.33% for the respective years. Despite a slight increase in the daily minimum temperature under the SSP1-2.6 scenario, there was no significant alteration in the annual count of frost days. Based on the above climate scenario, we simulated the mean DBH and biomass of Norway spruce at the Bily Kriz site for the 2020–2100 period (Figure 6).

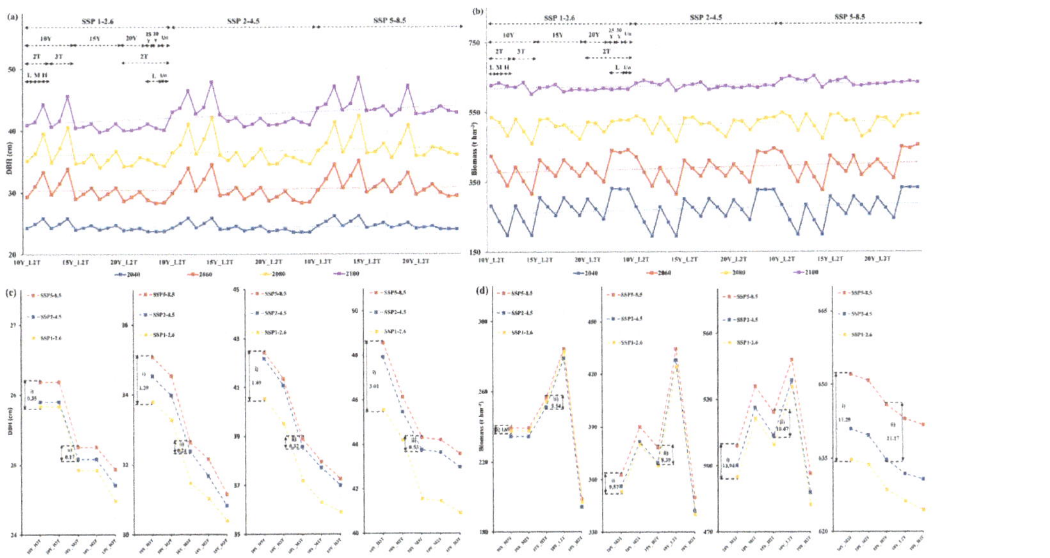

Figure 6. The future Norway spruce stand's mean diameter at breast height and biomass trends at the Bily Kriz site. (**a**,**b**) show the average diameter at breast height and biomass variation trend of the Norway spruce in 2040, 2060, 2080, and 2100 under several climate scenarios and 18 thinning strategies. Sections (**c**,**d**) identify the five most effective thinning techniques, based on the average diameter at breast height and biomass estimations of the forest stand in the year 2100. Where (i) represents the value difference between SSP1-2.6 and SSP5-8.5 scenarios, and (ii) Indicates the difference in values between SSP2-4.5 and SSP5-8.5 scenarios. The presented data display the simulated values of stand diameter at breast height and biomass for various thinning techniques in the years 2040, 2060, 2080, and 2100, arranged in a left-to-right formation.

The simulation results indicate that the productivity and DBH growth of spruce were most significant in the SSP5-8.5 scenario, regardless of the combination of thinning intensity and interval, when thinning from below was used (Figure 6a,b). Additionally, with the same thinning intensity and a climate scenario with a higher average temperature, the growth of Norway spruce varied more significantly. We selected the five thinning regimes with the highest biomass and DBH values for spruce in 2100 to verify our views (Figure 6c,d). The figures show the difference in spruce's response to the same thinning regimes and different climate scenarios at four time nodes in 2040, 2060, 2080, and 2100. Using the 10Y_H3T thinning approach as an example, the difference in DBH between the SSP5-8.5 and SSP1-2.6 scenarios increased from 0.35 cm in 2040 to 3.01 cm in 2100. The difference in biomass increased from 1.65 t hm^{-2} in 2040 to 17.28 t hm^{-2} in 2100. The same trend was present among the other thinning regimes (i.e., 10Y_H2T, 10Y_M3T, and other regimes) and climate scenarios. For stand diameter at breast height, with an increase in thinning intensity, the growth rate of stand diameter also increased. However, once the stand age reached 100a (2080), the growth of DBH per unit time began to slow. Furthermore, we discovered that when the thinning frequency was 10 years, the stand diameter was substantially greater than when the cutting frequency was 15 or 20 years. Light thinning at the middle-aged stage of spruce influenced stand diameter, but this effect depended on the thinning intensity at the young stage. For the light- and moderate-thinning plots, the impact on diameter at breast height was not obvious after light thinning in the middle-aged forest stage. However, on highly thinned plots, there was a greater potential for individual tree DBH growth after thinning at the middle-aged forest stage. For stand biomass, the number of Norway spruce trees in the stand fell constantly with increasing thinning severity during the continuous thinning era (before 2060). When compared to the un-thinned instance, the reduced number of trees resulted in lower total biomass. The number of trees was basically stable after the thinning period. When the trees reached the near-mature forest stage in 2080, the productivity of the 10Y_LT and 10Y_MT thinning forests were higher than that of the non-thinned stands. In 2100, The 10-year-cut and 15-year-cut stands' productivity was at a relatively high level, and the trees still had great growth potential. When Norway spruce is young, moderate-to-heavy thinning of the stand every 10 years makes it easier to produce large-diameter timber. Thinning at extended time intervals (more than 20 years) not only inhibits the growth of single-tree diameter at breast height but also limits the increase in total biomass.

4. Discussion

4.1. Estimation Accuracy and Interpretability of the Hybrid Model

The 3-PG model is designed to simulate key processes in the growth of forest stands. The model necessitates numerous parameters, and acquiring their values through direct observation is a challenging task [53]. Utilizing the LSTM approach driven by data, it is possible to effectively extract the spatio-temporal features of stand growth and their associations with climate, geography, and anthropogenic factors, based on historical data [54,55]. The process of training deep learning models typically necessitates a substantial amount of data. This study utilized the optimized 3-PG model to address the challenge of acquiring biomass data by boosting the available observation data. A substantial quantity of superior-grade data needs to be acquired to bolster the training of the LSTM model. The present study incorporates LSTM as a shared layer, multi-task learning, and biomass compatibility rules. The model adheres to the fundamental principle of mass conservation. Improvements in estimation accuracy, reliability, and generalization ability were also observed. The 3PG-MT-LSTM model more accurately captures the impact of thinning and climate change on stand biomass. The reasonable use of models for prediction and estimation can reduce uncertainty in forest management.

The density of tree stands is a crucial factor in determining the level of competition for resources among trees, ultimately impacting their growth [56,57]. The Norway spruce at the Solling site transitions from near-mature forest to mature forest. Thinning alters the structure of the canopy, allowing more light to reach the understory, which promotes forest growth. However, according to the studies described in Section 3.1, the age of the forest is a constraint that causes the photosynthetic efficacy of trees (or AlphaCx) to fluctuate and decline. The decrease in the PFS20 parameter value can also reflect the change in forest photosynthetic efficiency. After thinning, the leaf area index decreases, the gap increases, and each tree is allocated more appropriate growing space. The light resource is no longer the main object of tree competition. Whilst there was a gradual increase in leaf biomass, there was a decrease in the proportion of biomass that trees allocated to leaves. This discovery is in line with the research conducted by Deng et al. [58] and Chen et al. [59]. On the other hand, soil resources are finite. As forest grows, there is a concomitant decrease in soil fertility for a given plot. At this time, there is a tendency for forest stands to allocate a greater amount of carbon towards their root system, resulting in an increase in the PRn parameter. This phenomenon has the potential to enhance the competitive advantage of vegetation in acquiring soil resources. Competition for biomass allocation among stand organs is considered to be an adaptive feature of forests [60]. Similar findings were reported by Wang et al. [61] and Subedi et al. [62] in their respective investigations of the effects of thinning on tree biomass distribution.

It is challenging to adequately depict forest tree growth using a "white box" model because it is a complex process. For instance, the spatiotemporal dependence of vegetation and physiological and ecological characteristics needs to be considered in the prediction process. The LSTM regulates the flow and output of information by means of its distinctive gating mechanism. In contrast to Artificial Neural Networks and Recurrent Neural Networks, LSTM exhibits superior capacity in acquiring long-term dependencies within a sequence [63]. The temporal correlation between climate-induced effects on forests remains unclear. This is difficult to explain and quantify using known mechanisms [64,65]. Because of its gating mechanism, LSTM is better suited to processing and predicting a long-time series with relatively large intervals and delays [66,67]. LSTM can accurately capture the effect of events on forest biomass, whether it be forest thinning with extended gap intervals or climatic change with lag effects. While LSTM models can achieve high levels of accuracy during training, it is important to remain cautious of their inherent uncertainty. The constraints of physical processes must also be considered while maximizing knowledge learning from data. The 3-PG model exhibited more accuracy in simulating the allometric growth of the tree. The output of the process model was employed as the training sample for deep learning for this paper, and underlying physical law restrictions were introduced in the training process. The aforementioned approach not only satisfies the data prerequisites of Long Short-Term Memory (LSTM) and enhances the precision of the model but also upholds a degree of conformity with the principles governing plant growth. Through multi-task learning and penalty terms in loss function, the 3PG-MT-LSTM model integrates biomass compatibility rules. In forest ecosystems, the biomass compatibility rule is a rare and strong physical limitation. The principles of mass conservation-based compatibility rules are universally applicable to trees of all species, regardless of the age of the forest or its geographical location. During our research, we also considered models of volume compatibility (i.e., biomass = plant volume * plant density). However, after discussion, we discovered that the density of wood will alter slightly with the expansion of forest age, even for the same tree species. Additionally, obtaining the volume also presents challenges. This could cause the model to contain more mistakes and uncertainties. Because of this, for this study, we ultimately decided to solely employ the biomass compatibility constraint as the model's additional penalty. In deep learning algorithms, this reduces the build-up of unjustified errors. The accuracy of the 3PG-MT-LSTM model remains unaffected by variations in forest age or external environmental factors, irrespective of whether the model is trained on a single site or multiple sites. The integration of process modeling, and deep learning techniques enhances the predictive capabilities of the model in predicting future changes, thereby introducing a novel avenue for precision forestry research.

4.2. Synergistic Effects of Thinning and Climate on Forest Growth

Examining the relationship between tree growth and environmental and management practices is a crucial aspect of the advancement of precision forestry. In the young forest stage, stand density is generally higher. Thinning reduces competition among trees for nutrient space and resource constraints on tree growth. Based on the 3PG-MT-LSTM model proposed in this paper, we performed long-term predictions of Norway spruce biomass at the Bily Kriz site. The results showed that, when using the identical thinning strategy, SSP5-8.5's simulation of Norway spruce had the highest overall biomass and single-tree diameter. After combining the differences between different climate scenarios in Section 3.3, we can conclude that solar radiation and temperature are the most important driving variables for Norway spruce growth without considering the impact of extreme climates. This agrees with conclusions from earlier studies examining the environmental factors that influence spruce growth [68–70]. The observed variations in DBH and biomass across different scenarios indicate that the beneficial impact of forest management is particularly pronounced in warmer climates. The differences observed in SSP5-8.5 compared to other scenarios have evidenced that the Norway spruce species exhibits favorable ecological adaptability and thrives in areas with abundant sunlight and moderate humidity levels. The Norway spruce species can tolerate a variety of environmental conditions but prefers a climate with moderate humidity and lots of sunlight [71]. Temperature and solar radiation will exponentially increase after the 2070s, according to the SSP5-8.5 climate change scenario (Figure A1). The amount of precipitation exhibited a further decrease. As a result, in the SSP5-8.5 scenario, Norway spruce growth continued to increase. But compared to other situations, the growth rate was lower.

Different thinning intensities and intervals had a substantial impact on stand growth, in addition to climate change. The regimes taken during the young-aged forest stage for thinning played a decisive role in the growth of the stand's average diameter at breast height. For even-aged spruce forests, we used low thinning to reduce stand density and intra-stand competition. Under the low-thinning method, the heavy thinning of stands in young forests is conducive to the production of large-diameter timber. On this basis, further light thinning of stands in the middle-aged forest stage can increase the growth potential of stand diameter at breast height. Over time, the growth rate of a single tree in the light-to-moderate thinning stand slowed down significantly (at the same time interval, the average increase in diameter at breast height decreased). This is due to the limitations of light resources and soil resources [72]. Managers can improve ventilation and light conditions in the forest at this time by pruning branches and increasing forest gaps. In the thinning stage, there was an inverse relationship between thinning intensity and stand biomass. The higher the thinning intensity, the smaller the stand number and remaining biomass. This is consistent with the findings of Simon and Ameztegui [73]. After thinning, the stand entered a rapid growth period. In Section 3.3, we found that the growth rate of biomass in moderate- and heavy-thinning forests is much higher than that of light-thinning forests. In 2100, the biomass of moderate-thinning stands was the highest compared with other thinning regimes. This is a consequence of the regulation of multiple conditions, such as the total number of trees, soil resources, and competition within the forest. Rimal et al. examined the effects of different thinning intensities and intervals on the biomass growth of a 75-year-old Norway spruce stand [13]. They found that light thinning applied at longer intervals (greater than 25 years) resulted in higher stand biomass. However, according to our research, conducting moderate-to-heavy thinning 1–2 times during the young-aged Norway spruce forest stage and light thinning 0–2 times during the middle-aged forest stage are more favorable for biomass accumulation. The 10Y_H2T thinning regime can better balance the relationship between DBH and production. We attributed the differences in Rimal et al.'s results to the difference in the ages of the subjects. Specifically, our study was conducted on a 34-year-old forest stand, whereas Rimal et al. conducted their research on a 75-year-old one. Consequently, variations exist in the choice of thinning techniques and the resultant findings.

In the context of forest stand management, it may be advisable to conduct heavy thinning during the early stages of stand development in instances where there is a high demand for large-diameter timber. This approach is particularly relevant for operators in this field. To enhance the stand's overall productivity within a limited timeframe, opting for light thinning as a stand management strategy is advisable. To maximize the stand's long-term carbon sink capacity, we recommend implementing periodic moderate thinning with intervals of 10 to 15 years during the young-aged forest stage. Using the 3PG-MT-LSTM model has the potential to aid forest managers in promptly revising stand growth information. Before thinning, operators can formulate suitable thinning measures based on the results of model simulations. This can provide a direction to help them achieve a balance between producing large-diameter timber and economic benefits while taking forest carbon storage into account. After thinning, the model can be used to continuously monitor the growth of forest stands. By evaluating the real-time impact of management measures on the value of forest carbon sinks, operators can promote the development of sustainable forest management.

4.3. Limitations of Modeling Methods

The 3-PG model provides a larger training sample for deep learning algorithms. At the same time, the cost of data collection is reduced. However, limited by the output of the 3-PG model, we can only constrain the compatibility rules among stem, leaf, root, aboveground biomass, and total biomass. The biomass of bark and branches cannot be estimated. Due to constraints in data acquisition during the variable selection process, variables such as soil fertility were not included. Although the 3-PG sub-model has the capability to simulate alterations in soil fertility. To mitigate the potential for heightened ambiguity within the model, we refrained from incorporating its simulated value as a feature variable. Furthermore, we lack thinning regime designs for different thinning methods. Our model currently only provides forest growth estimates for a single thinning method. In the future, we hope to obtain more relevant data support, combined with feature coding technology, to design a more complete thinning experiment. This provides a reliable technical means of clarifying the long-term impacts of thinning on forests.

For the 3PG-MT-LSTM model, in addition to increasing the type and amount of data, the coupling mechanism is also an important avenue for improvement. Despite the fact that the interpretability of the hybrid model has improved, it fundamentally remains a model driven by data. The model exhibits a high degree of precision in its response to alterations in both climatic and human management factors. However, the challenge lies in accurately measuring the individual impact of each variable on the simulation outcomes. During the process of training a model, there is a relatively high demand for professional expertise and hardware capabilities, resulting in a relatively high computational cost. Researchers will be able to build genuine data–mechanism hybrid driving models in the future using methods such as mechanism model parameter optimization, model sub-module proxy, and model system error correction. This could further improve the robustness of forestry model and provides technical support for forestry accurate management.

5. Conclusions

Data gaps are a common problem in forestry modeling. In this paper, combined with the existing data parameterizing the 3-PG model, the stand diameter and biomass data were successfully filled using the monthly scale simulation results. The LSTM is trained based on the augmented data, and the 3PG-MT-LSTM model is constructed by adding MTL and compatibility rules. In contrast to the conventional LSTM architecture, the coupling model demonstrates enhanced precision in biomass estimation following thinning operations while also circumventing the accumulation of errors across individual components. There is no significant difference in the applicability of the model when it is applied to spruce forest stands in different regions and forest ages, and the generalization ability has improved. Utilizing the hybrid model, the present study showcases the impact of diverse thinning techniques on the growth of Norway spruce trees situated at the Bily Kriz location in light of anticipated climate change scenarios. According to our analysis, the periodic light-to-moderate thinning of Norway spruce stands at intervals of 10 to 15 years had a favorable impact on the accumulation of biomass over an extended period.

The heavy thinning technique is deemed more appropriate to produce timber with large diameters. Additionally, thinning management during the young-aged stages of forest growth is imperative for any production objective. The decision to implement additional management practices during the middle-aged stage of forest growth can be appropriately modified based on the current condition of the stand. The application of a hybrid model can help managers balance large-diameter timber production with economic benefits. The carbon sink value of a stand is determined through comprehensive research to provide scientific guidance for sustainable forest development. In conclusion, the in-depth application of deep learning presents an opportunity for the field of forestry management, but it also poses new challenges for forest managers.

Author Contributions: Conceptualization, J.Q. and X.S.; methodology, J.Q.; validation, J.Q., X.S. and M.M.; formal analysis, J.Q.; data curation, Y.Z.; writing—original draft preparation, J.Q.; writing—review and editing, X.S. and B.W.; visualization, J.Q., M.M. and X.S. All authors have read and agreed to the published version of the manuscript.

Funding: This research was funded by National Natural Science Foundation of China, grant number 32101516.

Institutional Review Board Statement: Not applicable for studies not involving humans or animals.

Informed Consent Statement: Not applicable for studies not involving humans.

Data Availability Statement: The data that support the findings of this study are openly available in the "PROFOUND database" (https://doi.org/10.5880/PIK.2020.006, accessed on 12 December 2022) and "The Coupled Model Intercomparison Project Phase 6" (https://esgf-node.llnl.gov/projects/cmip6/, accessed on 14 February 2023).

Acknowledgments: We thank the makers of the PROFOUND database for providing free access to their data. We acknowledge the World Climate Research Program's Working Group on Coupled Modeling, which is responsible for CMIP6, and thank the climate modeling groups for producing and making available their model outputs. We also thank the anonymous reviewers for their valuable suggestions.

Conflicts of Interest: The authors declare no conflict of interest.

Appendix A

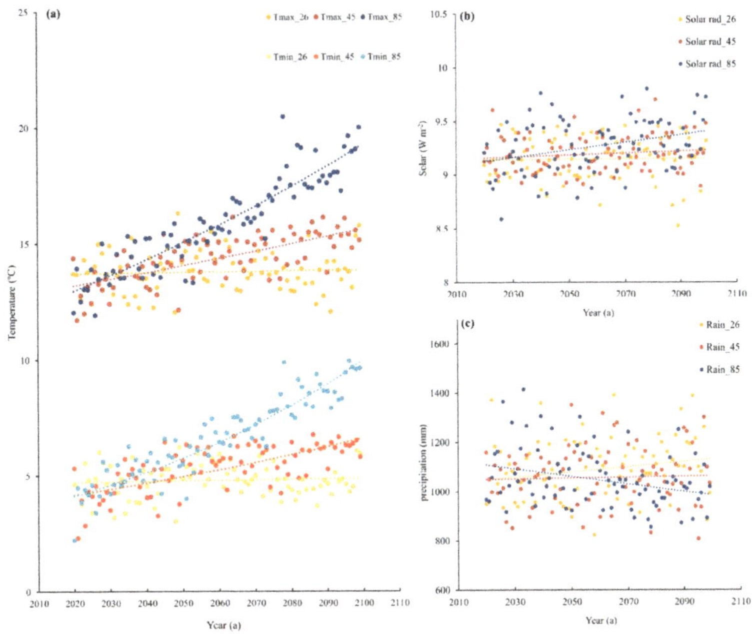

Figure A1. Change trends of annual average maximum and minimum temperature (**a**), annual average solar radiation (**b**), and annual total precipitation (**c**) under SSP1-2.6, SSP2-4.5, and SSP5-8.5 scenarios. The values shown here are from the same source as the future meteorological factors in the text. They are taken from the ensemble averages of three climate models: IPSL-CM6A-LR, MPI-ESM1-2-HR, and BCC-CSM2-MR.

References

1. Kramer, P.J. Carbon Dioxide Concentration, Photosynthesis, and Dry Matter Production. *BioScience* **1981**, *31*, 29–33. [CrossRef]
2. Reichstein, M.; Carvalhais, N. Aspects of Forest Biomass in the Earth System: Its Role and Major Unknowns. *Surv. Geophys.* **2019**, *40*, 693–707. [CrossRef]
3. Yadav, V.S.; Yadav, S.S.; Gupta, S.R.; Meena, R.S.; Lal, R.; Sheoran, N.S.; Jhariya, M.K. Carbon Sequestration Potential and CO_2 Fluxes in a Tropical Forest Ecosystem. *Ecol. Eng.* **2022**, *176*, 106541. [CrossRef]
4. Hansen, W.D.; Schwartz, N.B.; Williams, A.P.; Albrich, K.; Kueppers, L.M.; Rammig, A.; Reyer, C.P.O.; Staver, A.C.; Seidl, R. Global Forests Are Influenced by the Legacies of Past Inter-Annual Temperature Variability. *Environ. Res. Ecol.* **2022**, *1*, 011001. [CrossRef]
5. Augusto, L.; Boča, A. Tree Functional Traits, Forest Biomass, and Tree Species Diversity Interact with Site Properties to Drive Forest Soil Carbon. *Nat. Commun.* **2022**, *13*, 1097. [CrossRef] [PubMed]
6. Erdozain, M.; Bonet, J.A.; Martínez de Aragón, J.; de-Miguel, S. Forest Thinning and Climate Interactions Driving Early-Stage Regeneration Dynamics of Maritime Pine in Mediterranean Areas. *For. Ecol. Manag.* **2023**, *539*, 121036. [CrossRef]
7. Yu, J.; Zhang, X.; Xu, C.; Hao, M.; Choe, C.; He, H. Thinning Can Increase Shrub Diversity and Decrease Herb Diversity by Regulating Light and Soil Environments. *Front. Plant Sci.* **2022**, *13*, 948648. [CrossRef]
8. Zhang, X.; Guan, D.; Li, W.; Sun, D.; Jin, C.; Yuan, F.; Wang, A.; Wu, J. The Effects of Forest Thinning on Soil Carbon Stocks and Dynamics: A Meta-Analysis. *For. Ecol. Manag.* **2018**, *429*, 36–43. [CrossRef]
9. D'Amato, A.W.; Bradford, J.B.; Fraver, S.; Palik, B.J. Effects of Thinning on Drought Vulnerability and Climate Response in North Temperate Forest Ecosystems. *Ecol. Appl.* **2013**, *23*, 1735–1742. [CrossRef]
10. Gong, C.; Tan, Q.; Liu, G.; Xu, M. Forest Thinning Increases Soil Carbon Stocks in China. *For. Ecol. Manag.* **2021**, *482*, 118812. [CrossRef]
11. Olajuyigbe, S.; Tobin, B.; Saunders, M.; Nieuwenhuis, M. Forest Thinning and Soil Respiration in a Sitka Spruce Forest in Ireland. *Agric. For. Meteorol.* **2012**, *157*, 86–95. [CrossRef]
12. Melikov, C.H.; Bukoski, J.J.; Cook-Patton, S.C.; Ban, H.; Chen, J.L.; Potts, M.D. Quantifying the Effect Size of Management Actions on Aboveground Carbon Stocks in Forest Plantations. *Curr. For. Rep.* **2023**, *9*, 131–148. [CrossRef] [PubMed]

13. Rimal, S.; Djahangard, M.; Yousefpour, R. Forest Management under Climate Change: A Decision Analysis of Thinning Interventions for Water Services and Biomass in a Norway Spruce Stand in South Germany. *Land* **2022**, *11*, 446. [CrossRef]
14. Heimann, M.; Reichstein, M. Terrestrial Ecosystem Carbon Dynamics and Climate Feedbacks. *Nature* **2008**, *451*, 289–292. [CrossRef] [PubMed]
15. Higgins, S.I.; Conradi, T.; Muhoko, E. Shifts in Vegetation Activity of Terrestrial Ecosystems Attributable to Climate Trends. *Nat. Geosci.* **2023**, *16*, 147–153. [CrossRef]
16. Beer, C.; Reichstein, M.; Tomelleri, E.; Ciais, P.; Jung, M.; Carvalhais, N.; Rödenbeck, C.; Arain, M.A.; Baldocchi, D.; Bonan, G.B.; et al. Terrestrial Gross Carbon Dioxide Uptake: Global Distribution and Covariation with Climate. *Science* **2010**, *329*, 834–838. [CrossRef]
17. Medlyn, B.E.; Pepper, D.A.; O'Grady, A.P.; Keith, H. Linking Leaf and Tree Water Use with an Individual-Tree Model. *Tree Physiol.* **2007**, *27*, 1687–1699. [CrossRef]
18. Sturtevant, B.R. Forest Processes from Stands to Landscapes: Exploring Model Forecast Uncertainties Using Cross-Scale Model Comparison. *Can. J. For. Res.* **2010**, *40*, 2345–2359. [CrossRef]
19. Landsberg, J.J.; Waring, R.H. A Generalised Model of Forest Productivity Using Simplified Concepts of Radiation-Use Efficiency, Carbon Balance and Partitioning. *For. Ecol. Manag.* **1997**, *95*, 209–228. [CrossRef]
20. Restrepo, H.I.; Montes, C.R.; Bullock, B.P.; Mei, B. The Effect of Climate Variability Factors on Potential Net Primary Productivity Uncertainty: An Analysis with a Stochastic Spatial 3-PG Model. *Agric. For. Meteorol.* **2022**, *315*, 108812. [CrossRef]
21. Running, S.W.; Coughlan, J.C. A General Model of Forest Ecosystem Processes for Regional Applications I. Hydrologic Balance, Canopy Gas Exchange and Primary Production Processes. *Ecol. Model.* **1988**, *42*, 125–154. [CrossRef]
22. Yan, M.; Mao, F.; Du, H.; Li, X.; Chen, Q.; Ni, C.; Huang, Z.; Xu, Y.; Gong, Y.; Guo, K.; et al. Spatiotemporal Dynamic of Subtropical Forest Carbon Storage and Its Resistance and Resilience to Drought in China. *Front. Plant Sci.* **2023**, *14*, 1067552. [CrossRef] [PubMed]
23. Dang, A.T.N.; Nandy, S.; Srinet, R.; Luong, N.V.; Ghosh, S.; Senthil Kumar, A. Forest Aboveground Biomass Estimation Using Machine Learning Regression Algorithm in Yok Don National Park, Vietnam. *Ecol. Inform.* **2019**, *50*, 24–32. [CrossRef]
24. Huy, B.; Truong, N.Q.; Khiem, N.Q.; Poudel, K.P.; Temesgen, H. Deep Learning Models for Improved Reliability of Tree Aboveground Biomass Prediction in the Tropical Evergreen Broadleaf Forests. *For. Ecol. Manag.* **2022**, *508*, 120031. [CrossRef]
25. Kraft, B.; Jung, M.; Körner, M.; Requena Mesa, C.; Cortés, J.; Reichstein, M. Identifying Dynamic Memory Effects on Vegetation State Using Recurrent Neural Networks. *Front. Big Data* **2019**, *2*, 31. [CrossRef]
26. De Oliveira Neto, R.R.; Leite, H.G.; Gleriani, J.M.; Strimbu, B.M. Estimation of Eucalyptus Productivity Using Efficient Artificial Neural Network. *Eur. J. For. Res.* **2022**, *141*, 129–151. [CrossRef]
27. Xu, Q.; Lei, X.; Zhang, H. A Novel Method for Approaching the Compatibility of Tree Biomass Estimation by Multi-Task Neural Networks. *For. Ecol. Manag.* **2022**, *508*, 120011. [CrossRef]
28. Gu, Y.; Li, B.; Meng, Q. Hybrid Interpretable Predictive Machine Learning Model for Air Pollution Prediction. *Neurocomputing* **2022**, *468*, 123–136. [CrossRef]
29. Narayanan, H.; Cruz Bournazou, M.N.; Guillén Gosálbez, G.; Butté, A. Functional-Hybrid Modeling through Automated Adaptive Symbolic Regression for Interpretable Mathematical Expressions. *Chem. Eng. J.* **2022**, *430*, 133032. [CrossRef]
30. Perry, G.L.W.; Seidl, R.; Bellvé, A.M.; Rammer, W. An Outlook for Deep Learning in Ecosystem Science. *Ecosystems* **2022**, *25*, 1700–1718. [CrossRef]
31. Irrgang, C.; Boers, N.; Sonnewald, M.; Barnes, E.A.; Kadow, C.; Staneva, J.; Saynisch-Wagner, J. Towards Neural Earth System Modelling by Integrating Artificial Intelligence in Earth System Science. *Nat. Mach. Intell.* **2021**, *3*, 667–674. [CrossRef]
32. Reichstein, M.; Camps-Valls, G.; Stevens, B.; Jung, M.; Denzler, J.; Carvalhais, N. Prabhat Deep Learning and Process Understanding for Data-Driven Earth System Science. *Nature* **2019**, *566*, 195–204. [CrossRef] [PubMed]
33. Konapala, G.; Kao, S.-C.; Painter, S.L.; Lu, D. Machine Learning Assisted Hybrid Models Can Improve Streamflow Simulation in Diverse Catchments across the Conterminous US. *Environ. Res. Lett.* **2020**, *15*, 104022. [CrossRef]
34. Xu, T.; Longyang, Q.; Tyson, C.; Zeng, R.; Neilson, B.T. Hybrid Physically Based and Deep Learning Modeling of a Snow Dominated, Mountainous, Karst Watershed. *Water Resour. Res.* **2022**, *58*, e2021WR030993. [CrossRef]
35. Jiang, S.; Zheng, Y.; Solomatine, D. Improving AI System Awareness of Geoscience Knowledge: Symbiotic Integration of Physical Approaches and Deep Learning. *Geophys. Res. Lett.* **2020**, *47*, e2020GL088229. [CrossRef]
36. Li, M.; Yang, Y.; He, Z.; Guo, X.; Zhang, R.; Huang, B. A Wind Speed Forecasting Model Based on Multi-Objective Algorithm and Interpretability Learning. *Energy* **2023**, *269*, 126778. [CrossRef]
37. Wang, T.; Lin, Q. Hybrid Predictive Models: When an Interpretable Model Collaborates with a Black-Box Model. *J. Mach. Learn. Res.* **2021**, *22*, 6085–6122.
38. Pradhan, K.; Ettinger, A.K.; Case, M.J.; Hille Ris Lambers, J. Applying Climate Change Refugia to Forest Management and Old-Growth Restoration. *Glob. Chang. Biol.* **2023**, *29*, 3692–3706. [CrossRef]
39. Reyer, C.P.; Silveyra Gonzalez, R.; Dolos, K.; Hartig, F.; Hauf, Y.; Noack, M.; Lasch-Born, P.; Rötzer, T.; Pretzsch, H.; Meesenburg, H.; et al. The PROFOUND Database for Evaluating Vegetation Models and Simulating Climate Impacts on European Forests. *Earth Syst. Sci. Data* **2020**, *12*, 1295–1320. [CrossRef]
40. Sands, P.J.; Landsberg, J.J. Parameterisation of 3-PG for Plantation Grown Eucalyptus Globulus. *For. Ecol. Manag.* **2002**, *163*, 273–292. [CrossRef]

41. Song, X.; Song, Y. Introducing 3-PG2Py, an Open-Source Forest Growth Model in Python. *Environ. Model. Softw.* **2022**, *150*, 105358. [CrossRef]
42. Wallach, D.; Makowski, D.; Jones, J.W.; Brun, F. Chapter 3—The R Programming Language and Software. In *Working with Dynamic Crop Models*, 2nd ed.; Wallach, D., Makowski, D., Jones, J.W., Brun, F., Eds.; Academic Press: San Diego, CA, USA, 2014; pp. 71–117, ISBN 978-0-12-397008-4.
43. Trotsiuk, V.; Hartig, F.; Forrester, D.I. R3PG—An r Package for Simulating Forest Growth Using the 3-PG Process-Based Model. *Methods Ecol. Evol.* **2020**, *11*, 1470–1475. [CrossRef]
44. Morris, M.D. Factorial Sampling Plans for Preliminary Computational Experiments. *Technometrics* **1991**, *33*, 161–174. [CrossRef]
45. Caruana, R. Multitask Learning. *Mach. Learn.* **1997**, *28*, 41–75. [CrossRef]
46. Wang, Y.; Zhang, D.; Wulamu, A. A Multitask Learning Model with Multiperspective Attention and Its Application in Recommendation. *Comput. Intell. Neurosci.* **2021**, *2021*, 8550270. [CrossRef] [PubMed]
47. Qi, Y.; Li, Q.; Karimian, H.; Liu, D. A Hybrid Model for Spatiotemporal Forecasting of PM2.5 Based on Graph Convolutional Neural Network and Long Short-Term Memory. *Sci. Total Environ.* **2019**, *664*, 1–10. [CrossRef]
48. Wang, S.; Ren, Y.; Xia, B.; Liu, K.; Li, H. Prediction of Atmospheric Pollutants in Urban Environment Based on Coupled Deep Learning Model and Sensitivity Analysis. *Chemosphere* **2023**, *331*, 138830. [CrossRef]
49. Seltzer, M.L.; Droppo, J. Multi-Task Learning in Deep Neural Networks for Improved Phoneme Recognition. In Proceedings of the 2013 IEEE International Conference on Acoustics, Speech and Signal Processing, Vancouver, BC, Canada, 26–31 May 2013; pp. 6965–6969.
50. Zhang, Y.; Yang, Q. An Overview of Multi-Task Learning. *Natl. Sci. Rev.* **2018**, *5*, 30–43. [CrossRef]
51. Carvalho, D.; Rafael, S.; Monteiro, A.; Rodrigues, V.; Lopes, M.; Rocha, A. How Well Have CMIP3, CMIP5 and CMIP6 Future Climate Projections Portrayed the Recently Observed Warming. *Sci. Rep.* **2022**, *12*, 11983. [CrossRef]
52. Wang, Y.; Wei, X.; del Campo, A.D.; Winkler, R.; Wu, J.; Li, Q.; Liu, W. Juvenile Thinning Can Effectively Mitigate the Effects of Drought on Tree Growth and Water Consumption in a Young Pinus Contorta Stand in the Interior of British Columbia, Canada. *For. Ecol. Manag.* **2019**, *454*, 117667. [CrossRef]
53. Zhou, F.; Chai, F.; Huang, D.; Xue, H.; Chen, J.; Xiu, P.; Xuan, J.; Li, J.; Zeng, D.; Ni, X.; et al. Investigation of Hypoxia off the Changjiang Estuary Using a Coupled Model of ROMS-CoSiNE. *Prog. Oceanogr.* **2017**, *159*, 237–254. [CrossRef]
54. Willard, J.; Jia, X.; Xu, S.; Steinbach, M.; Kumar, V. Integrating Scientific Knowledge with Machine Learning for Engineering and Environmental Systems. *ACM Comput. Surv.* **2022**, *55*, 66. [CrossRef]
55. Zhang, Q.; Wang, R.; Qi, Y.; Wen, F. A Watershed Water Quality Prediction Model Based on Attention Mechanism and Bi-LSTM. *Environ. Sci. Pollut. Res.* **2022**, *29*, 75664–75680. [CrossRef]
56. Geng, Y.; Yue, Q.; Zhang, C.; Zhao, X.; von Gadow, K. Dynamics and Drivers of Aboveground Biomass Accumulation during Recovery from Selective Harvesting in an Uneven-Aged Forest. *Eur. J. For. Res.* **2021**, *140*, 1163–1178. [CrossRef]
57. Zhou, H.; Meng, S.; Liu, Q. Long-Term Response of Living Forest Biomass to Extensive Logging in Subtropical China. *J. For. Res.* **2019**, *30*, 1679–1687. [CrossRef]
58. Deng, C.; Ma, F.; Xu, X.; Zhu, B.; Tao, J.; Li, Q. Allocation Patterns and Temporal Dynamics of Chinese Fir Biomass in Hunan Province, China. *Forests* **2023**, *14*, 286. [CrossRef]
59. Chen, R.; Ran, J.; Hu, W.; Dong, L.; Ji, M.; Jia, X.; Lu, J.; Gong, H.; Aqeel, M.; Yao, S.; et al. Effects of Biotic and Abiotic Factors on Forest Biomass Fractions. *Natl. Sci. Rev.* **2021**, *8*, nwab025. [CrossRef]
60. Zhang, W.-P.; Jia, X.; Morris, E.C.; Bai, Y.-Y.; Wang, G.-X. Stem, Branch and Leaf Biomass-Density Relationships in Forest Communities. *Ecol. Res.* **2012**, *27*, 819–825. [CrossRef]
61. Wang, Z.C.; Li, Y.X.; Meng, Y.B.; Wang, C. Effect of Tending and Thinning on Spatial and Carbon Distribution Patterns of Natural Mixed Broadlesf-Conifer Secondary Forest in Xiaoxing'an Mountains, Pr China. *Appl. Ecol. Environ. Res.* **2021**, *19*, 4751–4764. [CrossRef]
62. Subedi, S.; Kane, M.; Zhao, D.; Borders, B.; Greene, D. Cultural Intensity and Planting Density Effects on Aboveground Biomass of 12-Year-Old Loblolly Pine Trees in the Upper Coastal Plain and Piedmont of the Southeastern United States. *For. Ecol. Manag.* **2012**, *267*, 157–162. [CrossRef]
63. Graves, A. Long Short-Term Memory. In *Supervised Sequence Labelling with Recurrent Neural Networks*; Graves, A., Ed.; Studies in Computational Intelligence; Springer: Berlin/Heidelberg, Germany, 2012; pp. 37–45, ISBN 978-3-642-24797-2.
64. Li, C.; Liu, Y.; Zhu, T.; Zhou, M.; Dou, T.; Liu, L.; Wu, X. Considering Time-Lag Effects Can Improve the Accuracy of NPP Simulation Using a Light Use Efficiency Model. *J. Geogr. Sci.* **2023**, *33*, 961–979. [CrossRef]
65. Wu, D.; Zhao, X.; Liang, S.; Zhou, T.; Huang, K.; Tang, B.; Zhao, W. Time-Lag Effects of Global Vegetation Responses to Climate Change. *Glob. Change Biol.* **2015**, *21*, 3520–3531. [CrossRef] [PubMed]
66. Cao, M.; Liang, Y.; Zhu, Y.; Lü, G.; Ma, Z. Prediction for Origin-Destination Distribution of Dockless Shared Bicycles: A Case Study in Nanjing City. *Front. Public Health* **2022**, *10*, 849766. [CrossRef] [PubMed]
67. Greff, K.; Srivastava, R.K.; Koutnik, J.; Steunebrink, B.R.; Schmidhuber, J. LSTM: A Search Space Odyssey. *IEEE Trans. Neural. Netw. Learn. Syst.* **2017**, *28*, 2222–2232. [CrossRef] [PubMed]

68. Leuchner, M.; Hertel, C.; Rötzer, T.; Seifert, T.; Weigt, R.; Werner, H.; Menzel, A. Solar Radiation as a Driver for Growth and Competition in Forest Stands. In *Growth and Defence in Plants: Resource Allocation at Multiple Scales*; Matyssek, R., Schnyder, H., Oßwald, W., Ernst, D., Munch, J.C., Pretzsch, H., Eds.; Ecology Research; Springer: Berlin/Heidelberg, Germany, 2012; pp. 175–191, ISBN 978-3-642-30645-7.
69. Reitz, O.; Bogena, H.; Neuwirth, B.; Sanchez-Azofeifa, A.; Graf, A.; Bates, J.; Leuchner, M. Environmental Drivers of Gross Primary Productivity and Light Use Efficiency of a Temperate Spruce Forest. *J. Geophys. Res. Biogeosci.* **2023**, *128*, e2022JG007197. [CrossRef]
70. Zhao, J.; Lange, H.; Meissner, H. Estimating Carbon Sink Strength of Norway Spruce Forests Using Machine Learning. *Forests* **2022**, *13*, 1721. [CrossRef]
71. Honkaniemi, J.; Rammer, W.; Seidl, R. Norway Spruce at the Trailing Edge: The Effect of Landscape Configuration and Composition on Climate Resilience. *Landsc. Ecol.* **2020**, *35*, 591–606. [CrossRef]
72. Aiba, S.; Kitayama, K. Light and Nutrient Limitations for Tree Growth on Young versus Old Soils in a Bornean Tropical Montane Forest. *J. Plant. Res.* **2020**, *133*, 665–679. [CrossRef]
73. Simon, D.-C.; Ameztegui, A. Modelling the Influence of Thinning Intensity and Frequency on the Future Provision of Ecosystem Services in Mediterranean Mountain Pine Forests. *Eur. J. For. Res.* **2023**, *142*, 521–535. [CrossRef]

forests

MDPI

Article

Paleo Distribution and Habitat Risks under Climate Change of *Helleborus thibetanus*

Xiaohua Shi [1,*], Lihui Mao [1], Miao Sun [2], Guangying Ma [1] and Kaiyuan Zhu [1]

1 Zhejiang Academy of Agricultural Sciences, Zhejiang Institute of Landscape Plants and Flowers, Hangzhou 310000, China
2 College of Landscape Architecture, Beijing Forestry University, Beijing Key Laboratory of Ornamental Plants Germplasm Innovation and Molecular Breeding, National Engineering Research Center for Floriculture, Beijing 100083, China
* Correspondence: shxh2021@aliyun.com

Abstract: As an endemic species and the only *Helleborus* species in China, *Helleborus thibetanus* is highly valued in medicinal and ornamental applications, and basic research is needed for its further resource conservation and utilization. Considering the interesting disjunct distribution of the genus *Helleborus*, we focus on the distribution pattern of *H. thibetanus* in this research. Based on species distribution models using three different algorithms (MaxEnt, RF, and FDA), we constructed a robust ensemble model and predicted potential distributions under different scenarios: current situation, paleo periods since the Last Glacial Maximum, and simulations of climate change in the 2070s. The habitat suitability of *H. thibetanus* across geography and scenarios was further analyzed by calculating regional areas and centroids. The results showed that *H. thibetanus* is currently distributed in southern Shaanxi and northern Sichuan, while central and southern Sichuan used to be suitable 14 thousand years ago but gradually became unsuitable, which may reflect the population decrease in Sichuan and the population expansion in Shaanxi over the last 14 thousand years. Our results showed that current populations are under limited extinction pressure in the soft climate change scenario (ssp126), but most populations in Shaanxi are under extinction pressure in the hardy situation scenario (ssp585). Fortunately, northern Sichuan is predicted to be relatively stable under climate change (both ssp126 and ssp585), and regions in western Sichuan and eastern Qinghai are predicted to become newly suitable for *H. thibetanus*. These findings should be helpful for the further conservation and utilization of *H. thibetanus* and also help us understand the history of the conjunct distribution pattern of the *Helleborus* genus.

Keywords: *Helleborus thibetanus*; species distribution; climate change; bioclimatic variables

Citation: Shi, X.; Mao, L.; Sun, M.; Ma, G.; Zhu, K. Paleo Distribution and Habitat Risks under Climate Change of *Helleborus thibetanus*. *Forests* **2023**, *14*, 630. https://doi.org/10.3390/f14030630

Academic Editors: Daniela Dalmonech, Alessio Collalti and Gina Marano

Received: 8 February 2023
Revised: 12 March 2023
Accepted: 15 March 2023
Published: 20 March 2023

1. Introduction

Belonging to Ranunculaceae, *Helleborus* comprises about 22 species in Eurasia [1]. However, all species are located in Europe, except a single species (*H. thibetanus*) found in Asia [1]. This disjunct and uneven diversity pattern suggests a complex evolutionary history of *Helleborus*, which may connect to historical climate change, geological events, and dispersal limitations. This evolution is ongoing [2]. Questions may arise, including why a single species is left in Asia and what historical process explains its current distribution pattern. Such questions are valuable in revealing the full evolutionary history of *Helleborus* and other genera with disjunct distribution patterns. This history suggests that *H. thibetanus* holds great value in basic scientific research [2]. Meanwhile, *H. thibetanus* has a long tradition in ethnomedicinal and ethnoveterinary applications [3]. Modern pharmacological studies have suggested that they possess antitumor, antibacterial, immune-regulation, and cytotoxic properties, and many components have been disclosed [4,5]. In addition to its extensive medicinal usage throughout history, *H. thibetanus* is widely used in the ornamental industry, given its complex flower structure and long fluorescence [6,7]. The frequent and disordered introduction of natural germplasms due to their value in medical and ornamental applications, along with other pressures such as habitat fragmentation and global climate change, is creating extinction pressure on *H. thibetanus* [7,8]. Therefore, it is important to perform distribution modeling on *H. thibetanus* to better understand its distribution patterns and its evolutionary history, as well as to aid in conservation efforts.

Current species distribution patterns can depend on a variety of factors, including environmental influences, historical influences, and competition between species [2]. Among them, climate factors often play important roles, as they can affect the growth, development, and survival of plants in a variety of ways. Warmer temperatures often lead to increased growth rates, while colder temperatures can lead to decreased growth rates or the death of the plant [9]. In most cases, plants remain in balance with local climate conditions. However, extreme temperatures and changes in weather patterns due to climate change can lead to increased stress on plants, resulting in decreased growth and susceptibility to disease [10]. To respond to dramatic changes in climate, plants can shift their distributions to follow changing environments, shift their physiology to adapt to changing conditions while remaining in place, remain in isolated pockets of the unchanged environment (climate refugia), or, more often, become extinct [11]. Considering dramatic climate fluctuations in the Quaternary, the current distribution pattern of the *Helleborus* species should result from a series of events, including range shifts and species extinction [11]. Resolving the Quaternary biogeographic history of *Helleborus* is important for fundamental ecological and evolutionary science, and it is also crucial for addressing applied scientific questions about species' responses to contemporary climate change.

Species distribution models (SDMs) are often used to understand environmental relationships and predict species distributions in both environmental and geographic spaces. SDMs are an important tool for conservation planners, ecologists, and natural resource managers [12]. SDM methods span from purely correlative methods (i.e., statistical assessments of relationships between species presence and a set of environmental variables) to purely process-based methods (i.e., explicit ecological relationships between environmental conditions and organism performance) [12]. Considering the limitation of demanding parameterization in process-based approaches, correlation-based SDMs have been widely applied because of their simplicity and the accessibility of software such as Maxent [13]. Meanwhile, different methods may have their own strengths and weaknesses. Ensemble modeling is reported to produce a more accurate and reliable prediction by combining the results of multiple individual models [14]. One of the most popular platforms for ensemble modeling is BIOMOD [15]. With the rapid development of Species Distribution Modeling (SDM) theory and technology, BIOMOD has become a powerful tool in several applications, particularly in ecology and conservation biology [12]. One of the main functions of SDM is to help us understand the niche of specific species. By modeling the relationships between species distribution and environmental variables, SDM can provide insights into the fundamental niche of a species or the set of environmental conditions under which it can persist [16]. Understanding the niche of a species can be crucial for conservation efforts, particularly in managing and restoring degraded habitats or identifying potential areas for reintroduction [16]. Another function of SDM is to identify suitable habitats for wildlife and assist in conservation efforts [12]. By projecting potential distribution across regions and periods, SDM can identify areas where species may be at risk of habitat loss or fragmentation [7]. This information can be used to prioritize areas for conservation and restoration efforts, as well as to inform land-use planning and decision-making [17]. Meanwhile, the utilization of SDM can facilitate the identification of regions that have undergone less pressure in the face of both past and present climatic changes, referred to as refugia [11,18]. The reconstruction of glacial refugial history has greatly contributed to our comprehension of the mechanisms underlying species' responses to environmental changes, which can, in turn, provide valuable insights into their evolutionary capacity [19]. Furthermore, predictions of future refugia under conditions of climate change hold significant implications for wildlife conservation and reintroduction efforts [18,20].

The purpose of this study is to predict the distribution dynamics of *H. thibetanus* from the Last Glacial Maximum to the near future under climate change, which is helpful for its conservation and utilization. To achieve this objective, (1) we conducted a systematic review of occurrences of *H. thibetanus* and constructed an ensemble species model based on three different algorithms. (2) We detected the proximal climatic variables that affected the distribution of *H. thibetanus* and explored their optimal limits. (3) We explored the climatically suitable habitats under current climate conditions and projected the paleo-distribution of *H. thibetanus* after the Last Glacial Maximum. (4) We predicted the future potential distribution of *H. thibetanus* under different climate change scenarios and evaluated the extinction risks among current distribution regions.

2. Materials and Methods

2.1. Occurrence Data and Distribution Range

A total of 302 specimen records of *H. thibetanus* were collected from 3 Chinese specimen platforms of plants, including the Chinese Virtual Herbarium (www.cvh.ac.cn, accessed on 31 July 2022), the Chinese Teaching Specimens Platform (http://mnh.scu.edu.cn/, accessed on 31 July 2022), and the China Nature Reserve Specimen Resources (www.papc.cn, accessed on 31 July 2022). Records without images or with uncertain locality were filtered, and then each record was identified manually to filter for those with possible taxonomy mistakes [1]. A total of 104 clean records were generated. Considering the possible bias caused by spatial autocorrelation [21,22], we thinned our occurrence records using the thin function of R package spThin [23], which generated a thinned dataset through a randomization algorithm and ensured distances among all records were above a minimum distance (10 km in this research). Finally, a total of 66 occurrences records were subjected to further analysis.

A distribution range was generated to sample pseudoabsence or background records. Firstly, the mean pairwise geographical distance of 66 thinned occurrences records was calculated by sp packages [24] in R. Then, buffers of each thinned occurrence record with a radium of the mean distance were generated and merged by raster package [25], which was treated as the current possible distribution range of *H. thibetanus*.

2.2. Environmental Data and Correlation Analysis

Considering our object of tracing dynamic distribution history and estimating future distribution loss risks under climatic change, a total of 19 bioclimatic data [26] (Table 1) were initially selected and downloaded from the Chelsa Climatic Database, Version 2, accessed on 31 July 2022 (the download path is listed in Supplementary File S1). Data from the period of 1981~2010 was used to construct species distribution models, in which most specimens were recorded. To estimate the risks under climate change, 2 shared socioeconomic pathways (ssp) scenarios (ssp126 and ssp585) from CMIP6 (https://esgf-node.llnl.gov/projects/cmip6/, accessed on 31 July 2022) were selected, which represent two contrasting scenarios of future greenhouse gas emissions, with ssp126 assuming a rapid and sustained reduction in emissions and ssp585 assuming a continued increase in emissions throughout the 21st century [27]. Specifically, datasets of 3 global climatic models (GFDL-ESM4, PSL-CM6A-LR, and MRI-ESM2-0, the download path is listed in Supplementary File S1) in the period of 2071~2100 under each scenario were downloaded to balance the possible bias generated by global climatic models. To trace historical distribution dynamics, modeled bioclimatic data of 6 historical periods (BC22K, BC18K, BC14K, BC10K, BC6K, and BC2K), ranging from 22 thousand years before the current period (BC22K) to 2 thousand years before the current period (BC2K), were also downloaded from the Chelsa Climatic Database (https://chelsa-climate.org/chelsa-trace21k/, accessed on 31 July 2022) [28]. All of the climatic data above are downloaded with a resolution of 30 arcsecs.

Table 1. Bioclimatic variables and importance in modeling.

ID	Bioclimatic Variables	Unit	Mean Importance	Standard Deviation
bio01	Mean temperature	°C	-	-
bio02	Diurnal air temperature range	°C	-	-
bio03	Isothermality (bio02/bio07 ×100)	\	2.20	0.08
bio04	Temperature seasonality (standard deviation ×100)	°C	15.06	0.40
bio05	Max temperature of warm month	°C	-	-
bio06	Min temperature of cold month	°C	-	-
bio07	Annual temperature range	°C	-	-
bio08	Mean temperature of wet quarter	°C	23.41	0.60
bio09	Mean temperature of dry quarter	°C	-	-
bio10	Mean temperature of warm quarter	°C	-	-
bio11	Mean temperature of cold quarter	°C	-	-
bio12	Annual precipitation	mm	-	-
bio13	Precipitation of wet month	mm	-	-
bio14	Precipitation of dry month	mm	-	-
bio15	Precipitation seasonality (coefficient of variation)	\	11.56	0.37
bio16	Precipitation of wet quarter	mm	-	-
bio17	Precipitation of dry quarter	mm	-	-
bio18	Precipitation of warm quarter	mm	14.68	0.46
bio19	Precipitation of cold quarter	mm	33.09	0.62

The function vifstep of R package usdm [29] was applied to reduce the multicollinearity between bioclimatic variables, in which the maximum number of observations within the training area was set to 10,000, and the thresholds of the correlation coefficients and vif were set to 0.8 and 5. Consequently, 6 bioclimatic variables (Table 1) were selected for further analysis.

2.3. Species Distribution Model Tuning and Construction

We applied 3 different algorithms to the following modeling, including maximum entropy (MaxEnt) [13], random forests (RF) [30], and flexible discriminant analysis (FDA) [31]. All were implemented in the R package of biomod2 (v4.2.3) [15]. The 70 thinned occurrence records and 6 bioclimatic variables obtained above were subject to tuning and constructing species distribution models. The best parameters of each algorithm were tuned through the function of BIOMOD_Tuning, which ran numerous models with different parameters and evaluated each with True Skill Statistics (TSS). In the process of model construction, 3 sets of pseudoabsence records were sampled from the distribution range, and 10 duplicate models were constructed for each data set and each algorithm; that is to say, we generated 90 single models total. In every single model, 70 percent of the input data were used to calibrate the model, while the remaining 30 percent was used to evaluate the model. Other model parameters were set according to the result of model tuning [12,32]. To balance the possible bias produced by a single run or algorithm, we only maintained models with a TSS value above 0.8 and then ensembled left models through the weight of the TSS value with the prob.mean.weight option [32]. The importance of each bioclimatic variable in modeling was calculated based on the ensembled model through the bm_VariablesImportance function in BIOMOD2 with a replicate time of 30 and summarized by mean value and standard deviation value, and then scaled to ensure the sum of the mean values is 1. The responsive curve of the proximal bioclimatic variables was plotted through the bm_plotresponcse function based on the training data [33] and the MaxEnt algorithm.

2.4. Potential Distribution Prediction and Geographical Analysis

The potential distribution under different scenarios of *H. thibethanus* was projected using the ensembled model through the BIOMOD_Forecasting function in the Biomod2 package. Firstly, we predicted the current potential distribution of *H. thibetanus* with the bioclimatic training data. For an exploratory analysis of the historical distribution dynamic of *H. thibethanus*, we projected its potential distribution in 6 historical periods (BC22K, BC18K, BC14K, BC10K, BC6K, and BC2K). Finally, projections under future scenarios were made for the purpose of evaluating its habitat risk under climate change.

Values below the threshold maximizing the TSS were masked by NA in each prediction result, and then a fitness map under each scenario was plotted. To further compare geographical changes of potential distributions under different scenarios, continuous probability maps were transformed into binary maps through a threshold that maximized the TSS. Regional areas and distribution centroids were then calculated with the help of the SDMTools package [34]. All of the figures in this research were plotted using QGIS (www.qgis.org, version 2.6.2) and the ggplot2 package [35].

3. Results

3.1. Single Model Accuracy and Ensembled Models

Single model accuracy based on AUC and TSS is illustrated in Figure 1, which refers to MaxEnt obtaining the highest mean AUC (0.944) and TSS (0.810). Most single models were shown to be high quality, which indicates that the model's prediction effect was excellent and that the prediction results had high accuracy and reliability. To further ensure high accuracy and reliability, we only maintained single models with TSS > 0.8 and ensembled them for further prediction analysis.

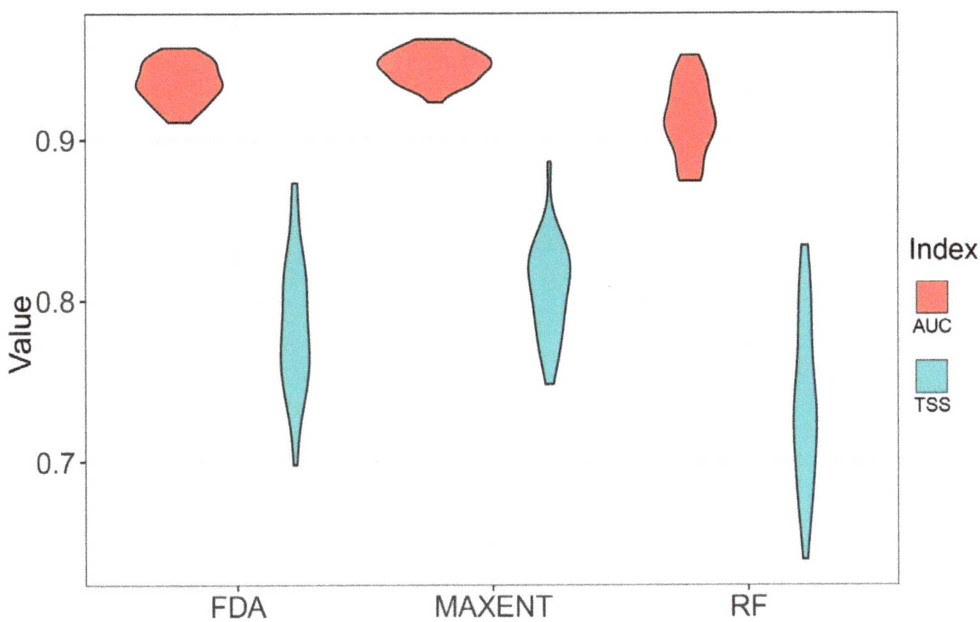

Figure 1. Model accuracy of single models applying three different algorithms.

3.2. Climatic Niche and Proximal Variables

The importance values of the six modeled climatic variables are listed in Table 1. As the table shows, three bioclimatic variables (bio19, bio08, bio04, and bio18) contributed the most to the prediction model. Bio19 had the highest importance (33.09%) in the prediction model, while bio03 had the least importance (2.20%) in the prediction model.

The bioclimatic niche of *H. thibetanus* can be described using a suitable range of multiple climatic variables, which can be displayed using the model's response curve. As

Figure 2 shows, most proximal bioclimatic variables (bio19, bio08, and bio04) are single-peak curves. Specifically, the most suitable range (distribution probability above 0.5) of bio19 (precipitation of the coldest quarter) is between 5.48 and 35.32 mm, and the peak at 20.40 mm. The most suitable range of bio08 (mean temperature of the wettest quarter) is between 13.59 and 19.96 °C, and the peak at 16.62 °C. The most suitable range of bio04 (temperature seasonality) is between 7.23 and 8.65 °C, while the peak is 7.94 °C. The response curve of bio18 (precipitation of warmest quarter) is not a peaked curve, which reached a distribution probability of 0.5 at 452.70 mm, followed by a maximum probability of 0.53 at 768.70 mm, and then remained stable as precipitation rose.

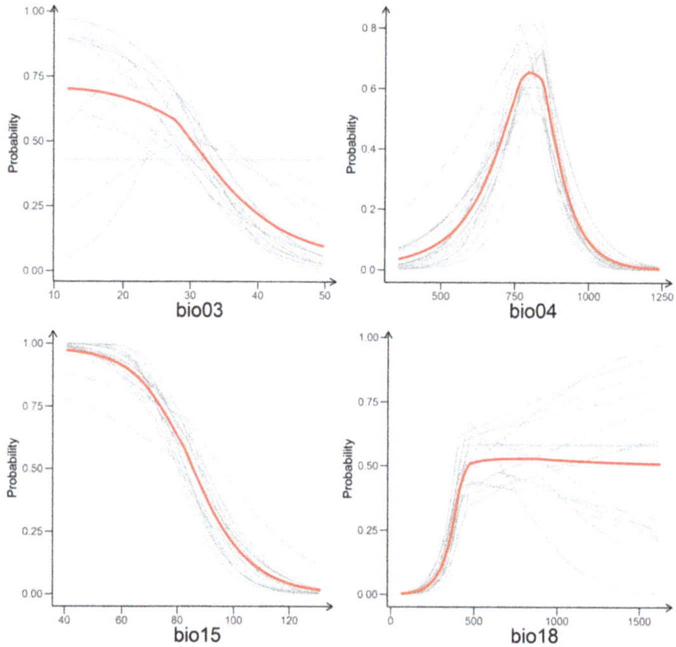

Figure 2. Response curves of 6 bioclimatic variables. We choose 18 MaxEnt models (with TSS > 0.8) and the ensemble model based on the 18 individual models to generate variable curves. The curve revealed by each individual model was drawn in grey, while a curve revealed by the ensemble model was drawn in red.

3.3. Current Potential Distribution of *H. thibetanus*

The current potential distribution of *H. thibetanus* was predicted using ensembled species distribution models. As Figure 3 shows, *H. thibetanus* is mainly located in three provinces (Shaanxi, Gansu, and Sichuan), although fragments can be found in some provinces (Xizang, Shanxi, Hubei, and Henan) that are also predicted to be suitable. The most highly suitable regions (which have higher prediction values) are located in Southern Gansu, Southern Shaanxi, and Northern Sichuan, corresponding to the Qinling mountains, Daba mountains, and northern Hengduan mountains.

According to the binary results, the total area of the current potential distribution of *H. thibetanus* is 287,556 km^2. Sichuan had the largest area (85,678 km^2) of suitable habitats for *H. thibetanus*, followed by Shaanxi (84,828 km^2) and Gansu (72,488 km^2). These three provinces accounted for a cumulative 84.5% of all suitable regions.

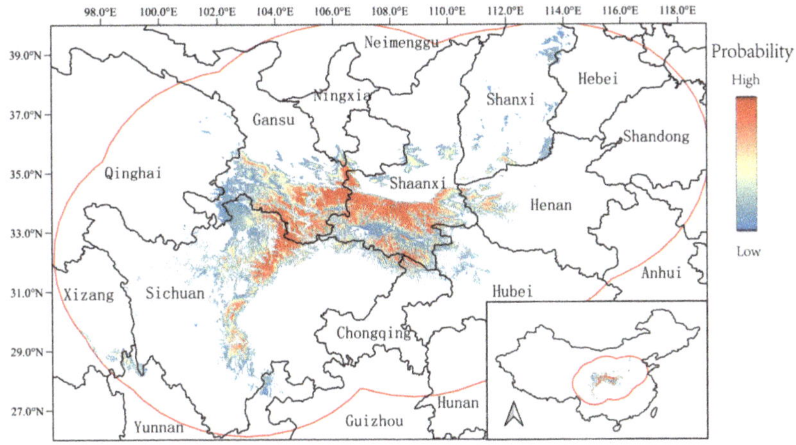

Figure 3. The current potential distribution of *H. thibetanus*.

3.4. Paleo Potential Distribution and Historical Dynamics

The paleo potential distribution of *H. thibetanus* (Figure 4) was predicted based on paleo bioclimatic data. It predicted a wider potential distribution of *H. thibetanus* in the period BC22K (22 thousand years before the current year) than the current scenario, which covers most regions in southern Shaanxi and northern Sichuan and extends to the southern Hengduan mountains. Four thousand years later, the predicted potential distribution in BC18K is mainly located in southern Sichuan, and most current potential regions in Shaanxi were predicted to have low suitability. This trend persisted at least until BC10K, and it was predicted that nearly all current potential regions in Shaanxi were unsuitable for *H. thibetanus* between BC14K and BC10K. In the next 10 thousand years, the suitability of Sichuan would decrease, while the suitability of Shaanxi and Gansu would increase, resulting in the current potential distribution pattern of *H. thibetanus*.

The areas of the three types of suitable regions and the centroid of potential distribution were calculated. As shown in Figure 5, the area of the suitable regions in BC22K is 256,836 km^2, and the following area dynamics can be divided into two periods. In the period of BC22K~BC14K, the area decreases until it reaches 121,549 km^2. In the period of BC14K~today, the area increases. The spatial dynamics of the paleo potential distribution are shown by the movement of the centroid (Figure 6). As the results show, the centroid in BC22K and BC2K was near the current centroid, while two significant movements occurred between BC22K and BC2K. The centroid moves southeastwards between BC22K and BC14K and then northeastern between BC14K and BC2K. The most southern centroid (with a latitude value of 29.29 degrees) is found in BC14K, while the most western centroid (with a longitude value of 103.30 degrees) is also found in BC14K.

3.5. Future Potential Distribution and Ranges under Risks

Two climate change scenarios were selected to predict the future potential distribution of *H. thibetanus* and evaluate the risks of habitats disappearing. As Figure 7 shows, a weaker suitable region change is found in the moderate climate change scenario (ssp126), *H. thibetanus* is predicted to lose a limited number of regions in eastern and southern Shaanxi and also some fragments in southern Sichuan. On the other hand, *H. thibetanus* is predicted to expand its current distribution westwards and cover wide regions in eastern Qinghai and western Sichuan. A greater suitable region change is predicted in the high climate change scenario (ssp585). *H. thibetanus* is predicted to lose nearly all suitable regions in Shaanxi in ssp585, but gain new suitable regions in eastern Qinghai and western Sichuan. Overall, the area of the suitable region is predicted to increase in both scenarios (by 63.6 in ssp126 and 74.2% in ssp585) (Figure 5). The centroids of suitable regions in both scenarios are predicted to move westwards (Figure 6) while they reach a longitude value of 103.4 in ssp126 and 101.0 in ssp585.

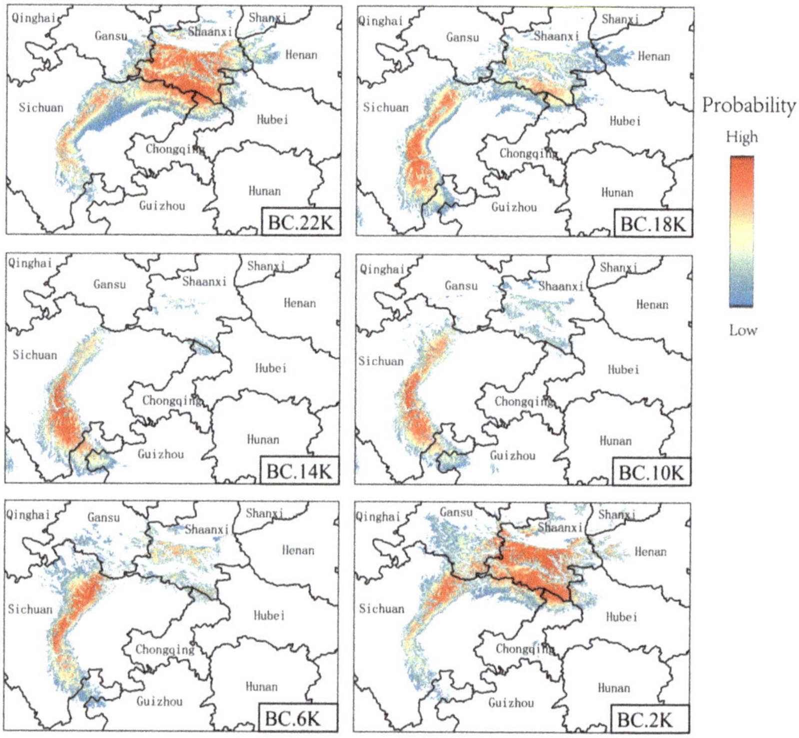

Figure 4. Paleo potential distribution of *H. thibetanus*.

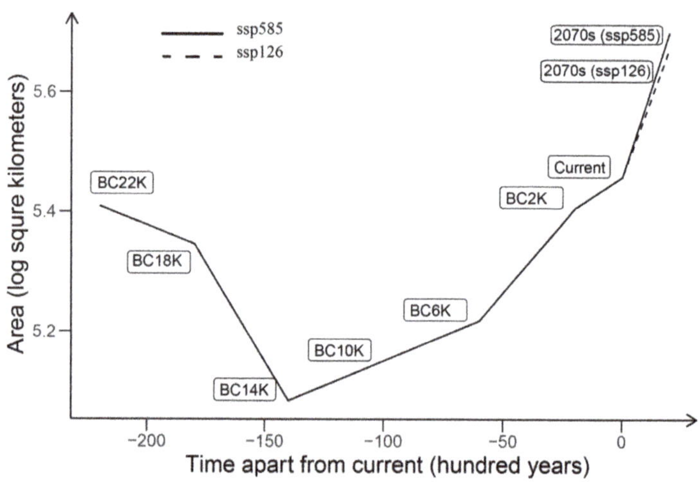

Figure 5. Area of different regions since the Last Glacial Maximum to the 2070s.

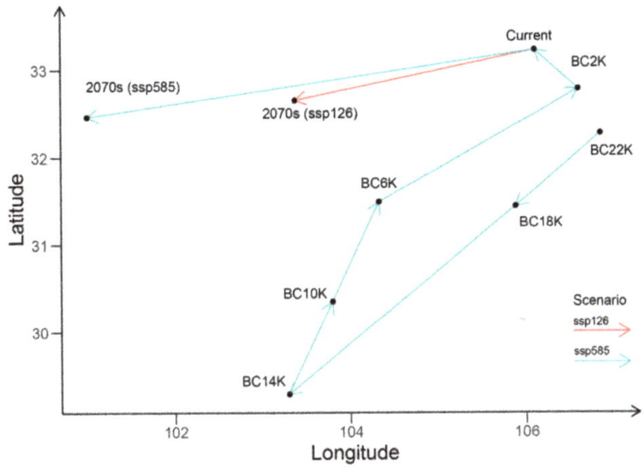

Figure 6. Distribution centroids of *H. thibetanus* from the Last Glacial Maximum to the 2070s.

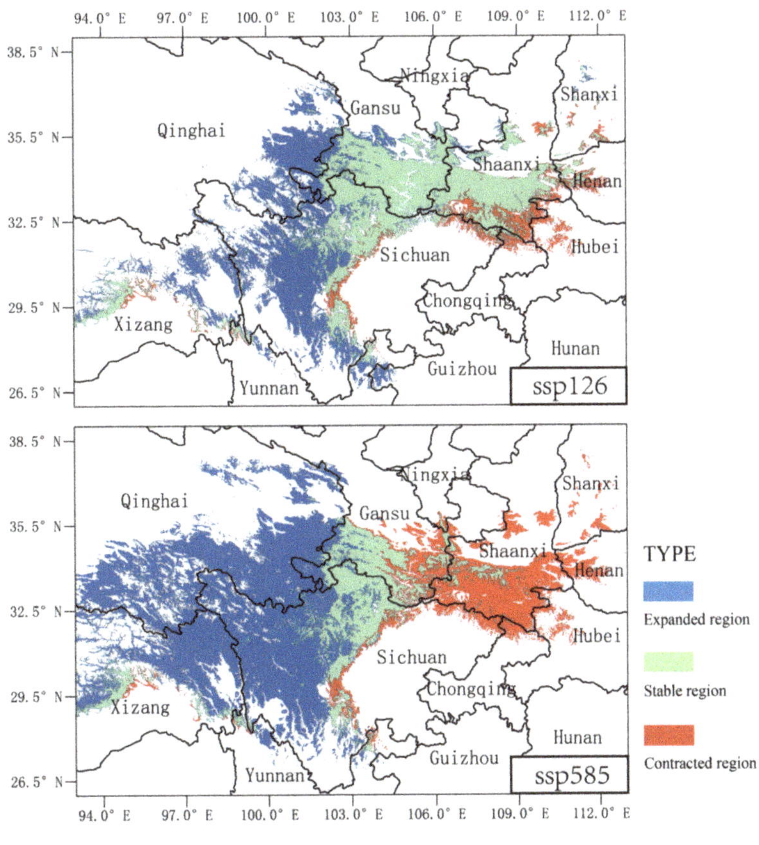

Figure 7. Changes of the potential distribution of *H. thibetanus* in the 2070s in two scenarios (ssp126 and ssp585).

4. Discussion

4.1. Climatic Niche and Proximal Variables

As a temperate perennial plant, *H. thibetanus* is reported to prefer cold, semi-shady and humid environments [6]. To describe the climatic niche of *H. thibetanus* quantitatively, six bioclimatic variables were selected and modeled. Given the high quality index (AUC and TSS) value and the consistency between a predicted potential distribution and experienced knowledge, the species distribution model of *H. thibetanus* is proved to be accurate and robust. According to the species distribution model, the total contribution rate related to temperature (bio03 and bio04) is 17.26%, the rate related to precipitation (bio15) is 11.56%, and the rate related to both temperature and precipitation (bio08, bio18, and bio19) is 71.18%. Herein, temperature contributes more compared to precipitation, although both variables are important for delimitating the distribution of *H. thibetanus*. Temperature is a key factor that influences plants' growth, development, and reproduction [36]. It is reported that *H. thibetanus* has strong adaptability to low temperatures and can flower in snowy weather [1,6]. However, *H. thibetanus* seems to have limited adaptability to hot temperatures, especially when accompanied by a wet environment (bio08). The optimal limit was 13.59~19.96 °C. Another important temperate restriction is temperature seasonality, which corresponds to demands in the process of seed dormancy and germination of *H. thibetanus* [1,6]. The optimal limit of temperature seasonality is 7.23~8.65 °C, which reflects a moderate temperature change and a limited dormancy in *H. thibetanus* seeds. Precipitation is another important factor for plant life, which is also correlated with many environmental factors influencing the physiological and biochemical processes of plants. For example, soil moisture is the main factor affecting the plant assimilation rate and root breath [36]. As a perennial plant that is dormant during the winter, *H. thibetanus* requires low precipitation in the winter, which reflects a low optimal limit of precipitation in the coldest quarter of 5.48~35.32 mm. On the other hand, *H. thibetanus* has grassy leaves and requires high air moisture during its growth periods [1], which reflects a high optimal limit of precipitation in the warmest quarter above 452 mm. Although both precipitation variables are important in delimitating the distribution of *H. thibetanus*, low precipitation during dormancy periods (33.09%) is more important than high precipitation during growth periods (14.68%). Overall, our research indicates that the species distribution model is powerful for helping us understand the niche of research objects. Besides bioclimatic variables, other environmental variables (including soil characteristics, land cover, and biological variables) [12,32] may also influence the distribution of *H. thibetanus*; thus, further research that considers more environmental variables may provide a more comprehensive understanding of the niche of *H. thibetanus*. Meanwhile, the current distribution pattern of *H. thibetanus* may not fully be conducted by the species niche. Geographical events and anthropogenic disturbances may also play significant roles, and species distribution models concerning these parameters will produce more details.

4.2. The Current Distribution Range and Climate Refugia

Global climate has fluctuated greatly during the past three million years, leading to the recent major ice ages [10,19]. An inescapable consequence for most living organisms is the great changes in their distribution, which are expressed differently in different zones and among different taxon [19]. To understand species' biogeographic histories, we modeled the potential distribution of *H. thibetanus* from the period of the Last Glacial Maximum to the current period, and the distribution area and centroids were compared in this research. The results show that current distribution regions of *H. thibetanus* are also suitable in the period of LGM, reflecting a wide paleodistribution. In the following 22 thousand years, climate fluctuations resulted in a round-trip movement of the distribution centroids of *H. thibetanus*. In particular, the distribution area in Shaanxi decreases and then increases, while the area in Sichuan increases and decreases. The extreme situation occurred near the maximum of the Bølling-Allerød warming (BC 14K), in which nearly all suitable regions in Shaanxi disappeared, while the area of the suitable region in Sichuan reached a peak.

Being a slow migrator, another *Helleborus* species (*H. niger*) is proven to survive the LGM in local refugia [8]. In our research, it is hard to prove that the current distribution area in Shaanxi is the result of postglacial translation and colonization from Sichuan, as complex terrain in the Qingling mountains may also provide local climate refugia for *H. thibetanus* to survive. On the other hand, current population dynamics seem clear, as the results reflect. The population located in Sichuan contracted over the last 14 thousand years, which is consistent with the fact that limited accessions of wild *H. thibetanus* were recorded in central and southern Sichuan, and most southern accession (a latitude of 30.6°) located in the Dengchi valley of Baoxing was recorded in 1954 (PE00428129). Meanwhile, populations in Shaanxi grew, which is consistent with the growing specimen records in Shaanxi and Gansu in recent years. Resolving biogeographic history is important for the fundamental ecological and evolutionary science of *H. thibetanus*, but a single method is limited to addressing a full story; thus, more methods (including phylogeographic and demographic methods) are needed in further research [11].

4.3. Habitat Risks under Climate Change

It has been shown that global climate change has led to a continuous rise in global temperature, precipitation mode (time and space), and precipitation intensity, which is threatening the habitats of plants and animals [22,37,38]. So far, according to our predictions, *H. thibetanus* revealed a wide area (287,556 km^2) of current potential distribution, which is much higher than the threshold (20,000 km^2) of the vulnerable (VU) conservation status. This indicates that *H. thibetanus* has a relatively low risk of extinction in the near future when neglecting climatic change influences. To identify the habitat changes of *H. thibetanus* as the global climate changes, potential distributions in two different scenarios were modeled in this research. All of the changes can be divided into three catalogs: expanded, stable, and contracted. Stable habitats refer to standing suitable regions in climate change for *H. thibetanus* and could be considered basic preserved stations; such regions include the highlands of southern Shaanxi and most regions of northern Sichuan in both scenarios. Contracted habitats refer to threatened regions of the current distribution of *H. thibetanus*. In ssp126, contracted habitats are located in the southern and eastern edges of current distribution regions, but it covers most regions in Shaanxi in ssp585. Thus, preservation actions are needed in all southern and eastern edges of current distribution regions, while continuous monitoring is needed for populations in most of Shaanxi. Specifically, populations in the Daba mountains are under higher climate pressure than those in the Qingling mountains, although both may be extinct in the hardy climate change scenario (ssp585). A similar situation is also reported in another *Helleborus* species (*H. odorus* subsp. *cyclophyllus*), which is projected to lose a significant portion of its current distribution by 2070 [7]. In our research, wide expanded regions are also predicted. Though naturally expanded areas should be restricted given the limited ability of dispersal of *H. thibetanus* [39], those regions could be optional for ex situ preservation. Such regions are located west of the current distribution and cover regions in northwestern Sichuan and eastern Qinghai. However, considering the limited dispersal ability of *H. thibetanus*, expanded suitable habitat may not be available for natural populations. In that case, *H. thibetanus* will lose 18.1% and 53.9% of its current habitat in ssp126 and ssp585, respectively. Overall, global climate change significantly affects the potential distribution of *H. thibetanus* even in a moderate change scenario, and preservation is urgent. On the other hand, a more comprehensive understanding of habitat risks calls for research on issues including genetic diversity [7], local adaptation [40], and plasticity [41], which will also improve the preservation and utilization of *H. thibetanus*.

5. Conclusions

As the single *Helleborus* species in Asia and an endemic species to China, *H. thibetanus* is in urgent need of further basic research in order to understand its evolution and ecological characteristics. Using three different algorithms, we constructed an ensemble model based on current bioclimatic data and occurrence records. We described the bioclimate niche of *H. thibetanus,* including properties like moderate temperature and wet precipitation in growth periods and dry precipitation in dormancy periods. Based on the ensembled species distribution model, we projected a map of the current potential distribution of *H. thibetanus* in high resolution, which displays continuous distribution in northern Sichuan and southern Shaanxi. With the help of paleo-bioclimatic data, we projected the potential distribution of *H. thibetanus* since the Last Glacial Maximum, and the result shows a round-trip movement of the distribution centroids of *H. thibetanus* corresponding to the climate fluctuation. Furthermore, we predicted the potential distribution of *H. thibetanus* in the 2070s under two different climate change scenarios, and the result shows that the current distribution region will be threatened by global climate change, but regions located west of the current distribution are predicted to be suitable for *H. thibetanus* under climate change. Our results provide an important scientific basis for the conservation, introduction, and utilization of *H. thibetanus* in China.

Supplementary Materials: The following supporting information can be downloaded at: https://www.mdpi.com/article/10.3390/f14030630/s1, File S1: Download Path of Climatic Data.

Author Contributions: X.S. conceived and designed the experiments; L.M., M.S. and G.M. performed the experiments and analyzed the data; X.S., M.S. and K.Z. wrote the paper. All authors have read and agreed to the published version of the manuscript.

Funding: Research and demonstration of key technologies of new varieties breeding and industrialization of *Helleborus* under the scientific and technological cooperation plan of "three agriculture and nine parties" in Zhejiang Province(2023SNJF032).

Data Availability Statement: Not applicable.

Conflicts of Interest: None of the authors have any actual or potential conflicts of interest that could inappropriately influence, or be perceived to influence, this work.

References

1. Li, L.; Michio, T. (Eds.) *Flora of China*; Missouri Botanical Garden Press: Saint Louis, MO, USA; Science Press: Beijing, China, 2013; Volume 6, p. 148.
2. Song, H.; Ordonez, A.; Svenning, J.; Qian, H.; Yin, X.; Mao, L.; Deng, T.; Zhang, J. Regional disparity in extinction risk: Comparison of disjunct plant genera between eastern Asia and eastern North America. *Glob. Change Biol.* **2021**, *27*, 1904–1914. [CrossRef] [PubMed]
3. Balázs, V.L.; Filep, R.; Ambrus, T.; Kocsis, M.; Farkas, Á.; Stranczinger, S.; Papp, N. Ethnobotanical, historical and histological evaluation of *Helleborus* L. genetic resources used in veterinary and human ethnomedicine. *Genet. Resour. Crop Evol.* **2020**, *67*, 781–797. [CrossRef]
4. Song, X.; Li, Y.; Zhang, Z.; Huang, W.; Zhang, H.; Jiang, Y.; Liu, J.; Zhang, D. Two New Spirostanol Glycosides from the Roots and Rhizomes of *Helleborus thibetanus* Franch. *Rec. Nat. Prod.* **2023**, *17*, 318–328. [CrossRef]
5. Li, Y.; Zhang, H.; Liang, X.; Song, B.; Zheng, X.; Wang, R.; Liu, L.; Song, X.; Liu, J. New cytotoxic bufadienolides from the roots and rhizomes of *Helleborus thibetanus* Franch. *Nat. Prod. Res.* **2020**, *34*, 950–957. [CrossRef]
6. Deng, W.; Zhao, L.; Shi, X.; Du, L. Progress in the development of germplasme resource in genus *Helleborus*. *Xiandai Hortic.* **2022**, *45*, 1–3. (In Chinese)
7. Fassou, G.; Kougioumoutzis, K.; Iatrou, G.; Trigas, P.; Papasotiropoulos, V. Genetic Diversity and Range Dynamics of *Helleborus odorus* subsp. *cyclophyllus* under Different Climate Change Scenarios. *Forests* **2020**, *11*, 620. [CrossRef]
8. Záveská, E.; Kirschner, P.; Frajman, B.; Wessely, J.; Willner, W.; Gattringer, A.; Hülber, K.; Lazić, D.; Dobeš, C.; Schönswetter, P. Evidence for Glacial Refugia of the Forest Understorey Species *Helleborus niger* (Ranunculaceae) in the Southern as Well as in the Northern Limestone Alps. *Front. Plant Sci.* **2021**, *12*, 683043. [CrossRef]
9. Sparey, M.; Cox, P.; Williamson, M.S. Bioclimatic change as a function of global warming from CMIP6 climate projections. *Biogeosciences* **2023**, *20*, 451–488. [CrossRef]
10. Hoban, S.; Dawson, A.; Robinson, J.D.; Smith, A.B.; Strand, A.E. Inference of biogeographic history by formally integrating distinct lines of evidence: Genetic, environmental niche and fossil. *Ecography* **2019**, *42*, 1991–2011. [CrossRef]
11. Mestre, F.; Barbosa, S.; Garrido-García, J.A.; Pita, R.; Mira, A.; Alves, P.C.; Paupério, J.; Searle, J.B.; Beja, P. Inferring past refugia and range dynamics through the integration of fossil, niche modelling and genomic data. *J. Biogeogr.* **2022**, *49*, 2064–2076. [CrossRef]

12. Zurell, D.; Franklin, J.; König, C.; Bouchet, P.J.; Dormann, C.F.; Elith, J.; Fandos, G.; Feng, X.; Guillera-Arroita, G.; Guisan, A.; et al. A standard protocol for reporting species distribution models. *Ecography* **2020**, *43*, 1261–1277. [CrossRef]
13. Phillips, S.J.; Anderson, R.P.; Schapire, R.E. Schapire, Maximum entropy modeling of species geographic distributions. *Ecol. Model.* **2006**, *190*, 231–259. [CrossRef]
14. Hao, T.; Elith, J.; Guillera-Arroita, G.; Lahoz-Monfort, J.J. A review of evidence about use and performance of species distribution modelling ensembles like BIOMOD. *Divers. Distrib.* **2019**, *25*, 839–852. [CrossRef]
15. Thuiller, W.; Lafourcade, B.; Engler, R.; Araújo, M.B. BIOMOD—A platform for ensemble forecasting of species distributions. *Ecography* **2009**, *32*, 369–373. [CrossRef]
16. Smith, A.B.; Godsoe, W.; Rodriguez-Sanchez, F.; Wang, H.-H.; Warren, D. Niche Estimation Above and Below the Species Level. *Trends Ecol. Evol.* **2019**, *34*, 260–273. [CrossRef] [PubMed]
17. Zhang, X.; Ci, X.; Hu, J.; Bai, Y.; Thornhill, A.H.; Conran, J.G.; Li, J. Riparian areas as a conservation priority under climate change. *Sci. Total Environ.* **2023**, *858*, 159879. [CrossRef] [PubMed]
18. Brambilla, M.; Rubolini, D.; Appukuttan, O.; Calvi, G.; Karger, D.N.; Kmecl, P.; Mihelič, T.; Sattler, T.; Seaman, B.; Teufelbauer, N.; et al. Identifying climate refugia for high-elevation Alpine birds under current climate warming predictions. *Glob. Change Biol.* **2022**, *28*, 4276–4291. [CrossRef]
19. Roces-Díaz, J.V.; Jiménez-Alfaro, B.; Chytrý, M.; Díaz-Varela, E.R.; Álvarez-Álvarez, P. Glacial refugia and mid-Holocene expansion delineate the current distribution of *Castanea sativa* in Europe. *Palaeogeogr. Palaeoclimatol. Palaeoecol.* **2018**, *491*, 152–160. [CrossRef]
20. Tang, C.Q.; Matsui, T.; Ohashi, H.; Dong, Y.F.; Momohara, A.; Herrando-Moraira, S.; Qian, S.; Yang, Y.; Ohsawa, M.; Luu, H.T.; et al. Identifying long-term stable refugia for relict plant species in East Asia. *Nat. Commun.* **2018**, *9*, 4488. [CrossRef]
21. Duan, X.; Li, J.; Wu, S. MaxEnt Modeling to Estimate the Impact of Climate Factors on Distribution of *Pinus densiflora*. *Forests* **2022**, *13*, 402. [CrossRef]
22. Li, Y.; Shao, W.; Huang, S.; Zhang, Y.; Fang, H.; Jiang, J. Prediction of Suitable Habitats for Sapindus delavayi Based on the MaxEnt Model. *Forests* **2022**, *13*, 1611. [CrossRef]
23. Aiello-Lammens, M.E.; Boria, R.A.; Radosavljevic, A.; Vilela, B.; Anderson, R.P. spThin: An R package for spatial thinning of species occurrence records for use in ecological niche models. *Ecography* **2015**, *38*, 541–545. [CrossRef]
24. Edzer, J.; Roger, S.B. Classes and methods for spatial data in R. *R J.* **2005**, *5*, 9–13.
25. Robert, J.H. raster: Geographic Data Analysis and Modeling. *R Package Version* **2021**, *2*, 1–49.
26. Karger, D.N.; Conrad, O.; Böhner, J.; Kawohl, T.; Kreft, H.; Soria-Auza, R.W.; Zimmermann, N.E.; Linder, H.P.; Kessler, M. Climatologies at high resolution for the earth's land surface areas. *Sci. Data* **2017**, *4*, 170122. [CrossRef]
27. Castillo, A.E.; Peña, L.S.; Delgado, S.G. Trayectorias Socioeconómicas Compartidas (SSP): Nuevas maneras de comprender el cambio climático y social. *Estud. Demográficos Urbanos* **2017**, *32*, 669–693. [CrossRef]
28. Karger, D.N.; Nobis, M.P.; Normand, S.; Graham, C.H.; Zimmermann, N.E. CHELSA-TraCE21k v1.0. Downscaled transient temperature and precipitation data since the last glacial maximum. *Clim. Past Discuss.* **2021**, *1*, 1–27.
29. Naimi, B.; Hamm, N.A.; Groen, T.A.; Skidmore, A.K.; Toxopeus, A.G. Where is positional uncertainty a problem for species distribution modelling? *Ecography* **2014**, *37*, 191–203. [CrossRef]
30. Breiman, L. Random Forests. *Mach. Learn.* **2001**, *45*, 5–32. [CrossRef]
31. Trevor, H.; Robert, T.; Jerome, F. *The Elements of Statistical Learning: Data Mining, Inference, and Prediction*, 2nd ed.; Springer: Berlin/Heidelberg, Germany, 2016.
32. Sillero, N.; Arenas-Castro, S.; Enriquez-Urzelai, U.; Vale, C.G.; Sousa-Guedes, D.; Martínez-Freiría, F.; Real, R.; Barbosa, A. Want to model a species niche? A step-by-step guideline on correlative ecological niche modelling. *Ecol. Model.* **2021**, *456*, 109671. [CrossRef]
33. Elith, J.; Ferrier, S.; Huettmann, F.; Leathwick, J. The evaluation strip: A new and robust method for plotting predicted responses from species distribution models. *Ecol. Model.* **2005**, *186*, 280–289. [CrossRef]
34. Van Der Wal, J.; Falconi, L.; Januchowski, S.; Shoo, L.; Storlie, C. SDMTools: Species Distribution Modelling Tools: Tools for processing data associated with species distribution modelling exercises. *R Package Version* **2014**, *1*, 1–221.
35. Hadley, W. *ggplot2: Elegant Graphics for Data Analysis*; Springer: New York, NY, USA, 2016; p. 260.
36. Metzger, M.J.; Bunce, R.G.H.; Jongman, R.H.G.; Sayre, R.; Trabucco, A.; Zomer, R. A high-resolution bioclimate map of the world: A unifying framework for global biodiversity research and monitoring. *Glob. Ecol. Biogeogr.* **2013**, *22*, 630–638. [CrossRef]
37. Chen, Y.; Shan, X.; Ovando, D.; Yang, T.; Dai, F.; Jin, X. Predicting current and future global distribution of black rockfish (*Sebastes schlegelii*) under changing climate. *Ecol. Indic.* **2021**, *128*, 107799. [CrossRef]
38. Zhao, G.; Cui, X.; Sun, J.; Li, T.; Wang, Q.; Ye, X.; Fan, B. Analysis of the distribution pattern of Chinese Ziziphus jujuba under climate change based on optimized biomod2 and MaxEnt models. *Ecol. Indic.* **2021**, *132*, 108256. [CrossRef]
39. Zhu, Y.; Wang, D.; Codella, S.G. Seed re-dispersal of four myrmecochorous plants by a keystone ant in central China. *Ecol. Res.* **2017**, *32*, 387–393. [CrossRef]

40. Chen, Q.; Yin, Y.; Zhao, R.; Yang, Y.; Da Silva, J.A.T.; Yu, X. Incorporating Local Adaptation into Species Distribution Modeling of *Paeonia mairei*, an Endemic Plant to China. *Front. Plant Sci.* **2019**, *10*, 1717–1731. [CrossRef]
41. Benito Garzón, M.; Robson, T.M.; Hampe, A. ΔTraitSDMs: Species distribution models that account for local adaptation and phenotypic plasticity. *New Phytol.* **2019**, *222*, 1757–1765. [CrossRef]

forests

Article

Predicting Spruce Taiga Distribution in Northeast Asia Using Species Distribution Models: Glacial Refugia, Mid-Holocene Expansion and Future Predictions for Global Warming

Kirill Korznikov [1,2], Tatyana Petrenko [2], Dmitry Kislov [2], Pavel Krestov [2] and Jiří Doležal [1,3,*]

1 Department of Functional Ecology, Institute of Botany CAS, 379 01 Třeboň, Czech Republic
2 Department of Geobotany, Botanical Garden-Institute FEB RAS, 690024 Vladivostok, Russia
3 Department of Botany, Faculty of Science, University of South Bohemia, 370 05 České Budějovice, Czech Republic
* Correspondence: jiriddolezal@gmail.com

Abstract: Spruce taiga forests in Northeast Asia are of great economic and conservation importance. Continued climate warming may cause profound changes in their distribution. We use prognostic and retrospective species distribution models based on the Random Forest machine learning method to estimate the potential range change of the dominant taiga conifer Jezo spruce (*Picea jezoensis* (Siebold & Zucc.) Carrière) for the year 2070 climate warming scenarios and for past climate epochs–the Last Glacial Maximum (LGM) (~21,000 years before present) and the mid-Holocene Climatic Optimum (MHO) (~7000 years before the present) using the MIROC-ESM and CCSM4 climate models. The current suitable climatic conditions for *P. jezoensis* are estimated to be 500,000 km^2. Both climatic models show similar trends in past and future ranges but provide different quantitative areal estimates. During the LGM, the main part of the species range was located much further south than today at 35–45° N. Projected climate warming will cause a greater change in the distributional range of *P. jezoensis* than has occurred since the MHO. Overlapping climatic ranges at different times show that the Changbai Mountains, the central parts of the Japanese Alps, Hokkaido, and the Sikhote-Alin Mountains will remain suitable refugia for Jezo spruce until 2070. The establishment of artificial forest stands of *P. jezoensis* and intraspecific taxa in the future climate-acceptable regions may be important for the preservation of genetic diversity.

Keywords: climate change; boreal forest; spruce forest; *Picea jezoensis*; species distribution modeling; Last Glacial Maximum; Northeast Asia

Citation: Korznikov, K.; Petrenko, T.; Kislov, D.; Krestov, P.; Doležal, J. Predicting Spruce Taiga Distribution in Northeast Asia Using Species Distribution Models: Glacial Refugia, Mid-Holocene Expansion and Future Predictions for Global Warming. *Forests* **2023**, *14*, 219. https://doi.org/10.3390/f14020219

Academic Editors: Daniela Dalmonech, Alessio Collalti and Gina Marano

Received: 20 December 2022
Revised: 17 January 2023
Accepted: 20 January 2023
Published: 24 January 2023

1. Introduction

Current climate changes in the boreal zone of Eurasia have led to visible changes in vegetation cover due to increased fire frequency, the proliferation of insect pests, desiccation, and wind disturbances, which alter the structure of the vegetation cover and the distribution of plants and whole biomes [1]. However, it remains unclear how current changes in the distribution of boreal forests and their dominant species deviate from long-term dynamics and what the prospects are. Studying the spatial distribution of dominant species in changing boreal forests can therefore help us better understand the factors behind their past and present occurrence, assess possible climate-induced range shifts, and predict future forest dynamics [2,3].

Recently, species distribution modeling (SDM) methods have been widely used to study the effects of climate change on species ranges [4–7]. Using data on the current distribution of climatic indicators characterizing species range, it is possible to predict climatically suitable areas under current climatic conditions, under climatic conditions of the past, or predicted climatic conditions of the future [8–11]. Identifying areas and climates that have been able to sustain relict populations of dominant boreal species from the Last Glacial Maximum (LGM) to the present and will be able to sustain them in the future are important for the establishment of protected reserves, as these areas have the potential to contain populations with a continuous history of several tens of thousands of years. This is important for maintaining genetic diversity [12] and further opens the possibility of adaptive management in areas such as forestry and agriculture as well as ex situ species conservation [13,14].

One of the most important taiga forest species for the ecosystem's functioning and timber industry in Northeast Asia is *Picea jezoensis* (Siebold & Zucc.) Carrière (Jezo or Yezo spruce, by the old name of Hokkaido Island), also known as *Picea ajanensis* Fisch. ex Carrière (Ajan spruce) [15]. *P. jezoensis* is a coniferous evergreen tree up to 35 m tall and 120 cm in diameter at breast height. The life expectancy of the trees is 300–400 years; the maximum age is 520 years [16]. Phylogenetically and ecologically, *P. jezoensis* is close to the North American *P. sitchensis* (Bong.) Carrière. This conclusion is also supported by the fact that both *P. jezoensis* and *P. sitchensis* have flattened leaves and loosely arranged seed scales [17]. *P. jezoensis* occurs in the sub-maritime and maritime areas of Northeastern Asia between 40° N and 55° N, where it forms zonal forests [18].

P. jezoensis forms mono-dominant or mixed spruce–fir stands (with *Abies nephrolepis* (Trautv.) Maxim. in the mainland area and *A. sachalinensis* (F.Schmidt) Mast. in the insular part of the region) from the elevation of sea level to 1500–2000 m, depending on latitude. Forests formed by *P. jezoensis* occupy almost the whole range of ecologically different sites in this belt, except only in mires and on rock outcrops. The optimal climatic parameters for this species include a mean annual temperature from −1 to 0 °C, a vegetative period of 145–155 days, and a mean summer precipitation of 370–590 mm. Mean annual air humidity within the range of species does not fall below 60%. The climatic optimum of *P. jezoensis* is thus much more severe than that of European species *Picea abies* (L.) H.Karst. [15,16].

In the most productive spruce forests, timber stock reaches 1000 m^3 with over 500 Mg ha^{-1} of aboveground biomass [15,16]. Due to its high economic potential, *P. jezoensis* is one of the region's main objects of logging and timber production. Besides logging, spruce forests are also affected by natural disturbances. Since the middle of the 20th century, particularly active processes of natural desiccation of *P. jezoensis* primeval forests have been recorded in the continental part of the Far East, with a total area of 55,000 km^2 already in the 1970s [16].

The critically important ecosystem function of *P. jezoensis* and the high economic value of this species in the context of current trends in boreal forest degradation due to climate change [19–21] make understanding the *P. jezoensis* range changes by projected climate change scenarios particularly important. Retrospective modeling over time of significant milestones of evolutionary vegetation dynamics, i.e., the LGM (~21,000 years before the present) and the mid-Holocene Climatic Optimum (MHO) (~7000 years before the present), is interesting for assessing the movement of *P. jezoensis* climate optimum in Northeast Asia. There is a clear relevance to paleoenvironmental and phylogenetic studies as well as the understanding of current trends in biodiversity and biome distributions [22].

In this study, we build the SDMs of *P. jezoensis* using the distribution data of the species in its natural habitats (not under culture or plantation conditions) and the WorldClim 1.4 climate dataset [23] associated with the downscaled paleoclimate data [24]. Using paleoclimate data MIROC-ESM [25,26] and CCSM4 [27], we reconstruct the spatial distribution of areas with climatic conditions suitable for *P. jezoensis* in the LGM and MHO. We also construct prognostic models of the potential distribution for 2070 under the RCP2.6 (representative concentration pathway) climate change scenarios [28] and RCP8.5 [29]. We used the MIROC-ESM and CCSM4 models because they cover paleoclimates and predicted future climates among other climatic models. Both prognostic and retrospective SDMs were developed using ensembles of decision trees. Tree-based supervised learning algorithms are quite efficient tools for handling complicated decision boundaries in multifactor spaces [30]. Another advantage of decision trees and their ensembles for SDMs is the ability to estimate the importance of climatic variables and the impacts on the observed species distribution. We focused on building SDMs using the RF classifier (a versatile machine learning algorithm) and the investigation of overlapping ranges of potential species occurrence under the LGM and projected future climates (RCP2.6, RCP8.5) which could be considered long-term *P. jezoensis* refugia.

2. Materials and Methods

2.1. Study Area

The dark coniferous forests of *P. jezoensis* are widespread from the mountainous regions of central Japan and South Korea in the south to central Kamchatka in the north. The species range extends in latitude from 40° N to more than 55° N [18]. Details of the species biology and ecology are described in previous works [15,16,18,31]. Clarification of intraspecific taxonomy was performed based on allozyme analysis [32,33] and nuclear and organelle DNA [34,35]. We suppose that the single subspecies of *P. jezoensis* subsp. *jezoensis* exists throughout the entire range, except for isolated relict populations in the mountains of central Honshu, attributed to the taxon *P. jezoensis* subsp. *hondoensis* (Mayr) P.A. Schmidt.

The range of *P. jezoensis* includes part of the Primorye Region, the southeast coast of the Sea of Okhotsk, Sakhalin Island, the southern islands of the Kuril Archipelago, part of Northeast China, central Kamchatka (the so-called "coniferous forest island"), the north of the Korean Peninsula, Hokkaido Island, and the central part of Honshu Island (Figure 1) [15,18]. The range of *P. jezoensis* lies in a deeply rugged mountain area. The climate in the area of the *P. jezoensis* range is controlled by the seasonally alternating maritime and continental air masses brought about by monsoon circulation. Annual precipitation ranges from 460 mm in central Kamchatka to 1250 mm in the southern Kuril Islands, and average annual temperatures range from −3 °C in the continental part to +7.5 °C in Hokkaido. The influence of the East Asian monsoon decreases with distance inland. In the continental part of the range, winters are sparsely snowy and cold, while summers are cool and humid. Mean annual air humidity within the range of *P. jezoensis* does not fall below 60%, although it varies considerably from region to region. Summarizing climatic conditions in the areas where *P. jezoensis* forms pure and fir-mixed stands, the climatic optimum is much harsher than for European *P. abies*. [15].

The presence of refugia both in the northern part of the range (isolated populations in Kamchatka) and in the southern part (isolated populations in the mountains of South Korea and *P. jezoensis* ssp. *hondoensis* in the Japanese Alps) testifies to an extensive shifting of the species range in the past associated with periods of warming and cooling in the Pleistocene. Thus, the explanation of the modern range of *P. jezoensis* lies not only in the current climatic conditions of the present but also in its changes over the past millennia. This is confirmed, among other things, by modern population genetics data [34,35].

2.2. Presence Points

Georeferenced occurrence points of *P. jezoensis* were taken from different sources: 1—local herbarium collections of the Botanical Garden-Institute FEB RAS (herbarium acronym VBGI) and Institute of Biology and Soil Science FEB RAS (VLA); 2—Global Biodiversity Information Facility database (GBIF) [36]; 3—own archival data of geobotanical relevés and occurrences points sampled in the field research. To create the models, we used only those points of *P. jezoensis* presence that belong to *P. jezoensis* subsp. *jezoensis*, i.e., not including *P. jezoensis* subsp. *hondoensis*, whose relict populations are isolated from the contiguous range of this species in the mountains of central Honshu.

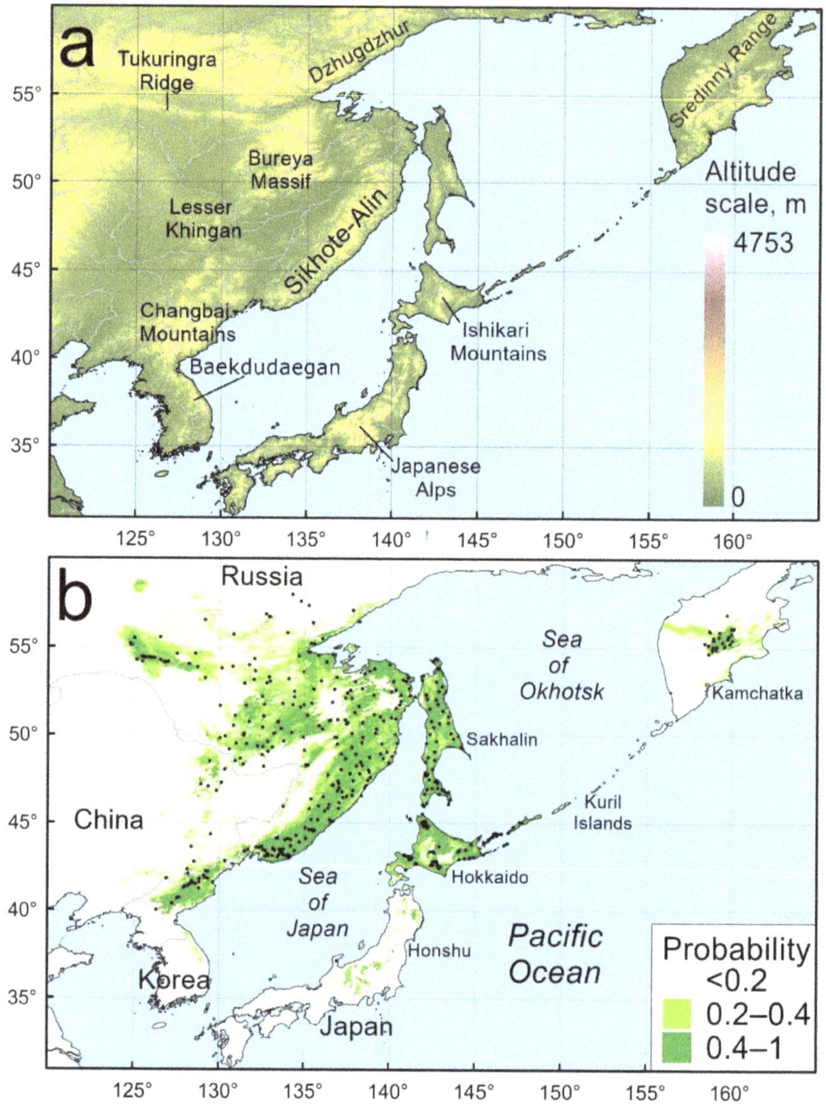

Figure 1. Topography map of the region (**a**), the built species distribution model of *Picea jezoensis* (Siebold & Zucc.) Carrière represented as a probability map; black dots indicate presence points in the model train (n = 479) (**b**).

The presence of points outside the natural distribution area (forest plantations on the islands of Hokkaido and Sakhalin) and in urbanized areas (gardens, parks) was excluded using high-resolution satellite images. A filtering algorithm was then applied to remove presence points located closer to each other than 2 km apart. The algorithm was implemented using the geopy package [37]. We then calculated the average nearest neighbor index implemented in the ArcMap 10.8 program [38] for the remaining data which compares the observed average distance between all presence points to the expected distance for a set of evenly distributed points. If the index is less than 1, the pattern exhibits clustering; if the index is greater than 1, the trend is toward dispersion or competition. Thus, we managed to avoid significant data imbalance effects when one region could be represented by a disproportionately large number of presence points. As a result, 479 unique points of species presence were used in the modeling (Figure 1b). The number of pseudo absence points randomly placed throughout the simulation area was estimated to be 2 times greater than the number of presence points.

2.3. Climatic Data

To model the distribution of *P. jezoensis*, we used the 5 most informative bioclimatic indices from a set of more than 30 indices provided by [39,40]: Kira's warmth index (WKI, the sum of average monthly temperatures above +5 °C), which showed the highest contributions to the distribution of *P. jezoensis* vegetation in Northeast Asia; Kira's coldness index (CKI, the sum modulo of average monthly temperatures below +5 °C); the index of continentality (IC, difference between annual maximum and minimum average monthly temperatures); the rain precipitation index (Pp, the amount of precipitation during the period with positive average monthly temperatures); and the snow precipitation index (Pn, the amount of precipitation in the period with negative average monthly temperatures) [40]. We conclude that it is better to choose a few predictors with a clear biological interpretation than to select a slightly optimal subset of predictors that have an implicit or unclear impact on species distribution [11].

Preconditions checked before training the classifier included a multicollinearity check of selected bioclimatic indices using NumPy package for Python [41] and the "omcdiag" function from the mctest R-package [42]. The combination of selected bioclimatic indices led to significantly different than zero values of the determinant of the covariance matrix ($p < 0.05$). The index values were calculated from monthly mean temperatures and total precipitation data provided in the WorldClime v.1.4 [23] with a spatial resolution of 30 arc-seconds (~0.0083°), which were extracted from the source data files using the Geospatial Data Abstraction Library [43]. Similar data presented in the MIROC-ESM [25,26] and CCSM4 [27] climate models were used to reconstruct the climatic situation during the LGM and MHO and to forecast the climatic situation for the year 2070. Prognostic modeling was performed in accordance with two global climate change scenarios: 1—RCP2.6 implies an increase in the average planetary temperature of 0.3–1.7 °C by 2100 [28]; 2—RCP.5 implies an increase of 2.6–4.8 °C [29].

2.4. Model Building

The formal side of SDM consists in finding nonlinear relationships between species distribution and bioclimatic parameters. To handle this problem, we chose the Random Forest (RF) machine learning method implemented in the Python programming language in the Scikit-learn package [44]. We selected RF as a method to build the models following the results of several studies indicating that RF may be more applicable in predicting the native potential distribution of species with sufficient species occurrence data [45,46]. Scikit-learn is a general-purpose machine learning package focused on rapid prototyping, validating, and deploying supervised and unsupervised learning models. It is widely used in the data science world and allows researchers to formulate the process of building SDMs at a high level of abstraction. Using Scikit-learn, the SDM creation process is expressed as a piece of code in Python programming language, which efficiently performs all the necessary steps related to machine learning model development, such as feature engineering and feature selection, training, and model testing phases. We used a grid search cross-validation procedure to find the optimal subset of RF hyperparameters. As a result, optimal values for the configuration parameters of the RF algorithm were found to be equal to the values used in similar models [47]. The optimal number of random trees was found to be equal to 100 and the maximum tree depth was limited to 10. The remaining RF parameters were set to their default values.

The constructed model was evaluated using the continuous Boyce index [48], which is calculated using only species presence points, based on 100 iterations by randomly dividing the original spatial data set into training (3/4 points) and test (1/4 points) data sets. Using the continuous Boyce index to assess model quality is preferable to using ROC AUC because it is based solely on empirical data on the location of species sites, without reference to pseudo presence points [49].

We evaluated the contribution of each of the five predictors to the final model using the "feature_importances" attribute [50] implemented for the RF from Scikit-learn [44].

The result of applying the trained classifier to climatic data is a probability map (from 0—presence is unlikely to 1—the maximum probability of presence) of habitat suitability for *P. jezoensis*. For practical purposes, such as calculating the area of territory that a species can potentially occupy, we represented the probability maps in binary form, namely "species absent" (0) or "species present" (1). Binary probability maps require finding the optimal threshold value. If the probability in each point exceeds the threshold value, we convert it to 1 and treat it as a "species presence" point. Otherwise, the probability value is converted to 0 and the corresponding point is considered a "species absence" point.

To estimate the optimal threshold value, we considered the problem of maximizing the mean value of maxSSS [51], calculated based on 100 random splits of the original spatial data set into training (3/4) and test (1/4) data sets. A similar optimization issue was noted when compared to actual skill statistics and the F1 score metric (a measure of accuracy, the harmonic mean of precision and recall). To verify the obtained optimal threshold value, we used an expert approach [51,52]. Based on computational experiments, we concluded that the optimal maxSSS yield values for the *P. jezoensis* distribution maps are consistent with the expert evaluation. Binarization using an optimal threshold calculation is a convenient way of quantification, but this approach is not the only one possible; the overall interpretation of the ranges is also important [11]. To this end, we created potential distribution maps with probability levels of 0.4–1 and 0.2–0.4.

Thus, the process of creating the SDM using the RF classifier consisted of the following phases: (1) collection of *P. jezoensis* occurrence data; (2) data preprocessing (removal of duplicates, local equalization of point density, generation of pseudo-absence points); (3) applying recursive feature elimination and expert-based feature selection; (4) grid search for the best set of model parameters (number of trees, tree depth, available trees building criterion); (5) finding the best threshold value (by maximizing maxSSS and expert-based approach); and (6) applying the model to past, present, and future climatic data to result in interpretation.

Finally, we calculated response curves for each model predictor. Response curves are essentially smooth estimates of the modeled probability of species occurrence for a fixed value of a particular predictor. Higher values on the response curves correspond to a higher probability of species occurrence and suitability of climate.

All distribution maps were built in ArcMap 10.8. The relief map was drawn using elevation data from the Shuttle Radar Topography Mission (SRTM) [53].

3. Results

Verification of decision trees by cross-validation of the obtained models of the modern distribution of *P. jezoensis* showed high predictive accuracy. The continuous Boyce index value for all models was 0.99, indicating their high prognostic abilities; the AUC value was 0.89 ± 0.004; the maxSSS was 1.696 ± 0.179; and the accuracy was 0.932 ± 0.019. Using the five selected bioclimatic factors as predictors, the most important predictors are related to moisture rather than temperature. The absence of strong differences in the contribution of the factors generally indicates their common high importance in constructing the model (Table 1).

The binarization probability level according to the maxSSS optimal threshold value is 0.43. The climatic ranges of *P. jezoensis* distribution correspond well to the species distribution from ground-based data. A comparison of the distribution model of *P. jezoensis* (Figure 1) with expert range maps of the species shows a high degree of agreement, thus allowing the model of the current climatic range of the species to be used for retrospective and predictive modeling. The inferred climatic ranges of *P. jezoensis* distribution corresponded well to the distribution maps of the species derived from ground-based expert surveys (Figure S1) [16,18,35]. Cartographic models of the area potentially suitable for *P. jezoensis* for different climatic conditions are shown in Figure 2.

Table 1. Importance of the climatic predictors in the Random Forest model.

Model Predictor	Importance (Mean ± SE, n = 100)
Pp	0.234 ± 0.010
Pn	0.234 ± 0.009
WKI	0.210 ± 0.005
CKI	0.164 ± 0.003
IC	0.158 ± 0.001

Pp—annual precipitation in the months with the positive mean temperature; Pn—annual precipitation in the months with the negative mean temperature; WKI—Kira's warmth index; CKI—Kira's coldness index; IC—index of continentality.

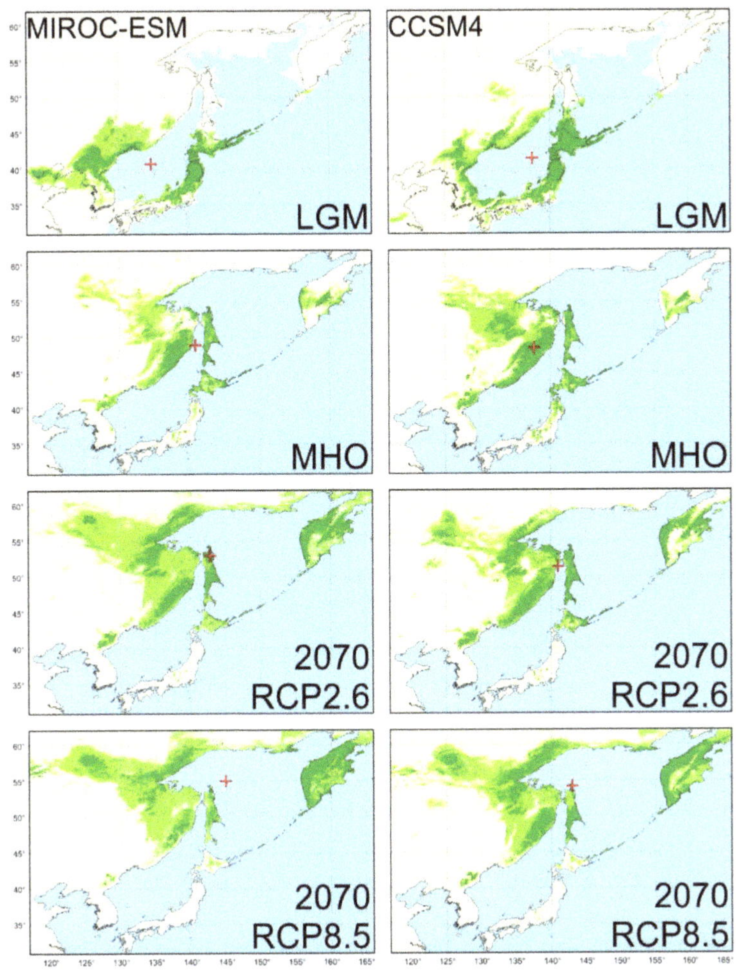

Figure 2. Potential distribution of *Picea jezoensis* (Siebold & Zucc.) Carrière built using the MIROC-ESM and CCSM4 climatic models under the Last Glacial Maximum (LGM), the mid-Holocene Climatic Optimum (MHO), and 2070 RCP2.6 and RCP8.5 scenarios; red crosses indicate the central geographical points of the predictable distributions.

The potential distribution area of *P. jezoensis* is estimated at 513,000 km^2 in current climate conditions. Quantification of the climatically suitable area of *P. jezoensis* in the LGM, the MHO, and climate projections for the year 2070 are shown in Table 2. Both the MIROC-ESM and CCSM4 climate models showed similar trends in past and future climatic range patterns but provide different quantitative areal estimates. Based on the MIROC-ESM climate model, the potential area of suitable climate conditions of *P. jezoensis* was predicted to be the highest in the 2070-year RCP2.6 scenario and lowest in the MHO (Table 3). Based on the CCSM4 climate model, the potential area of suitable climate conditions of *P. jezoensis* was predicted to be the highest in the 2070-year RCP8.5 scenario and lowest in the MHO. In addition to reflecting similar trends in bioclimatic ranges, the two models are different from each other in terms of quantitative areal estimates: the MIROC-ESM predicted area is significantly (~100,000 km^2) smaller than that of CCSM4 (Figure 2).

Table 2. The potential area (km^2) of highly suitable climate conditions of *Picea jezoensis* (Siebold & Zucc.) Carrière.

Scenario	CCSM4	MIROC-ESM
LGM	546,250 *	456,471
MHO	494,278	322,155
RCP2.6	614,347 *	581,760 *
RCP8.5	625,076 *	483,805

* Asterisks indicate an increase in the potential area in comparison to the current distribution.

Table 3. The estimation of the overlapped area with the LGM time to the 2070-year climate condition.

Climate Model	Scenario	MIROC-ESM
MIROC-ESM	RCP2.6	18,293
	RCP8.5	4480
CCSM4	RCP2.6	54,725
	RCP8.5	20,416

During the LGM, the main part of the species range was located much further south than today at 35–45° N. Projections for the MHO indicate a retreat from southern territories and northward expansion with a distribution center shifted to 45–55° N.

The SDMs of *P. jesoensis* from the LGM to the year 2070 superimposed on each other revealed the geographical locations where *P. jezoensis* always had favorable conditions. The intersection of potential areas occupied by this species in different periods showed the location of long-term stable refugia. Overlapping climatic ranges at different times showed that the Changbai Mountains, the central parts of the Japanese Alps, the Hokkaido mountains, and the Sikhote-Alin Mountains were the areas where *P. jezoensis* persisted over time. These mountain areas indicate the existence of long-term stable refugia (Figure 3) that deserve the highest priority in the conservation of the *P. jezoensis* gene pool and are expected to be represented by the most ancient populations.

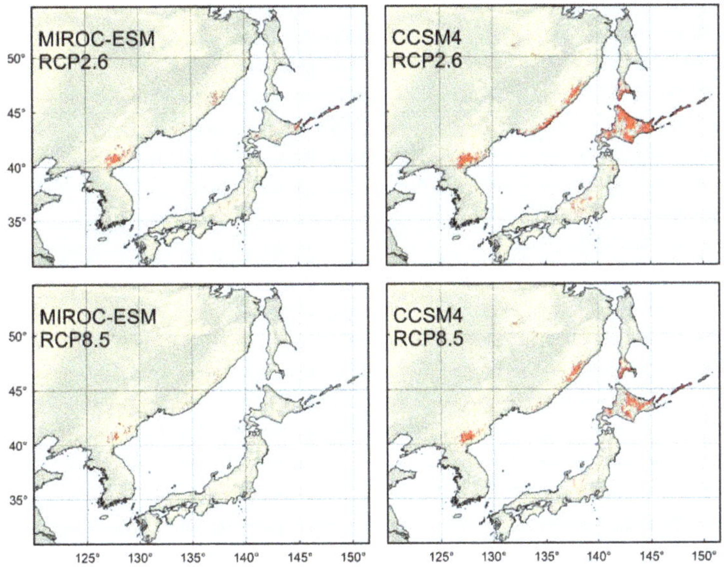

Figure 3. Overlay of *Picea jezoensis* (Siebold & Zucc.) Carrière potential distribution models of the four-time frames LGM-MHO-Current-Future follow the projected state of the MIROC-ESM and CCSM4 models and the RCP2.6 and RCP8.5 scenarios for the year 2070.

4. Discussion

4.1. Model of Current Distribution

In the continental part, *P. jezoensis* has an almost continuous range along the Pacific coast of Eurasia from 40 to 50° N. The ranges most distant from the seacoasts correspond to the Lesser Khingan Ranges and the Tukuringra Range. The climatic model predicted suitable areas for this species as relict isolated populations in the mountainous region of the Korean Peninsula, for the northern part of which there are only literature indications on the presence of *P. jezoensis* [54], but the presence points were absent in these locations according to our modeling protocols. Non-zero presence probabilities of *P. jezoensis* were obtained for several areas in South Korea, such as Mount Seorak (38.12° N, 128.46° E) and Mount Odae (37.80° N, 128.53° E), for which we had no presence points, but the *P. jezoensis* distribution is known from literature data [54]. On the other hand, the sensitivity of our model was not sufficient to predict the southernmost point of occurrence of *P. jezoensis* on Mount Jiri (35.33° N, 127.73° E).

Although we excluded the presence points of *P. jezoensis* subsp. *hondoensis* in central Honshu from the analysis, the mountainous areas where this taxon is commonly found were marked with a non-zero probability of presence. This region supports conditions for the existence of many other rare conifer taxa isolated from the main range besides *P. jezoensis*, for example, *Pinus koraiensis* Siebold & Zucc. [47].

4.2. Reconstructed Distribution in the LGM

The LGM climate in the region was characterized by lower temperatures and an arid environment [55]. Fundamentally different climatic conditions and different land contours during the sea level regression were reflected in a different distribution of biomes and their constituent species [56]. The MIROC-ESM and CCSM4 climate models provide generally similar climatic ranges of *P. jezoensis* during the LGM, with the main part of the species range located much further south than today, at 41–42° N. In addition to reflecting similar trends in bioclimatic extent, the two models are very different from each other: the MIROC-ESM predicted area is significantly smaller than that of the CCSM4. The climatically suitable area for the species according to the MIROC-ESM is somewhat smaller and the CCSM4 is somewhat larger than the area of the modern climatic range. It should be considered that the total land area during the LGM was higher due to marine regression. In particular, the CCSM4 model predicts the area between modern Hokkaido and Sakhalin (the Soya Landbridge) as a climatically suitable area. The seabed in the form of coastal plains represented additional habitats [57].

According to the CCSM4 model, areas with suitable bioclimatic conditions for *P. jezoensis* up to 50° N were preserved along the coast of the Sea of Japan, the eastern slopes of the Sikhote-Alin Mountains, and the western coast of Sakhalin Island. In the MIROC-ESM model, the northern limits of the climatically acceptable area for the species are located much more south. Although vegetation reconstructions [56] carried out for the area north of 45° N imply the presence of sparse larch forests (tundra-like vegetation) and the landscape complex supporting the mammoth tundra–steppe vegetation [58,59], *P. jezoensis* is recorded in the palynological spectra of Sakhalin and Northeastern Hokkaido during the LGM [60,61]. At the same time, *Picea* (probably *P. jezoensis*) pollen is recorded in the Lake Khanka area (45° N 132° E) [62]. The continental regions of Northeastern China had an arid climate and, judging by pollen surveys, supported forest-steppe and shrub-steppe vegetation. Herbs expanded rapidly, dominated mainly by *Poaceae*. During the LGM, the forest in the northern part of Northeast China was relatively open and dominated by larches. Forests composed of birch, pine, and alder have developed in the Changbai Mountains [63]. Vegetation structure in areas with less arid climates closer to seacoasts was clearly more complex than in homogeneous larch forests, and areas of light coniferous taiga were interspersed with refugia of dark coniferous taiga [64], and the overall vegetation heterogeneity was supported by ample populations of megafauna [65].

The range of *P. jezoensis* was located much farther south in the climatic conditions of the LGM than at present (Figure 2). On the contrary, the current isolated area of *P. jezoensis* on the mainland in the central part of Kamchatka corresponds with the minimal influence of the sea within the whole peninsula, and in this zone, the so-called "coniferous forest island" is formed. Populations of *P. jezoensis* in Kamchatka are thought to be relict and have been preserved in this area since past warm epochs when the distribution area of *P. jezoensis* included the entire coast of the northern Okhotsk Sea. The time of isolation of the Kamchatka population of *P. jezoensis* is estimated to be more than 400 thousand years [32]. However, both models do not predict the preservation of climatic refugia of *P. jezoensis* in central Kamchatka, even though there is no doubt about the preservation of the species in this area since the interglacial period. According to [34], the Kamchatka Peninsula population of *P. jezoensis* was part of the mainland Asian range and separated during the mid-Pleistocene. We explain this by the insufficient accuracy of model reproduction for remote and sparsely populated areas of Northeast Asia.

The genetic structure of *P. jezoensis* on the mainland is closer to the population in the northern part of Sakhalin Island [35]. The southern part of the island is closer to Hokkaido Island, which was settled from the mainland by land bridges in the mid-Pleistocene. Analysis of microsatellite loci indicates that *P. jezoensis* populations in southern Sakhalin and Hokkaido have passed through a series of bottlenecks [35]. In the context of our modeling data, this clearly signals the existence of isolated refugia in Sakhalin and Hokkaido during the LGM period, as confirmed by palynological studies [60,61,66].

4.3. Reconstructed Distribution in the MHO

The MHO in the region was characterized by a higher temperature compared to the present, which was reflected in the expansion of mixed stands of the main plant species, including more thermophilic taxa, as evidenced by palynospectrum imprints [67,68]. Changes in climatic conditions in the region from the LGM to the MHO were accompanied by the transformation of natural complexes and changes in the boundaries of the main vegetation types. Warming and increased precipitation were accompanied by the northward expansion of the forest-forming species of the dark coniferous forests from more southern latitudes and isolated refugia. Simultaneously with the poleward expansion, populations disappeared in the southern part of the range, where boreal and mixed forest ecosystems were replaced by more thermophilic vegetation [66,68,69].

The MIROC-ESM and CCSM4 models for the MHO climates predict smaller areas of *P. jezoensis* ranges than those under modern climates (Table 3). Both models show that in the MHO, the zone of a continuous distribution of *P. jezoensis* in mainland Northeast Asia was restricted to the southern Sikhote-Alin Mountains in the south, and the mountain ranges of North Korea and Northeast China (the Changbai Mountains) represented refugia separated from the main range. Palynological data from the southern Sikhote-Alin Mountains show that *P. jezoensis* did not disappear from plant communities during the LGM, but the proportion of pollen from this species was significantly lower than at present and was higher in broadleaved species [70,71]. In the island part of the region, the MHO was a period of a significant decrease in *P. jezoensis* pollen and an increase in the proportion of *Quercus mongolica* Fisch. ex Ledeb. s.l. pollen [61,66]. In the MHO time, relict populations of *P. jezoensis* subsp. *hondoensis* on Honshu and isolated populations of *P. jezoensis* in the mountains of the southern part of the Korean Peninsula have formed. At the same time, warmer climatic conditions in Kamchatka facilitated the spread of *P. jezoensis* from refugia preserved in the LGM [34,35].

4.4. Predicted Distribution in the Year 2070

RCP models ensure an increase in the temperature and precipitation balance depending on the concentration of greenhouse gases in the atmosphere. According to the optimistic scenario RCP2.6, the temperature increase by 2100 will be 2 °C, and according to the scenario RCP8.5 by 5 °C [28,29].

All scenarios and climate models for the year 2070 envisage more dramatic changes in the range of *P. jezoensis* than those that have occurred from the MHO to the present. In the RCP2.6 scenario, both the MIROC-ESM and CCSM4 project growth in areas of optimal climatic conditions. Areas in much of Kamchatka and along the coast of the Sea of Okhotsk will be suitable for the species. At the same time, a continuous area of climatically acceptable habitats in the southern part of the species range on the continent will disappear. A further reduction of potentially suitable areas will also occur in the extreme south of the species range, in the mountains of the southern Korean Peninsula [72,73]. Effects of climate change on coniferous tree species in the region have been observed [74–78]. Dendrochronological methods revealed a decline in the annual growth of *P. jezoensis* in China and Korea since 1980 in the lower elevations of the Changbai Mountains [79]. At the same time, an increase in the width of annual tree rings was observed in the higher elevations, as well as an extension of the length of the growing season.

Warming under the RCP8.5 scenario would result in an even more significant change in the contours of potentially suitable habitat for *P. jezoensis*, but while the projected area of the CCSM4 model would be higher than the current one, the MIROC-ESM model would reduce the final area of climatically suitable habitat.

Nevertheless, even the realization of the most pessimistic climate change scenarios will not cause the extinction of mainland populations in the Pektusan region (the southern face of the Changbai Mountains), which has an uninterrupted history since the LGM, and will not cause the complete disappearance of refugia in central Japan, although it will greatly reduce them.

The overlay of climatically acceptable areas for *P. jezoensis* from the LGM to the year 2070 shows that such areas are extremely small. Even the pessimistic RCP8.5 scenarios do not foresee the complete disappearance of *P. jezoensis* habitats from the Changbai Mountains, where populations of this species have existed continuously since the LGM. In the CCSM4 model, such areas of long-existing *P. jezoensis* include, in addition to the Changbai Mountains, central and southern parts of the Sikhote-Alin Mountains, partially, Hokkaido, southern Sakhalin, and southern Kurils (Figure 3).

4.5. Implications for Conservation and Management

It should be noted that modeling methods provide a probabilistic assessment of potential niches in terms of climate. Species distributions are affected by competition, dispersal, niche size, and environmental conditions in space and time [80]. Natural shifts in vegetation distribution may take longer because they depend on, among other things, the availability of diaspores, competitive relationships between plants, and local factors of a particular habitat [81,82]. At the same time, predictive models must always consider not only the extent of suitable habitats but also the rate of species distribution expansion, which is usually much slower than global climate change. The use of modeling techniques provides insight into trends in the general state of populations, allows planning of the conservation risks of *P. jezoensis* within the current range, and builds a systematic concept for creating forest crops and establishing forest plantations outside the current distribution of the species, with respect to expected climate changes [13,83,84].

To preserve the genetic diversity of *P. jezoensis* and intraspecific taxa, it is advisable to think about establishing plantations in places where the climate will be acceptable in the long term and in the context of projected changes. Forestry must take climate trends into account when establishing new plantations of *P. jezoensis*. Establishing artificial forest stands of this species in the southern part of its range against the background of a changing climate appears to be a bad decision, while a deeper introduction of this species into forestry practices could be a very prudent decision for areas of Northeast Asia where *P. jezoensis* does not currently grow in natural ecosystems.

The departure of *P. jezoensis* populations from the optimal climate zone will not cause their one-step extinction but will determine a trend towards gradual extinction by increased tree elimination due to bacterial diseases, fungal diseases, limitation of natural regeneration processes, drought, and fires accompanying drought. Within the study region, the previously unknown occurrence of bark beetle outbreaks took place in the Sakhalin and Kuril Islands as a result of massive windthrows in spruce and fir forests [85–87].

Due to the genetic diversity found in the populations of the species [32–35] and in order to preserve it, it is necessary to create stands of *P. jezoensis* from those places where the extinction of species is assumed. Such work cannot be carried out within one country and will require the consolidation of the efforts of all the states of Northeast Asia into a common project. The genetic structure of local populations of *P. jezoensis* in the mainland part of the species range has not been sufficiently studied, in contrast to detailed studies on the Japanese islands. First, it is of interest to collect materials from the boundaries of the modern distribution of the species on the Sikhote-Alin Mountains, the Lesser Khingan Mountains, the Tukuringra Range, and the southern Kurils.

5. Conclusions

SDMs of *P. jezoensis* built in this study are based on five bioclimatic factors and considered the distribution of climate continentality, heat balance throughout the year, and precipitation in warm and cold periods. The area of current suitable climatic conditions for *P. jezoensis* is estimated at more than 500,000 km^2. The MIROC-ESM and CCSM4 climate models for retrospective and predictive modeling provide slightly different estimates of potential range but describe similar trends in species range shifts.

We identify areas in the Changbai Mountains (China, North Korea) and the Sikhote-Alin Mountains (Russia) as long-term climatically stable *P. jezoensis* refugia from the LGM to projective climate conditions of the year 2070 under the scenario RCP8.5. These areas could be prioritized for the in situ conservation of species populations. In addition to its ecosystem role, *P. jezoensis* is also an economically important species, so the obtained results should also be applied in forestry planning. Potentially favorable climatic areas in the northern parts of Northeast Asia according to the obtained models should be considered and used as places for establishing artificial forest stands of *P. jezoensis* in the future. A reforestation process using *P. jezoensis* and commercial planting does not have long-term perspectives in more southern areas.

Supplementary Materials: The following supporting information can be downloaded at https://www.mdpi.com/article/10.3390/f14020219/s1, Figure S1: Distribution maps of *Picea jezoensis* (a) by Manko (1987); (b), Nakamura and Krestov (2005); (c) Aizawa et al., 2009.

Author Contributions: Conceptualization, K.K. and T.P.; methodology, D.K., software, D.K.; validation, K.K., T.P. and D.K.; formal analysis, T.P.; data curation, D.K.; writing—original draft preparation, T.P.; writing—review and editing, K.K., D.K., P.K. and J.D.; visualization, K.K., T.P. and D.K.; supervision, P.K. and J.D.; funding acquisition, K.K., P.K. and J.D. All authors have read and agreed to the published version of the manuscript.

Funding: This work was funded by the Ministry of Education, Youth and Sport of the Czech Republic (MŠMT), the project Mobility 2020 (CZ.02.2.69/0.0/0.0/18_053/0017850), K.K.; the Czech Science Foundation (20-05840Y), K.K.; (21-26883S), J.D.; long-term research development project of the Czech Academy of Sciences (RVO 67985939), K.K. and J.D.; the scientific research project of the Botanical Garden-Institute FEB RAS (FWFR-2022-0008) No. 122040800089-2, P.K.

Data Availability Statement: Data The data presented in this study are available in the article.

Conflicts of Interest: The authors declare no conflict of interest.

References

1. Chen, I.-C.; Hill, J.K.; Ohlemüller, R.; Roy, D.B.; Thomas, C.D. Rapid Range Shifts of Species Associated with High Levels of Climate Warming. *Science* **2011**, *333*, 1024–1026. [CrossRef]
2. Becknell, J.M.; Desai, A.R.; Dietze, M.C.; Schultz, C.A.; Starr, G.; Duffy, P.A.; Franklin, J.F.; Pourmokhtarian, A.; Hall, J.; Stoy, P.C.; et al. Assessing Interactions Among Changing Climate, Management, and Disturbance in Forests: A Macrosystems Approach. *Bioscience* **2015**, *65*, 263–274. [CrossRef]
3. Seidl, R.; Thom, D.; Kautz, M.; Martin-Benito, D.; Peltoniemi, M.; Vacchiano, G.; Wild, J.; Ascoli, D.; Petr, M.; Honkaniemi, J.; et al. Forest disturbances under climate change. *Nat. Clim. Chang.* **2017**, *7*, 395–402. [CrossRef] [PubMed]
4. Guisan, A.; Zimmermann, N.E. Predictive habitat distribution models in ecology. *Ecol. Model.* **2000**, *135*, 147–186. [CrossRef]
5. Elith, J.H.; Graham, C.P.H.; Anderson, R.P.; Dudík, M.; Ferrier, S.; Guisan, A.; Hijmans, R.J.; Huettmann, F.; Leathwick, J.R.; Lehmann, A.; et al. Novel methods improve prediction of species' distributions from occurrence data. *Ecography* **2006**, *29*, 129–151. [CrossRef]
6. Elith, J.; Leathwick, J.R. Species Distribution Models: Ecological Explanation and Prediction Across Space and Time. *Annu. Rev. Ecol. Evol. Syst.* **2009**, *40*, 677–697. [CrossRef]
7. Kearney, M.; Porter, W. Mechanistic niche modelling: Combining physiological and spatial data to predict species' ranges. *Ecol. Lett.* **2009**, *12*, 334–350. [CrossRef]
8. Pearson, R.G.; Dawson, T.P. Predicting the impacts of climate change on the distribution of species: Are bioclimate envelope models useful? *Glob. Ecol. Biogeogr.* **2003**, *12*, 361–371. [CrossRef]
9. Hijmans, R.J.; Graham, C.H. The ability of climate envelope models to predict the effect of climate change on species distributions. *Glob. Chang. Biol.* **2006**, *12*, 2272–2281. [CrossRef]
10. Booth, T.H. Species distribution modelling tools and databases to assist managing forests under climate change. *For. Ecol. Manag.* **2018**, *430*, 196–203. [CrossRef]
11. Santini, L.; Benítez-López, A.; Maiorano, L.; Čengić, M.; Huijbregts, M.A.J. Assessing the reliability of species distribution projections in climate change research. *Divers. Distrib.* **2021**, *27*, 1035–1050. [CrossRef]
12. Tang, C.Q.; Matsui, T.; Ohashi, H.; Dong, Y.-F.; Momohara, A.; Herrando-Moraira, S.; Qian, S.; Yang, Y.; Ohsawa, M.; Luu, H.T.; et al. Identifying long-term stable refugia for relict plant species in East Asia. *Nat. Commun.* **2018**, *9*, 4488. [CrossRef] [PubMed]
13. Janowiak, M.K.; Swanston, C.; Nagel, L.M.; Brandt, L.A.; Butler, P.R.; Handler, S.D.; Shannon, P.D.; Iverson, L.R.; Matthews, S.; Prasad, A.; et al. A Practical Approach for Translating Climate Change Adaptation Principles into Forest Management Actions. *J. For.* **2014**, *112*, 424–433. [CrossRef]
14. Schelhaas, M.-J.; Nabuurs, G.-J.; Hengeveld, G.; Reyer, C.; Hanewinkel, M.; Zimmermann, N.E.; Cullmann, D. Alternative forest management strategies to account for climate change-induced productivity and species suitability changes in Europe. *Reg. Environ. Chang.* **2015**, *15*, 1581–1594. [CrossRef]
15. Krestov, P.V. Forest Vegetation of Easternmost Russia (Russian Far East). In *Forest Vegetation of Northeast Asia. Geobotany*; Kolbek, J., Šrůtek, M., Box, E.O., Eds.; Springer: Dordrecht, The Netherlands, 2003; Volume 28, pp. 93–180.
16. Manko, Y.I. *El' Ajanskaya (Picea Ajanensis)*; Nauka: Leningrad, Russia, 1987. (In Russian)
17. Shao, C.-C.; Shen, T.-T.; Jin, W.-T.; Mao, H.-J.; Ran, J.-H.; Wang, X.-Q. Phylotranscriptomics resolves interspecific relationships and indicates multiple historical out-of-North America dispersals through the Bering Land Bridge for the genus Picea (Pinaceae). *Mol. Phylogenet. Evol.* **2019**, *141*, 106610. [CrossRef] [PubMed]

18. Nakamura, Y.; Krestov, P.V. Coniferous forests of the temperate zone of Asia. Coniferous forests. *Ser. Ecosyst. World* **2005**, *6*, 163–220.
19. Allen, C.D.; Macalady, A.K.; Chenchouni, H.; Bachelet, D.; McDowell, N.; Vennetier, M.; Kitzberger, T.; Rigling, A.; Breshears, D.D.; Hogg, E.H.; et al. A global overview of drought and heat-induced tree mortality reveals emerging climate change risks for forests. *For. Ecol. Manag.* **2010**, *259*, 660–684. [CrossRef]
20. Gauthier, S.; Bernier, P.; Kuuluvainen, T.; Shvidenko, A.Z.; Schepaschenko, D.G. Boreal forest health and global change. *Science* **2015**, *349*, 819–822. [CrossRef]
21. Kuuluvainen, T.; Gauthier, S. Young and old forest in the boreal: Critical stages of ecosystem dynamics and management under global change. *For. Ecosyst.* **2018**, *5*, 26. [CrossRef]
22. Svenning, J.-C.; Eiserhardt, W.L.; Normand, S.; Ordonez, A.; Sandel, B. The Influence of Paleoclimate on Present-Day Patterns in Biodiversity and Ecosystems. *Annu. Rev. Ecol. Evol. Syst.* **2015**, *46*, 551–572. [CrossRef]
23. Hijmans, R.J.; Cameron, S.E.; Parra, J.L.; Jones, P.G.; Jarvis, A. Very high resolution interpolated climate surfaces for global land areas. *Int. J. Climatol.* **2005**, *25*, 1965–1978. [CrossRef]
24. WorldClim. 1.4 Downscaled Paleo Climate. Available online: https://www.worldclim.org/data/v1.4/paleo1.4.html (accessed on 17 January 2023).
25. Watanabe, S.; Hajima, T.; Sudo, K.; Nagashima, T.; Takemura, T.; Okajima, H.; Nozawa, T.; Kawase, H.; Abe, M.; Yokohata, T.; et al. MIROC-ESM 2010: Model description and basic results of CMIP5-20c3m experiments. *Geosci. Model Dev.* **2011**, *4*, 845–872. [CrossRef]
26. Kawamiya, M.; Hajima, T.; Tachiiri, K.; Watanabe, S.; Yokohata, T. Two decades of Earth system modeling with an emphasis on Model for Interdisciplinary Research on Climate (MIROC). *Prog. Earth Planet. Sci.* **2020**, *7*, 64. [CrossRef]
27. Gent, P.R.; Danabasoglu, G.; Donner, L.J.; Holland, M.M.; Hunke, E.C.; Jayne, S.R.; Lawrence, D.M.; Neale, R.B.; Rasch, P.J.; Vertenstein, M.; et al. The Community Climate System Model Version 4. *J. Clim.* **2011**, *24*, 4973–4991. [CrossRef]
28. Van Vuuren, D.P.; Stehfest, E.; den Elzen, M.G.J.; Kram, T.; Van Vliet, J.; Deetman, S.; Isaac, M.; Goldewijk, K.K.; Hof, A.; Beltran, A.M.; et al. RCP2.6: Exploring the possibility to keep global mean temperature increase below 2 °C. *Clim. Chang.* **2011**, *109*, 95–116. [CrossRef]
29. Riahi, K.; Rao, S.; Krey, V.; Cho, C.; Chirkov, V.; Fischer, G.; Kindermann, G.E.; Nakicenovic, N.; Rafaj, P. RCP 8.5—A scenario of comparatively high greenhouse gas emissions. *Clim. Chang.* **2011**, *109*, 33–57. [CrossRef]
30. Rokach, L.; Maimon, M.O. Classification Trees. In *Data Mining and Knowledge DisCovery Handbook*; Maimon, O., Rokach, L., Eds.; Springer: Boston, MA, USA, 2009; pp. 149–174.
31. Krestov, P.V.; Nakamura, Y. Phytosociological study of thePicea jezoensis forests of the far east. *Folia Geobot.* **2002**, *37*, 441–473. [CrossRef]
32. Potenko, V.V.; Knysh, Y.D. Genetic variation of Yeddo spruce populations in Russia. *For. Genet.* **2003**, *10*, 55–64.
33. Potenko, V.V. Allozyme Variation and Phylogenetic Relationships in Picea jezoensis (Pinaceae) Populations of the Russian Far East. *Biochem. Genet.* **2007**, *45*, 291–304. [CrossRef]
34. Aizawa, M.; Yoshimaru, H.; Saito, H.; Katsuki, T.; Kawahara, T.; Kitamura, K.; Shi, F.; Kaji, M. Phylogeography of a northeast Asian spruce, Picea jezoensis, inferred from genetic variation observed in organelle DNA markers. *Mol. Ecol.* **2007**, *16*, 3393–3405. [CrossRef]
35. Aizawa, M.; Yoshimaru, H.; Saito, H.; Katsuki, T.; Kawahara, T.; Kitamura, K.; Shi, F.; Sabirov, R.; Kaji, M. Range-wide genetic structure in a north-east Asian spruce (*Picea jezoensis*) determined using nuclear microsatellite markers. *J. Biogeogr.* **2009**, *36*, 996–1007. [CrossRef]
36. GBIF. Global Biodiversity Information Facility. Available online: https://www.gbif.org/ (accessed on 9 October 2022).
37. GeoPy's Documentation. Available online: https://geopy.readthedocs.io/en/stable/ (accessed on 9 October 2022).
38. Average Nearest Neighbor, ArcMap 10.8. Available online: https://desktop.arcgis.com/en/arcmap/latest/tools/spatial-statistics-toolbox/average-nearest-neighbor.htm (accessed on 9 October 2022).
39. Nakamura, Y.; Krestov, P.V.; Omelko, A.M. Bioclimate and zonal vegetation in Northeast Asia: First approximation to an integrated study. *Phytocoenologia* **2007**, *37*, 443–470. [CrossRef]
40. Noce, S.; Caporaso, L.; Santini, M. A new global dataset of bioclimatic indicators. *Sci. Data* **2020**, *7*, 398. [CrossRef] [PubMed]
41. NumPy. The Fundamental Package for Scientific Computing with Python. Available online: https://numpy.org/ (accessed on 9 October 2022).
42. Imdadullah, M.; Aslam, M.; Altaf, S. mctest: An R Package for Detection of Collinearity among Regressors. *R J.* **2016**, *8*, 495–505. [CrossRef]
43. GDAL Documentation. Available online: https://gdal.org/ (accessed on 9 October 2022).
44. Pedregosa, F.; Varoquaux, G.; Gramfort, A.; Michel, V.; Thirion, B.; Grisel, O.; Blondel, M.; Prettenhofer, P.; Weiss, R.; Dubourg, V.; et al. Scikit-learn: Machine learning in Python. *J. Mach. Learn. Res.* **2011**, *12*, 2825–2830. [CrossRef]
45. Čengić, M.; Rost, J.; Remenska, D.; Janse, J.H.; Huijbregts, M.A.J.; Schipper, A.M. On the importance of predictor choice, modelling technique, and number of pseudo-absences for bioclimatic envelope model performance. *Ecol. Evol.* **2020**, *10*, 12307–12317. [CrossRef]
46. Zhao, Z.; Xiao, N.; Shen, M.; Li, J. Comparison between optimized MaxEnt and random forest modeling in predicting potential distribution: A case study with Quasipaa boulengeri in China. *Sci. Total Environ.* **2022**, *842*, 156867. [CrossRef] [PubMed]

47. Petrenko, T.Y.; Korznikov, K.A.; Kislov, D.E.; Belyaeva, N.G.; Krestov, P.V. Modeling of cold-temperate tree Pinus koraiensis (Pinaceae) distribution in the Asia-Pacific region: Climate change impact. *For. Ecosyst.* **2022**, *9*, 100015. [CrossRef]
48. Boyce, M.S.; Vernier, P.R.; E. Nielsen, S.; Schmiegelow, F.K. Evaluating resource selection functions. *Ecol. Model.* **2002**, *157*, 281–300. [CrossRef]
49. Lobo, J.M.; Jiménez-Valverde, A.; Real, R. AUC: A misleading measure of the performance of predictive distribution models. *Glob. Ecol. Biogeogr.* **2007**, *17*, 145–151. [CrossRef]
50. Breiman, L. Random forests. *Mach. Learn.* **2001**, *45*, 5–32. [CrossRef]
51. Liu, C.; White, M.; Newell, G. Selecting thresholds for the prediction of species occurrence with presence-only data. *J. Biogeogr.* **2013**, *40*, 778–789. [CrossRef]
52. Konowalik, K.; Nosol, A. Evaluation metrics and validation of presence-only species distribution models based on distributional maps with varying coverage. *Sci. Rep.* **2021**, *11*, 1482. [CrossRef]
53. USGS EROS Archive—Digital Elevation—Shuttle Radar Topography Mission (SRTM) 1 Arc-Second Global. Available online: https://www.usgs.gov/centers/eros/science/usgs-eros-archive-digital-elevation-shuttle-radar-topography-mission-srtm-1 (accessed on 9 October 2022).
54. Černý, T.; Kopecký, M.; Petřík, P.; Song, J.-S.; Šrůtek, M.; Valachovič, M.; Altman, J.; Doležal, J. Classification of Korean forests: Patterns along geographic and environmental gradients. *Appl. Veg. Sci.* **2014**, *18*, 5–22. [CrossRef]
55. Herzschuh, U.; Birks, H.J.B.; Laepple, T.; Andreev, A.; Melles, M.; Brigham-Grette, J. Glacial legacies on interglacial vegetation at the Pliocene-Pleistocene transition in NE Asia. *Nat. Commun.* **2016**, *7*, 11967. [CrossRef]
56. Tsukada, M. *Vegetation in prehistoric Japan: The last 20,000 years, In Windows on the Japanese Past: Studies in Archeology and Prehistory*; Center for Japanese Studies, University of Michigan: Ann Arbor, MI, USA, 1986; pp. 11–56.
57. Sakaguchi, S.; Sakurai, S.; Yamasaki, M.; Isagi, Y. How did the exposed seafloor function in postglacial northward range expansion of Kalopanax septemlobus? Evidence from ecological niche modelling. *Ecol. Res.* **2010**, *25*, 1183–1195. [CrossRef]
58. Vereshchagin, N.K.; Baryshnikov, G.F. The ecological structure of the "Mammoth Fauna" in Eurasia. *Ann. Zool. Fenn.* **1991**, *28*, 253–259. Available online: https://www.jstor.org/stable/23735450 (accessed on 9 October 2022).
59. Markova, A.K.; Smirnov, N.G.; Kozharinov, A.V.; Kazantseva, N.E.; Simakova, A.N.; Kitaev, L.M. Late Pleistocene distribution and diversity of mammals in Northern Eurasia (PALEOFAUNA database). *Paleontol. Evol.* **1995**, *28–29*, 5–66.
60. Igarashi, Y.; Zharov, A.E. Climate and vegetation change during the late Pleistocene and early Holocene in Sakhalin and Hokkaido, northeast Asia. *Quat. Int.* **2011**, *237*, 24–31. [CrossRef]
61. Igarashi, Y. Vegetation and climate during the LGM and the last deglaciation on Hokkaido and Sakhalin Islands in the northwest Pacific. *Quat. Int.* **2016**, *425*, 28–37. [CrossRef]
62. Belyanin, P.S.; Belyanina, N.I. On the Prikhanka depression vegetation cover evolution and its mountain framing in the Late Neopleistocene-Holocene (from palynological data). *Russ. J. Pac. Geol.* **2012**, *31*, 96–108. (In Russian)
63. Li, X.; Zhao, C.; Zhou, X. Vegetation pattern of Northeast China during the special periods since the Last Glacial Maximum. *Sci. China Earth Sci.* **2019**, *62*, 1224–1240. [CrossRef]
64. Herzschuh, U. Legacy of the Last Glacial on the present-day distribution of deciduous versus evergreen boreal forests. *Glob. Ecol. Biogeogr.* **2019**, *29*, 198–206. [CrossRef]
65. Zimov, S.; Zimov, N.; Tikhonov, A.; Chapin, F. Mammoth steppe: A high-productivity phenomenon. *Quat. Sci. Rev.* **2012**, *57*, 26–45. [CrossRef]
66. Ooi, N. Vegetation history of Japan since the last glacial based on palynological data. *Jpn. J. Hist. Bot.* **2016**, *25*, 1–101. [CrossRef]
67. Cao, X.; Herzschuh, U.; Ni, J.; Zhao, Y.; Böhmer, T. Spatial and temporal distributions of major tree taxa in eastern continental Asia during the last 22,000 years. *Holocene* **2014**, *25*, 79–91. [CrossRef]
68. Binney, H.; Edwards, M.; Macias-Fauria, M.; Lozhkin, A.; Anderson, P.; Kaplan, J.O.; Andreev, A.; Bezrukova, E.; Blyakharchuk, T.; Jankovska, V.; et al. Vegetation of Eurasia from the last glacial maximum to present: Key biogeographic patterns. *Quat. Sci. Rev.* **2017**, *157*, 80–97. [CrossRef]
69. Zhao, C.; Li, X.; Zhou, X.; Zhao, K.; Yang, Q. Holocene vegetation succession and responses to climate change in the northern sector of Northeast China. *Sci. China Earth Sci.* **2016**, *59*, 1390–1400. [CrossRef]
70. Razjigaeva, N.; Ganzey, L.; Grebennikova, T.; Mokhova, L.; Kudryavtseva, E.; Arslanov, K.; Maksimov, F.; Starikova, A. Landscape and environmental changes along the Eastern Primorye coast during the middle to late Holocene and human effects. *J. Asian Earth Sci.* **2018**, *158*, 160–172. [CrossRef]
71. Razjigaeva, N.; Ganzey, L.; Lyaschevskaya, M.; Makarova, T.; Kudryavtseva, E.; Grebennikova, T.; Panichev, A.; Arslanov, K.; Maksimov, F.; Petrov, A.Y.; et al. Climatic and human impacts on landscape development of the Murav'ev Amursky Peninsula (Russian South Far East) in the Middle/Late Holocene and historical time. *Quat. Int.* **2019**, *516*, 127–140. [CrossRef]
72. Jang, W.; Park, P.S. Stand Structure and Maintenance of Picea jezoensis in a Northern Temperate Forest, South Korea. *J. Plant Biol.* **2010**, *53*, 180–189. [CrossRef]
73. Jang, W.; Keyes, C.R.; Running, S.W.; Lim, J.-H.; Park, P.S. Climate–growth relationships of relict *Picea jezoensis* at Mt. Gyebang, South Korea. *For. Sci. Technol.* **2014**, *11*, 19–26. [CrossRef]
74. Yu, D.; Wang, Q.; Wang, Y.; Zhou, W.; Ding, H.; Fang, X.; Jiang, S.; Dai, L. Climatic effects on radial growth of major tree species on Changbai Mountain. *Ann. For. Sci.* **2011**, *68*, 921–933. [CrossRef]

75. Wang, H.; Shao, X.-M.; Jiang, Y.; Fang, X.-Q.; Wu, S.-H. The impacts of climate change on the radial growth of Pinus koraiensis along elevations of Changbai Mountain in northeastern China. *For. Ecol. Manag.* **2013**, *289*, 333–340. [CrossRef]
76. Gai, X.; Wang, S.; Zhou, L.; Wu, J.; Zhou, W.; Bi, J.; Cao, L.; Dai, L.; Yu, D. Spatiotemporal evidence of tree-growth resilience to climate variations for Yezo spruce (Picea jezoensis var. komarovii) on Changbai Mountain, Northeast China. *J. For. Res.* **2018**, *31*, 927–936. [CrossRef]
77. Hiura, T.; Go, S.; Iijima, H. Long-term forest dynamics in response to climate change in northern mixed forests in Japan: A 38-year individual-based approach. *For. Ecol. Manag.* **2019**, *449*, 117469. [CrossRef]
78. Kim, E.S.; Lee, J.S.; Park, G.E.; Lim, J.-H. Change of subalpine coniferous forest area over the last 20 years. *J. Korean Soc. For. Sci.* **2019**, *108*, 10–20. [CrossRef]
79. Zhu, L.; Cooper, D.J.; Yang, J.; Zhang, X.; Wang, X. Rapid warming induces the contrasting growth of Yezo spruce (Picea jezoensis var. microsperma) at two elevation gradient sites of northeast China. *Dendrochronologia* **2018**, *50*, 52–63. [CrossRef]
80. Pulliam, H. On the relationship between niche and distribution. *Ecol. Lett.* **2000**, *3*, 349–361. [CrossRef]
81. Michalet, R.; Maalouf, J.-P.; Choler, P.; Clément, B.; Rosebery, D.; Royer, J.-M.; Schöb, C.; Lortie, C. Competition, facilitation and environmental severity shape the relationship between local and regional species richness in plant communities. *Ecography* **2014**, *38*, 335–345. [CrossRef]
82. Antúnez, P. Main environmental variables influencing the abundance of plant species under risk category. *J. For. Res.* **2021**, *33*, 1209–1217. [CrossRef]
83. Kolström, M.; Lindner, M.; Vilén, T.; Maroschek, M.; Seidl, R.; Lexer, M.J.; Netherer, S.; Kremer, A.; Delzon, S.; Barbati, A.; et al. Reviewing the Science and Implementation of Climate Change Adaptation Measures in European Forestry. *Forests* **2011**, *2*, 961–982. [CrossRef]
84. Jandl, R.; Spathelf, P.; Bolte, A.; Prescott, C.E. Forest adaptation to climate change—Is non-management an option? *Ann. For. Sci.* **2019**, *76*, 48. [CrossRef]
85. Vozmishcheva, A.S.; Bondarchuk, S.N.; Gromyko, M.N.; Kislov, D.E.; Pimenova, E.A.; Salo, M.A.; Korznikov, K.A. Strong Disturbance Impact of Tropical Cyclone Lionrock (2016) on Korean Pine-Broadleaved Forest in the Middle Sikhote-Alin Mountain Range, Russian Far East. *Forests* **2019**, *10*, 1017. [CrossRef]
86. Kislov, D.E.; Korznikov, K.A.; Altman, J.; Vozmishcheva, A.S.; Krestov, P.V. Extending deep learning approaches for forest disturbance segmentation on very high-resolution satellite images. *Remote Sens. Ecol. Conserv.* **2021**, *7*, 355–368. [CrossRef]
87. Korznikov, K.; Kislov, D.; Doležal, J.; Petrenko, T.; Altman, J. Tropical cyclones moving into boreal forests: Relationships between disturbance areas and environmental drivers. *Sci. Total Environ.* **2022**, *844*, 156931. [CrossRef]

forests MDPI

Article

Optimal Management Strategies to Maximize Carbon Capture in Forest Plantations: A Case Study with *Pinus radiata* D. Don

Alex Altamirano-Fernández [1], Alejandro Rojas-Palma [1] and Sergio Espinoza-Meza [2,*]

[1] Departamento de Matemática, Física y Estadística, Facultad de Ciencias Básicas, Universidad Católica del Maule, Talca 3460000, Chile
[2] Facultad de Ciencias Agrarias y Forestales, Universidad Católica del Maule, Talca 3460000, Chile
* Correspondence: espinoza@ucm.cl

Abstract: Plantations with fast-growing species play a crucial role in reducing global warming and have great carbon capture potential. Therefore, determining optimal management strategies is a challenge in the management of forest plantations to achieve the maximum carbon capture rate. The objective of this work is to determine optimal rotation strategies that maximize carbon capture in forest plantations. By evaluating an ecological optimal control problem, this work presents a method that manages forest plantations by planning activities such as reforestation, felling, thinning, and fire prevention. The mathematical model is governed by three ordinary differential equations: live biomass, intrinsic growth, and burned area. The characterization of the optimal control problem using Pontryagin's maximum principle is analyzed. The model solutions are approximated numerically by the fourth-order Runge–Kutta method. To verify the efficiency of the model, parameters for three scenarios were considered: a realistic one that represents current forestry activities based on previous studies for the exotic species *Pinus radiata* D. Don, another pessimistic, which considers significant losses in forest productivity; and a more optimistic scenario which assumes the creation of new forest areas that contribute with carbon capture to prevent the increase in global temperature. The model predicts a higher volume of biomass for the optimistic scenario, with the consequent higher carbon capture than in the other two scenarios. The optimal solution for the felling strategy suggests that, to increase carbon capture, the rotation age should be prolonged and the felling rate decreased. The model also confirms that reforestation should be carried out immediately after felling, applying maximum reforestation effort in the optimistic and pessimistic scenarios. On the other hand, the model indicates that the maximum prevention effort should be applied during the life cycle of the plantation, which should be proportional to the biomass volume. Finally, the optimal solution for the thinning strategy indicates that in all three scenarios, the maximum thinning effort should be applied until the time when the fire prevention strategy begins.

Keywords: ecological model; biomass volume; carbon dioxide; optimal control; numerical simulation

Citation: Altamirano-Fernández, A.; Rojas-Palma, A.; Espinoza-Meza, S. Optimal Management Strategies to Maximize Carbon Capture in Forest Plantations: A Case Study with *Pinus radiata* D. Don. *Forests* **2023**, *14*, 82. https://doi.org/10.3390/f14010082

Academic Editors: Daniela Dalmonech, Alessio Collalti and Gina Marano

Received: 25 November 2022
Revised: 26 December 2022
Accepted: 29 December 2022
Published: 1 January 2023

1. Introduction

Carbon dioxide (CO_2) is one of the main greenhouse gases (GHG) in the atmosphere. Multiple human activities in most industrialized countries have contributed to the increase in this gas and have exacerbated the negative effects of climate change. According to the latest report of the Intergovernmental Panel on Climate Change (IPCC), climate change is devastating today, in particular, because of the changes in the patterns of humidity, temperature, winds, snow, and ice, especially in coastal zones. These changes in climate conditions could have negative impacts on human health, agriculture, and the economy [1–3]. Under this worldwide situation, governments are making cooperative efforts agreements (e.g., the Paris Agreement and the Kyoto Protocol) to create new forest areas to help prevent the global average temperature rising more than 2 °C during the 21st century [4–6]. Forest ecosystems cover approximately 4100 billion hectares of the Earth's surface and have a huge potential for carbon capture [7]. Of this total area, approximately 45% are exotic plantations whereas the other 55% corresponds to native forests [8]. Because forest ecosystems can store the largest amounts of carbon [9], it has been suggested that expanding forest areas and prolonging the rotation age (i.e., the growth period required to derive maximum value from a stand of timber), especially in exotic forest plantations [10], are key strategies to maximize carbon capture and mitigate the negative effects of global climate change [11]. There is a large body of literature where carbon capture is estimated [12,13]. In a temperate forest in Southern Europe, the aboveground carbon capture in the species *Eucalyptus nitens* (Deane and Maiden), *Eucalyptus globulus* Labill, and *P. radiata*, with rotation ages ranging from 10 to 35 years, was estimated to be from 443 to 634 Tn C ha^{-1} [14]. The carbon sequestration with the same species established in Chile was 212 Tn C ha^{-1} for *P. radiata*, 180 Tn C ha^{-1} for *E. nitens*, and 117 Tn C ha^{-1} for *E. globulus* (age of 20–24 years for *P. radiata* and 10–14 years for Eucalyptus) [15]. On the other hand, in Panama the carbon stored in *Tectona grandis* E.L (Teca) plantations during 1 and 10 years was estimated to be 2.9 Tn C ha^{-1} and 40.7 Tn C ha^{-1}, respectively [16].

On the other hand, there are studies on the oil palm (*Elaeis guineensis* Jacq) which, due to its high biomass production and expansion dynamics, plays an important role in carbon capture [17]. By means of mathematical modelling the dynamics of both oil production and carbon capture have been studied [18]. In [19], they formulated an optimal control problem based on a system of ordinary differential equations that relate the dynamics of young and mature trees and considers felling as a control variable. The authors concluded that palm oil production and carbon capture increases with a controlled felling rate.

Notwithstanding, to increase CO_2 capture the trees must remain for longer periods in the field, which delays the rotation age [20,21]. However, in some situations, it is risky to prolong the rotation age in order to increase carbon capture, since it increases the probability of forest fires when there is more fuel in the field. More frequent forest fires will increase CO_2 levels in the atmosphere, causing extreme climate events and decreasing relative humidity in many regions of the world [22]. To model the probability of forest fire occurrence some authors have used the Faustmann model generalized to the stochastic Poisson process [23], whereas others have studied this phenomenon by using the Bellman equation to determine the optimal rotation age in a forest stand that produces timber and carbon benefits under fire risk [24]. The authors showed that higher fire risk will reduce the optimal rotation age due to a lack of fire prevention and low carbon prices, while a higher carbon price will increase the rotation age, thus obtaining a higher ecological benefit. It is known that fires contribute to the increase of CO_2 in the atmosphere. In [25] they developed a meteorological fire index to predict the risk of fire occurrence and help forest managers take appropriate preventive measures. The authors determined that relative humidity is a simple and feasible parameter to describe the occurrence of fires. Several mathematical models have been developed to describe the dynamics of CO_2 capture in reforestation projects [26–28]. The atmospheric CO_2 concentration decreases as the rate of reforestation increases. Also in [29], they presented a study to model the greenhouse effect caused by CO_2 emissions through the optimal control theory. In the model, the authors addressed the optimization of investments in reforestation and clean technologies associated with state variables such as CO_2 emissions, planted area, and Gross Domestic Product (GDP). They concluded that it is more efficient to invest in reforestation than in clean technologies.

Because forested areas can contribute to climate change mitigation, it is necessary to find optimal management strategies that maximize carbon capture. Strategies such as large-scale reforestations are efficient in capturing huge amounts of carbon [30], whereas the optimization of thinning, fire prevention, and harvesting strategies can also reduce CO_2 emissions in forest plantation management [31]. In [32] they applied a thinning strategy in Korean pine (*Pinus koraiensis* Sieb. et Zucc.) forest plantations and determined that the optimal rotation age that maximizes wood production and carbon capture was at the age of 86 years. In another study on oil palm [18], they applied the optimal control theory to model the dynamics of biomass growth and intrinsic biomass growth as state variables and considered felling as a control variable. The authors showed that the maximum oil production and carbon capture was reached at the age of 20 years. However, to our knowledge, no mathematical models have simultaneously modeled the relationship between the living biomass, the intrinsic biomass growth, the burned area, the reforestation, the felling and thinning, the fire prevention, and the relative humidity. Recently, [33] modeled the effects of the dynamics of living biomass, intrinsic growth, and burned area on carbon capture in forest plantations. The authors showed that biomass decreases in each cycle of regeneration because of forest fires, and suggested a strategy based on fire prevention in order to obtain maximum carbon capture. In this context, the objective of the present work is to determine optimal rotation strategies that maximize carbon sequestration in forest plantations. Based on the optimal control theory, a mathematical model is proposed to describe the dynamic relationship of carbon capture in forest plantations with control strategies such as reforestation, felling, fire prevention, and thinning, which are associated with state variables such as living biomass, intrinsic growth, and burned area. To verify the efficiency of the model, three scenarios are considered: realistic, pessimistic, and optimistic, using numerical methods to approximate its solution. In the case of the realistic scenario we tested with data of the species *P. radiata*.

2. Materials and Methods

2.1. The Mathematical Model

Referring to the models of [33,34], we created a mathematical model that studies optimal rotation strategies that maximize carbon sequestration in forest plantations. This model is based on a system of three ordinary differential equations (ODEs) that are governed by three state variables, $B(t)$, $r(t)$, and $I(t)$, that denote the amount of living biomass (considers aboveground and belowground biomass), intrinsic biomass growth, and burned area, respectively. From the state variables, four control strategies are associated: reforestation $R(t)$, felling $F(t)$, fire prevention $S(t)$, and thinning $T(t)$. In the following, the notation "dot" represents the derivative of a variable with respect to time t. The dynamics of the live biomass has a logistic growth with a carrying capacity K. The biomass also increases proportionally and indirectly with respect to reforestation activities and decreases proportionally due to the immediate effects of fire, felling, and thinning. The contribution of relative humidity is not considered in the biomass dynamics, since it exceeds the carrying capacity [33]. The differential equation governing biomass is as follows.

$$\dot{B}(t) = r(t)B(t)\left(1 - \frac{B(t)}{K}\right) + [\beta R(t)]B(t) - [\mu_1 I(t) + \sigma F(t) + \tau\, T(t)]B(t). \quad (1)$$

From Equation (1), β is the rate at which biomass increases with respect to reforestation, μ_1 is the rate at which biomass decreases due to fire effects, σ is the rate at which biomass decreases due to felling effects, and τ is the rate at which biomass decreases due to thinning (there is an instantaneous decrease in biomass).

Intrinsic growth was originally studied in [34]. Here, we consider that thinning has an indirect contribution with respect to individual growth in a linear manner, since, in the long term, it is related to biomass to control density and timber quality [35], such that

$$\dot{r}(t) = r_0 - \rho r(t) + \nu T(t). \quad (2)$$

In Equation (2), r_0 represents the maximum individual growth rate under ideal conditions [36] and ρ is the effect of the natural mortality rate on individual growth. In the case of thinning, in the long term, this silvicultural operation enhances the growth of the remaining trees, and ν is the parameter by which the intrinsic growth of the living biomass is increased by the effects of thinning.

Finally, the burned area increases as biomass increases, but this increasing behavior cannot be unlimited. Therefore, a relationship between burned area and biomass is considered that limits the growth of burned area without inhibiting fire. On the other hand, burned area decreases due to fire prevention, thinning, and relative humidity. Thus, the dynamics of the burned area are

$$\dot{I}(t) = \mu_2 I(t)\left(\frac{B(t)}{1+B(t)}\right) - \theta S(t) - \eta T(t) - hI(t). \tag{3}$$

From Equation (3), μ_2 is the fire rate, θ is the fire prevention rate, η is the thinning rate, and h is the relative humidity threshold at which fire occurs.

Considering Equations (1)–(3), the mathematical model is presented by means of a system of nonlinear ODEs, as follows

$$\begin{cases} \dot{B}(t) = r(t)B(t)\left(1 - \frac{B(t)}{K}\right) + [\beta R(t)]B(t) - [\mu_1 I(t) + \sigma F(t) + \tau\, T(t)]B(t) \\ \dot{r}(t) = r_0 - \rho r(t) + \nu T(t) \\ \dot{I}(t) = \mu_2 I(t)\left(\frac{B(t)}{1+B(t)}\right) - \theta S(t) - \eta T(t) - hI(t), \end{cases} \tag{4}$$

satisfying the initial conditions $B(t_0) = B_0,\ r(t_0) = r_0$, and $I(t_0) = I_0$. It is assumed that the amount of carbon stored in the biomass is proportional to the amount of biomass [37], therefore $C(t) = \alpha B(t)$ where α is the rate of carbon capture. The following set of assumptions allows building a model that simulates control strategies that maximize carbon capture in forest plantations:

- The model is formulated for fast-growing managed forest plantations;
- There are many plantations with different ages, thus biomass never goes to zero;
- The ambient humidity is considered constant for simplicity;
- No soil fertilization in each cycle of the forest regeneration is considered;
- The area burned per year is considered, but human intentionality is not taken into account;
- There are no incentives for reforestation or carbon capture;
- The harvesting method corresponds to clear-cutting;
- In the thinning, the thinner, lower quality and less commercially valuable trees will be removed. Two types of thinning effects are considered, which are explained below;
- The presence of artificial irrigation is neglected in the model;
- The budget for fire prevention is limited;
- Intensive management of forestry is not included in our model;
- Trees burned by fire are replaced by new plants and natural regeneration is not used;
- The mortality rate of extreme events is neglected in our model.

Where the variables $B(t) \geq 0, r(t) \geq 0, I(t) \geq 0,\ R(t) \geq 0, F(t) \geq 0, S(t) \geq 0$ and $T(t) \geq 0$, the parameters $(\mathrm{K}, \sigma, \mathrm{h}, \mu_1, \mu_2, \tau, \beta, r_0, \rho, \eta, \theta, \alpha, \nu) \in R^{13}_{+}$.

The relationships between the variables and parameters used in the controlled mathematical model (4) are schematically represented in Figure 1.

Their notations, definitions, and units for variables and parameters are described in Tables 1 and 2, respectively.

Table 1. Notation, definition, and units of each variable.

Notation	Definition	Unit
$B(t)$	Volume of living biomass	$m^3\ ha^{-1}$
$r(t)$	Intrinsic growth of biomass	$year^{-1}$
$I(t)$	Burned area per year	$m^2\ year^{-1}$
$C(t)$	Carbon capture	$Tn\ C\ ha^{-1}year^{-1}$
$R(t)$	Forest reforestation	$ha\ year^{-1}$
$F(t)$	Forest felling	$ha\ year^{-1}$
$S(t)$	Fire prevention	$US\$m^2\ ha^{-1}year^{-1}$
$T(t)$	Forest thinning	$ha\ year^{-1}$

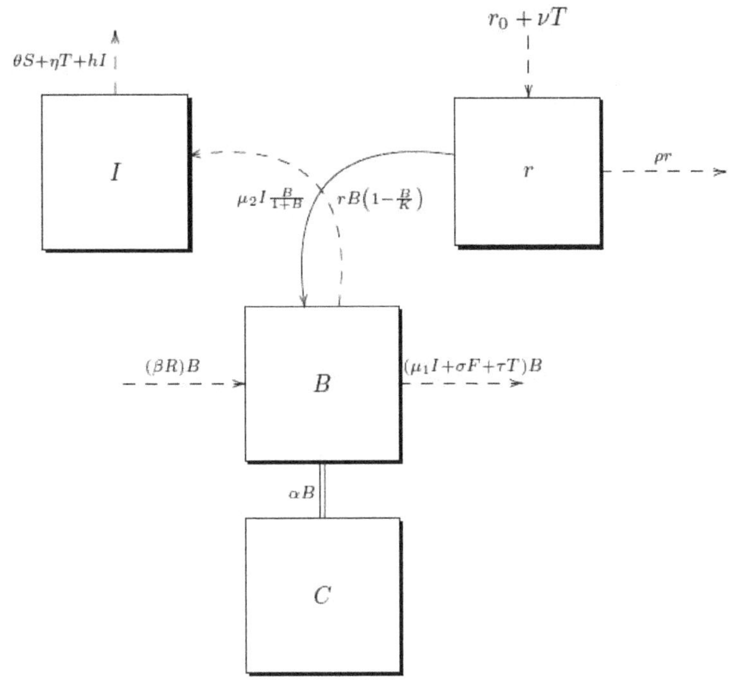

Figure 1. Mathematical model scheme. Dashed arrows indicate positive input and negative output values. The continuous arrow indicates that the term adds to the input variable but does not subtract the output variable. The double solid line represents the observable variable assumed to be proportional to biomass.

The parameters, their units, definition, and notation are as follows.

Table 2. Notation, definition, and units for each parameter.

Notation	Definition	Unit
β	Rate of increase in biomass due to the effects of reforestation	ha^{-1}
h	Relative humidity threshold to reduce fire	$year^{-1}$
μ_1	Rate at which biomass decreases due to fire effects	m^{-2}
μ_2	Fire parameter	$year^{-1}$
σ	Rate at which biomass decreases due to felling effects	ha^{-1}
τ	Rate at which biomass decreases due to thinning effects	ha^{-1}
r_0	Maximum growth rate	$year^{-2}$
ρ	Natural mortality rate	$year^{-1}$
ν	Rate of increase in thinning over individual growth	ha^{-1}
θ	Fire prevention rate	ha $US\$^{-1} year^{-1}$
η	Thinning rate	m^2 ha^{-1} $year^{-1}$
α	Carbon capture rate	Tn C m^{-3} $year^{-1}$

2.2. The Optimal Control Problem

An optimal ecological control problem is presented with the objective to maximize the objective function J representing the carbon capture in a fixed period $\left[0, T_f\right]$, such that

$$J(S, R, F, T) = \int_0^{T_f} \left[\omega_1 C(t) + \omega_2 S^2(t) + \omega_3 R^2(t) - \omega_4 F^2(t) - \omega_5 T^2(t)\right] dt, \quad (5)$$

subject to the state variables of the system (4). The state variables B, r, and I are assumed free at the final time. The initial conditions are the real and adjusted values for the species *P. radiata* from studies carried out in central Chile:

$$
\begin{gathered}
B(0) = B_0 \text{ (real)}, \quad r(0) = r_0\text{(real)}, \quad I(0) = I_0\text{(adjusted)} \\
B\left(T_f\right) = \text{free}, \; r\left(T_f\right) = \text{free}, \; I\left(I_f\right) = \text{free},
\end{gathered}
\tag{6}
$$

where the quantities ω_i, $i = 1, \cdots, 5$ are the weight parameters that balance the units of the terms of the objective function. It is considered $S^2(t)$, assuming that the contribution to carbon capture will be much higher if high prevention is applied and much lower if low prevention is applied. It is considered $R^2(t)$, assuming that the greater the reforestation, the greater the carbon capture. It is also considered $T^2(t)$, which implies that reducing the availability of fuel in the forest will contribute to decreasing the fire risk. Finally, it is considered $F^2(t)$, which increases the harvest rate to not compromise the owner's timber production. On the other hand, quadratic terms are introduced in the Lagrangian framework to avoid the problem being linear, thus usual techniques can be applied [38].

To prove the existence of the optimal control problem we will follow the results of [39,40], which is demonstrated in detail in Appendix A. So, now we are in the condition to characterize the optimal control problem by means of Pontryagin's maximum principle.

To solve the optimal control problem, Pontryagin's maximum principle is used to characterize the optimal controls $S^*,\ R^*,\ F^*, T^*$ [41]. To do this, it is necessary to determine the expressions for the adjoint variables using the Hamiltonian function [42].

$$
\begin{aligned}
H(B,r,I,R,F,T,S,\lambda_1,\ \lambda_2,\lambda_3,t) &= \left[\omega_1 C + \omega_2 S^2 + \omega_3 R^2 - \omega_4 F^2 - \omega_5 T^2\right] \\
&+ \lambda_1 \left[rB\left(1 - \tfrac{B}{K}\right) + [\beta R]B - [\mu_1 I + \sigma F + \tau T]B\right] \\
&+ \lambda_2[r_0 - \rho r + \nu T] + \lambda_3\left[\mu_2 I\left(\tfrac{B}{1+B}\right) - \theta S - \eta T - hI\right].
\end{aligned}
\tag{7}
$$

From Equation (7), the adjoint variables λ_1, λ_2, and λ_3 satisfy the following adjoint ODE.

$$
\begin{aligned}
\dot{\lambda_1}(t) &= -\tfrac{\partial H}{\partial B} = -\alpha\omega_1 - \lambda_1\left[r\left[1 - \tfrac{2B}{K}\right] + \beta R - [\mu_1 I + \sigma F + \tau T]\right] - \lambda_3 \tfrac{\mu_2 I}{(1+B)^2} \\
\dot{\lambda_2}(t) &= -\tfrac{\partial H}{\partial r} = -\lambda_1 B\left(1 - \tfrac{B}{K}\right) + \lambda_2 \rho \\
\dot{\lambda_3}(t) &= -\tfrac{\partial H}{\partial I} = \lambda_1 \mu_1 B - \lambda_3\left[\tfrac{\mu_2 B}{1+B} - h\right].
\end{aligned}
\tag{8}
$$

Since the initial conditions for the state variables given in Equation (6) are free, the transversality conditions for the adjoint variables given in Equation (8) are:

$$
\lambda_1\left(T_f\right) = 0, \;\; \lambda_2\left(T_f\right) = 0, \;\; \lambda_3\left(T_f\right) = 0. \tag{9}
$$

Using Pontryagin's maximum principle, the optimality of the control variables $S,\ R,\ F$ and T establishes the characterization of the optimal controls $S^*,\ R^*,\ F^*$, and T^* which satisfy the necessary first-order conditions

$$
\begin{aligned}
S^* &= min\left(max\left(0, \tfrac{\lambda_3\theta}{2\omega_2}\right), 1\right) \\
R^* &= min\left(max\left(0, -\tfrac{\lambda_1\beta B}{2\omega_3}\right), 1\right) \\
F^* &= min\left(max\left(0, -\tfrac{\lambda_1\sigma B}{2\omega_4}\right), 1\right) \\
T^* &= min\left(max\left(0, -\tfrac{(\lambda_1\tau B - \lambda_2\nu + \lambda_3\eta)}{2\omega_5}\right), 1\right).
\end{aligned}
\tag{10}
$$

For solving optimal control problems three approaches are feasible: direct methods, indirect methods, and dynamic programming [43,44]. We used the indirect methods because they are more robust than the base on the classical theory of Pontryagin's maximum principle that reduces the optimal control problem to the solution of a boundary value problem. We used the fourth-order Runge–Kutta scheme using Octave/MATLAB software [44,45]. In the following section, we performed numerical simulations to approximate the solutions to the optimal control problem. We simulate using the fourth-order forward Runge–Kutta method [38]. This iterative method consists of solving the controlled system (4) using a fourth-order forward Runge–Kutta scheme and the traversal or terminal conditions given in Equation (9) in a time interval $\left[0, T_f\right]$. Then, the adjoint system (8) is solved by a fourth-order backward Runge–Kutta scheme using the solution of the current iteration of the controlled system (4). The characterization of the controls of the Equation (10) is updated through a convex combination of the above controls. The procedure stalls if the values of the variables of the previous iteration are very close to the present iteration [44].

3. Results and Discussion

The main objective of this work is to determine the optimal strategies of reforestation $R(t)$, felling $F(t)$, fire prevention $S(t)$, and thinning $T(t)$ that maximize carbon capture in fast-growing forest plantations. To start with, the final time is set equal to $T_f = 200$ years. The initial conditions assumed for the biomass of *P. radiata* plantations are $B(0) = 7.5$ m^3 ha^{-1} and, for the intrinsic growth, a maximum growth $r(0) = 0.0725$ year^{-1}, given in [46], is considered. For the burned area, an average $I(0) = 0.07$ m^2 year^{-1} is adjusted. Then, the Runge-Kutta solver of order four is run for the following three scenarios:

Realistic Scenario. Real parameters from Table 3 (see Appendix B) were considered for the species *P. radiata* from studies carried out in central Chile.

Table 3. Real parameters for *P. radiata*.

K	μ_1	β	σ	τ	ρ	r_0	ν	μ_2	θ	η	h	α
400	0.055	2×10^{-5}	2×10^{-5}	1×10^{-5}	0.06	0.07275	0.159	0.4097	0.15	9.3×10^{-7}	0.135	0.5

Pessimistic Scenario. According to the IPCC, in recent decades global warming has increased GHG emissions and their accumulation is causing irreversible climatic changes. As a result, the Earth is experiencing an increase in temperature, which will lead to a sustained decrease in ambient relative humidity and a prolongation of the duration and frequency of drought events [47] and fires [48]. In this context, this scenario considers a decrease in ambient relative humidity, which will have a negative effect on biomass growth, and will also favor conditions for an increase in the frequency and intensity of fires which, in turn, will increase the burned area [49]. This scenario also considers a decrease in the budget dedicated to fire prevention, due to the effects that the global health and war situation will have on local and global economies [50,51]. For this scenario, the fixed values of the parameters shown in Table 4, which correspond to an artificial data set, are considered.

Table 4. Pessimistic parameters for forest plantations.

K	μ_1	β	σ	τ	ρ	r_0	ν	μ_2	θ	η	h	α
400	0.02	0.1	0.09	1×10^{-5}	0.06	0.061	0.03	0.47	0.052	9.3×10^{-7}	0.11	0.5

Optimistic Scenario. This scenario assumes that there will be a substantial increase in the annual rate of afforestation and reforestation, due to the growing concern to prevent the global average temperature from rising over the next century and the growth of new commodities based on environmental values, such as carbon capture [5,52,53]. The Paris Agreement and the Kyoto Protocol have designed economic instruments that provide financial incentives for governments to protect the environment using the private sector for strict pollution standards [29]. In this scenario, most of the parameters of the realistic scenario in Table 3 were considered, assuming an afforestation and reforestation rate higher than that of the realistic scenario ($\beta = 0.04$). By increasing the rate of afforestation and reforestation, carbon capture will increase, decreasing the greenhouse effect and regulating environmental temperature and humidity patterns. According to [54,55], they showed that trees have a positive effect on relative humidity increase and temperature reduction. Thus, this scenario assumes an increase in ambient relative humidity in the area where *P. radiata* plantations are concentrated ($h = 0.14$). On the other hand, the presence of financial incentives would increase the fire prevention budget by 120% ($\theta = 0.18$) in activities such as thinning, which will indirectly contribute positively to individual growth in the long term ($\nu = 0.42$) and decrease the continuity of fuel susceptible to fire.

The optimal trajectories for live biomass are similar for the three scenarios (Figure 2a), however, as expected, in the optimistic scenario there is greater biomass accumulation (the total area under the curve is 12 300) compared to the other scenarios (10 739 in the realistic and 5 125.4 in the pessimistic scenarios, respectively). In general, the minimum biomass volume coincides with the maximum burned area (Figure 2c), although in the pessimistic scenario this occurs at earlier ages as a result of lower ambient humidity and higher fire propensity [56], which shortens the rotation age. It is also observed that in the following plantation rotation cycles the biomass is lower than in the first cycle for the three scenarios. It could be due to the presence of fires in the three scenarios which causes the volatilization of the main soil nutrients [57]. Volatilization of some major soil nutrients, such as nitrogen and phosphorus, can affect tree growth and limit terrestrial carbon sequestration [58,59]. This model does not consider artificial fertilization of the soil and the plantation starts growing when the minimum burned area occurs. In [10], they argue that a higher reforestation rate, together with prolonging the rotation age, are key strategies to maximize carbon capture and mitigate the negative effects of global climate change. The model corroborates the above since the optimistic scenario considers a higher rate of afforestation and reforestation than the realistic scenario, which increases the volume of forest biomass and there is a prolongation of the rotation age to 29 years, as opposed to the earlier rotation ages determined by the pessimistic (23 years) and realistic (27 years) scenarios. The rotation age of the realistic scenario is within the rotation age range reported by [60] in operational plantations of *P. radiata* in central Chile. In the realistic scenario the model maintains the same rotation age as in [60], while for the pessimistic scenario the rotation age was reduced since the forest was affected by the fire, which forces early felling to avoid damage by new fires, with the consequent emission of CO_2 to the atmosphere. In the case of the optimistic scenario, the model prolongs the rotation age due to the positive impact of market incentives for environmental protection.

Carbon capture and burned area follow the same trend in all three scenarios (Figure 2b,c). As expected, the maximum burned area occurs years after the maximum biomass volume is produced. The realistic scenario shows a small burned area under the curve of 6258.8, which suggests that the model realistically reflects the current fire prevention and firefighting situation in Chile, which is more efficient [61,62]. It is also observed in the optimistic scenario that the burned area under the curve is 8313.2, which is greater than the realistic scenario. That scenario happens because in the optimistic scenario there is a greater volume of biomass (Figure 2a). However, in the pessimistic scenario, despite a low volume of biomass, there is a greater burned area under the curve of 10,974 compared to the other two scenarios. This situation is because the relative humidity threshold is the lowest of the three scenarios and the forest is more prone to burning, which decreases the biomass.

Finally, Figure 2d shows that the intrinsic growth variable is higher in the optimistic scenario, while the pessimistic scenario is lower and with a flat trend due to the negative effect of the higher number of fires, which affects individual plantation growth.

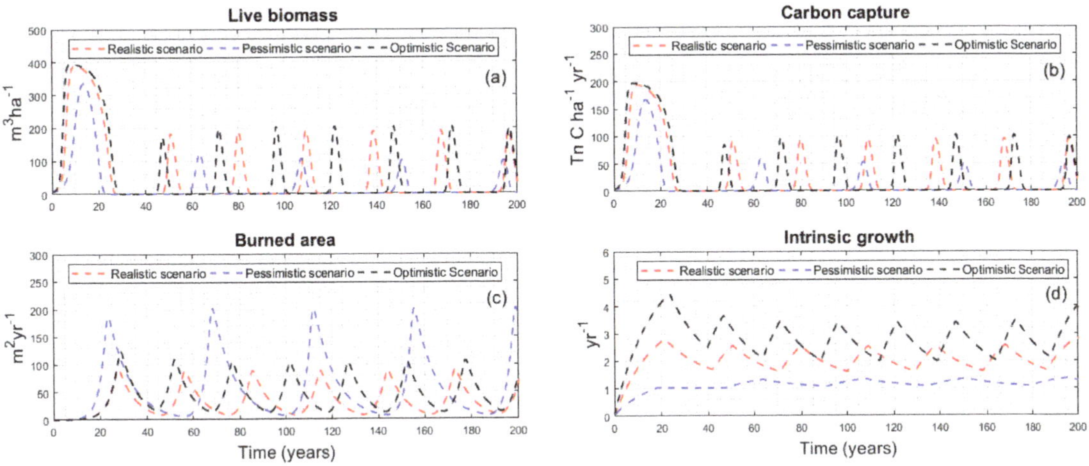

Figure 2. Optimal state trajectories for living biomass, intrinsic growth, burned area, and carbon capture for the three scenarios: (**a**) represents the dynamics of living biomass, (**b**) represents carbon sequestration, (**c**) represents the dynamics of burned area, and (**d**) represents intrinsic growth.

Figure 3 shows that the carbon accumulated in 200 years is 6050, 5240, and 2560 Tn C for the optimistic, realistic, and pessimistic scenarios, respectively. Overall, the optimistic scenario is 57.69% higher than the pessimistic scenario and 13.4% higher than the realistic scenario.

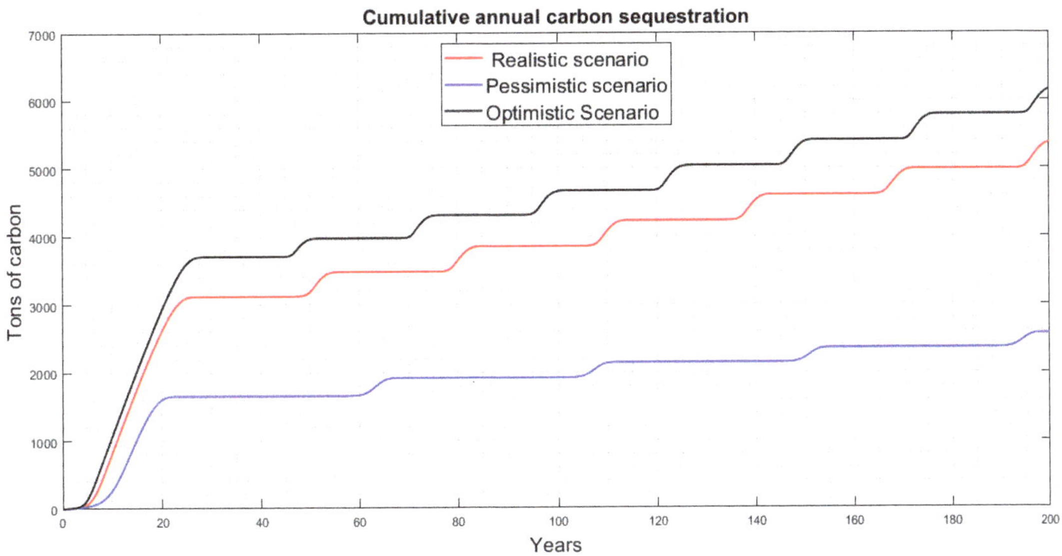

Figure 3. Cumulative carbon in the three scenarios for a period of 200 years.

As abovementioned, the optimal control theory was applied in [18,29] by independently assessing incentives that promote reforestation, the use of clean technologies, or other factors that modify the rotation age as control strategies to maximize carbon capture. However, in the present work, optimal control theory is applied with four strategies accordingly in three simulation scenarios (Figure 4). The results indicate that in the realistic and optimistic scenarios the felling strategy considers an optimal rotation age of 29 and 27 years, and no maximum felling effort is applied, while in the pessimistic scenario the optimal rotation age is shorter and maximum felling effort should be applied throughout the plantation life cycle, and the maximum effort has a limit value of one (Equation (10)). The previous facts are due to the presence of fire risk, which decreases the optimal rotation age [20]. Then, as fires increase, short rotations could be used and maximum felling effort applied as in the pessimistic scenario, so as not to increase the release of CO_2 to the atmosphere and not compromise the forest owner business. As previously mentioned, in the case of the optimistic and realistic scenarios, no maximum felling effort is applied. This coincides with the Chilean national reality since, despite the increase in demand for timber in the country in the last 15 years, the felling rate has dropped from 55,000 ha year^{-1} to 25,000 ha year^{-1} [63,64]. While this is positive from an environmental point of view, as more planted hectares are being maintained and contribute to mitigating climate change, it could lead to a shortage of wood for processing plants, which would increase the pressure to shorten the rotation age. However, this hypothesis needs further analysis. According to the IPCC, a significant percentage of GHGs (30%–40%) can be reduced by avoiding felling, forest degradation, and the recovery of forest areas.

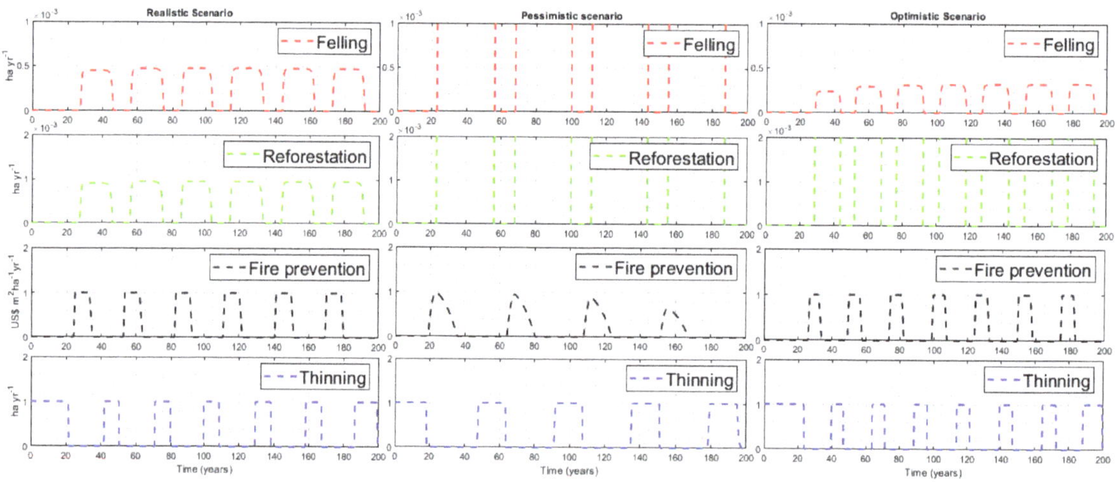

Figure 4. Optimal trajectories for the control variables in the three scenarios. The first column represents the realistic scenario, the second column the pessimistic scenario, and the third column the optimistic scenario.

On the other hand, in all three scenarios, the reforestation strategy is applied when felling begins. The model shows that maximum reforestation effort should be applied for the pessimistic and optimistic scenarios, while in the realistic scenario no maximum reforestation effort is applied. Thus, the model reports that today 100% of the reforestation rate is not reached, despite the Paris Agreement and the Kyoto Protocol that offer financial incentives to governments to increase forest area. According to Chilean Decree Law 701, between 1998 and 2015 36% of the area was forested, 39% of the area corresponds to reforestation, and more than 20,000 hectares are subsidized for forest management, which meant an investment of 388 million USD for the State in that period. According to [29] they suggest that it is more efficient to invest in reforestation than other practices such as clean technologies, since forested areas can contribute to climate change mitigation in a more efficient and environment friendly way. However, even though maximum reforestation effort is applied in the pessimistic scenario, the biomass volume is lower than in the other two scenarios (Figure 2a), which corroborates the negative effect of the larger burned area on biomass growth.

In the fire prevention strategy, the model suggests that in the three scenarios presented, the maximum prevention effort should be applied years after reaching the maximum biomass volume since the plantation in its adult state forms a continuous mass of fuel that is more difficult to control in case of fire. Therefore, the model based on this strategy, considering all plantations, suggests investing in fire prevention on average four years before and four years after felling (Figure 2c). Prevention applied before helps to minimize the fire risk in plantations ready to be felled, whereas prevention applied after helps to manage the forest residues in the soil (i.e., fuel) by felling operations. On the other hand, in the pessimistic scenario, it is observed that there is a decrease in the prevention strategy of the following rotation cycles because the biomass decreases (Figure 2a). Therefore, the model communicates that the owner should invest according to the amount of existing biomass.

Finally, for the thinning strategy in the three scenarios, it is observed that the maximum thinning effort should be applied from the initial moment to the moment of fire prevention. It is also observed that in the case of the pessimistic scenario the thinning periods are longer because the forest is more prone to burning.

In general, the optimal rotation age for the realistic scenario in the following rotation cycles turned out to be 29 years, whereas for the optimistic scenario in the following rotation cycles this is shortened to 24 years; this is because the presence of fires shortens the rotation age. Even though the rotation age is shortened there is a greater biomass and therefore greater carbon capture than in the other two scenarios (Figure 2a,b). However, in the pessimistic scenario in the following cycles, the rotation age is prolonged to ca. 45 years, because forests need a long time to recover from the fire. Therefore, the rotation age should be applied at the maximum time that the burned area reaches.

4. Conclusions

In this study, we determined optimal management strategies that maximize carbon capture in fast-growing forest plantations using the optimal control theory. The model effectively simulates the optimal dynamics of live biomass, intrinsic growth, and burned area to consider four strategies such as reforestation, felling, fire prevention, and thinning. To evaluate the effectiveness of the model, three scenarios have been considered: realistic, pessimistic, and optimistic. The parameters for modeling the realistic scenario were based on real data for the exotic species *P. radiata*.

The model predicts a higher biomass volume for the optimistic scenario, with a consequent higher carbon capture than in the other two scenarios. The optimal solution for the felling strategy suggests that, in order to increase carbon capture, the rotation age should be extended and the felling rate decreased. Likewise, the model corroborates that reforestation should be carried out immediately after felling, applying maximum reforestation effort in the optimistic and pessimistic scenarios. It is suggested that in order to increase CO_2 capture, large forestry companies and government agencies should increase investment in afforestation and reforestation. On the other hand, the model indicates that, although maximum prevention effort should be applied during the life cycle of the plantation, it should be proportional to the volume of biomass. Finally, the optimal solution for the thinning strategy indicates that in all three scenarios maximum thinning effort should be applied until the time when the fire prevention strategy is applied. Here, the optimistic scenario is considered as a possible alternative in forestry activities to maximize carbon capture. While this scenario yields the largest carbon capture, its implementation requires joint efforts between forest companies and the government to prolong the rotation age and economically incentivize reforestation, respectively.

Author Contributions: Conceptualization, A.A.-F., A.R.-P. and S.E.-M.; methodology, A.A.-F., A.R.-P. and S.E.-M.; software, A.A.-F.; experimental execution and validation, A.A.-F.; research, A.A.-F., A.R.-P. and S.E.-M.; resources, A.A.-F.; writing—revising and editing, A.A.-F. and S.E.-M.; scientific project coordination, A.R.-P. and S.E.-M.; funding acquisition, A.A.-F. All authors have read and agreed to the published version of the manuscript.

Funding: This research was partially funded by Universidad Católica del Maule Doctoral Studies Scholarship 2019 (Beca Doctoral Universidad Católica del Maule 2019).

Institutional Review Board Statement: Not applicable.

Informed Consent Statement: Not applicable.

Data Availability Statement: Not applicable.

Acknowledgments: A.A.-F. would like to thank the Vicerrectoría de Investigación y Postgrado at Universidad Católica del Maule, Chile. This work is part of A.A.-F.'s Ph.D. thesis in the program of Doctorado en Modelamiento Matemático Aplicado.

Conflicts of Interest: The authors declare no conflict of interest.

Appendix A. Existence of Solutions to the Optimal Control Problem

Then, from the system (4), we denote $u(t) = (R(t),\ F(t),\ T(t),\ S(t))$ four control variables and are associated with the three state variables $\mathrm{E}(t) = (B(t),\ r(t), I(t))$, they are bounded and measurable

$$\mathcal{U} = \{u : \text{ is measurable according to Lebesgue on } [0,1],\ 0 \leq F, R, T, S \leq 1\}, \tag{A1}$$

where $\mathcal{U}$ is the class of admissible controls. We will now mission the following theorem that guarantees the existence of solutions to the optimal control problem.

Theorem A1. *Assuming that* $\omega_4 + \omega_5 \geq \omega_2 + \omega_3$, $\omega_2 < 1 - \kappa_1$, $0 < \kappa_1 < 1$ *and* $\omega_3 < 1 - \kappa_2$, $0 < \kappa_2 < 1$, *there is a quadruple of optimal control* $(R^*,\ F^*,\ S^*,\ T^*) \in \mathcal{U}$, *such that*

$$J(R^*,\ F^*,\ S^*,\ T^*) = \max_{(S,R,F,T)\in\mathcal{U}} (J(S,R,F,T)) \tag{A2}$$

subject to the system of differential Equation (4) with non-negative initial conditions. If the following conditions are satisfied:

1. *The set of admissible controls and state variables of the problem is non-empty;*
2. *The admissible control class* $\mathcal{U}$ *is convex and bounded;*
3. *The right-hand side of the system Equation (4) is bounded by a nonlinear function that depends on the state and control variables;*
4. *The integral of the objective function is concave;*
5. *There exist positive constants* $d_1,\ d_2 > 0$ *and* $\gamma > 1$ *satisfying the integrating* J *of the objective functional, such that*

$$(R, F, S, T) = d_2 - d_1\left(S^2 + R^2 + F^2 + T^2\right)^{\gamma/2}$$

Proof of Theorem A1

1. The solutions of the system (4) are considered to be bounded in a finite time interval and making use of a result from [65], the existence of a solution for the controlled system can be assured;
2. From Equation (A1) the set of admissible controls $\mathcal{U}$ is known to be topologically closed and convex by definition;
3. For this point, let us represent the system (4) as follows:

$$\dot{\mathrm{E}}(t) = A\mathrm{E} + \mathcal{F}(\mathrm{E}) + Y(u) \tag{A3}$$

in its matrix form

$$\mathrm{E}=\begin{pmatrix} B \\ r \\ I \end{pmatrix},\ A = \begin{pmatrix} \beta R - (\sigma F + \tau T) & 0 & 0 \\ 0 & -\rho & 0 \\ 0 & 0 & h \end{pmatrix},$$

$$\mathcal{F}(\mathrm{E}) = \begin{pmatrix} rB\left(1-\frac{B}{K}\right) - \mu_1 IB \\ 0 \\ \mu_2 I\left(\frac{B}{1+B}\right) \end{pmatrix}, \; Y(u) = \begin{pmatrix} 0 \\ r_0 + \nu T \\ -\theta S - \eta T \end{pmatrix}$$

The system (A3) is a nonlinear system with a bounded coefficient. So, now the system (A3) is defined as follows:

$$G(\mathrm{E}) = A\mathrm{E} + \mathcal{F}(\mathrm{E}) + Y(u). \tag{A4}$$

From the second term of Equation (A4), we apply the inequality of Hölder we obtain

$$\begin{aligned} |\mathcal{F}(\mathrm{E}_1) - \mathcal{F}(\mathrm{E}_2)| &= \Big|\left(r_1B_1\left(1-\tfrac{B_1}{K}\right) - \mu_1 I_1 B_1\right) - \left(r_2B_2\left(1-\tfrac{B_2}{K}\right) - \mu_1 I_2 B_2\right) \\ &\quad + \left(\mu_2 I_1\left(\tfrac{B_1}{1+B_1}\right) - \left(\mu_2 I_2\left(\tfrac{B_2}{1+B_2}\right)\right)\right)\Big| \\ &\leq |r_1B_1 - r_2B_2| + \left|\tfrac{r_2B_2^2}{K} - \tfrac{r_1B_1^2}{K}\right| + \mu_1|I_1B_1 - I_2B_2| \\ &\quad + \mu_2|I_1B_1 - I_2B_2| \\ &\leq Z_1|B_1 - B_2| + Z_2|r_1 - r_2| + Z_3|I_1 - I_2| \\ &\leq \max(Z_1, Z_2, Z_3)(|B_1 - B_2| + |r_1 - r_2| + |I_1 - I_2|) \end{aligned}$$

where $Z_1 = \left[r_m + (\mu_1+\mu_2)\left(\frac{B_m(r_m+\beta R)}{\mu_1(1+B_m)} + r_m\right) + \frac{r_m}{K}(2B_m)\right]$, $Z_2 = \left(B_m + \frac{B_m^2}{K}\right)$, and $Z_3 = B_m(\mu_1 - \mu_2)$. In addition, by the constraints of the solutions of Equation (4) $r_m = \frac{r_0+\nu T}{\rho}$ and $B_m = K + K\frac{\beta R}{r_m}$ [33]. Then, the constant Z is positive, taking $Z = \max(Z_1, Z_2, Z_3, ||A|| < \infty$ which is independent of the state variables, we have that

$$|G(\mathrm{E}_1) - G(\mathrm{E}_2)| \leq Z|\mathrm{E}_1 - \mathrm{E}_2|. \tag{A5}$$

Following Equation (A5), it is stated that the function $G(\mathrm{E})$ is Lipschitz uniform continuous. From the definition of $\mathcal{U}$ and the restriction of B, r and I we can guarantee the existence of the solution of the controlled system [66,67].

4. To show the concavity of the integrand of the objective functional, let us denote as follows

$$\mathcal{N}(t, \mathrm{E}, u) = \omega_1 C(t) + \omega_2 S^2(t) + \omega_3 R^2(t) - \omega_4 F^2(t) - \omega_5 T^2(t),$$

for this, we must prove that

$$(1-q)\mathcal{N}(t, \mathrm{E}, u) + q\mathcal{N}(t, \mathrm{E}, v) \leq \mathcal{N}(t, \mathrm{E}, (1-q))u + qv) \tag{A6}$$

let u, v be two control vectors and $q \in (0,1)$. Applying Equation (A2), the definition of convex set, we obtain

$$\begin{aligned} &(1-q)\mathcal{N}(t, \mathrm{E}, u) + q\mathcal{N}(t, \mathrm{E}, v) - \mathcal{N}(t, \mathrm{E}, (1-q)u + qv) \geq \\ &\omega_2\left[\sqrt{q(1-q)}u_1 - \sqrt{q(1-q)}v_1\right]^2 + \omega_3\left[\sqrt{q(1-q)}u_2 - \sqrt{q(1-q)}v_2\right]^2 \\ &-\omega_4\left[\sqrt{q(1-q)}u_3 - \sqrt{q(1-q)}v_3\right]^2 - \omega_5\left[\sqrt{q(1-q)}u_4 - \sqrt{q(1-q)}v_4\right]^2 \end{aligned}$$

therefore if $\omega_4 + \omega_5 \geq \omega_2 + \omega_3$ one obtains from Equation (A6)

$$(1-q)\mathcal{N}(t, \mathrm{E}, u) + q\mathcal{N}(t, \mathrm{E}, v) - \mathcal{N}(t, \mathrm{E}, (1-q)u + qv) \leq 0$$

then the integrating of the objective function $\mathcal{N}(t, \mathrm{E}, u)$ is concave.

5. Finally, considering that $\omega_1 < 1 - \kappa_1,\ 0 < \kappa_1 < 1$ and $\omega_3 < 1 - \kappa_2,\ 0 < \kappa_2 < 1$ it follows that

$$\mathcal{N}(t, \mathrm{E}, u\ ) \leq \omega_1 C + (1 - \kappa_1)S^2 + (1 - \kappa_2)R^2 - \omega_4 F^2 - w_5 T^2$$
$$\mathcal{N}(t, \mathrm{E}, u\ ) \leq \omega_1 C + S^2 + R^2 - \kappa_1 S^2 - \kappa_2 R^2 - \omega_4 F^2 - \omega_5 T^2$$
$$\mathcal{N}(t, \mathrm{E}, u\ ) \leq 2 + \omega_1 \alpha B - \kappa_1 S^2 - \kappa_2 R^2 - F^2 - \omega_5 T^2$$
$$\mathcal{N}(t, \mathrm{E}, u\ ) \leq d_2 - d_1\left(R^2 + +S^2 + F^2 + T^2\right)$$

where d_2 depends on B, while for $d_1 = \min(\kappa_1, \kappa_2, \omega_3, \omega_5)$ and $\gamma = 2$ the required is obtained. It is satisfied that there exists an optimal control quadruple $J(R^*,\ F^*,\ S^*,\ T^*)$ such that $J(R, F, S, T)$ which was given in Equation (5) is maximized. □

Appendix B. Description of Parameters

The carrying capacity of live biomass (K) for a rotation age of 24 years in the species *P. radiata* reaches 400–450 m^3 ha^{-1} [68]. In this study, $K = 400$ m^3 ha^{-1} has been considered. Knowing that relative humidity is related to mean annual temperature and annual precipitation [69], and that relative humidity could contribute positively to reducing fire propagation [70], in [71] they evaluated the volume growth between sites in a range of six and seven years; the site with higher precipitation (1492 mm $year^{-1}$) and lower mean annual temperature compared to other sites (10.5 °C) obtained 47.3 m^3 ha^{-1} and the mean annual increase is 6.4 m^3 ha^{-1} $year^{-1}$. Then, the relative humidity rate is h = 6.4/47.3 = 0.135 $year^{-1}$. According to [72], a reforestation rate of 34304 ha was obtained in 2017, then $\beta = 1/34304$ = 0.00002 ha^{-1}. If the current situation is maintained, in which the annual planting rate does not increase and corresponds only to reforestation, which is the replacement of felled areas [72], then the felling rate is $\sigma = \beta$. In the case of the thinning, there is an instantaneous decrease in biomass because, with intensive silvicultural management, 50% of the trees per hectare are extracted [73,74]. Then, the rate of biomass declines due to the effects of thinning $\tau = \sigma/2 =$ 0.00001 ha^{-1}. On the other hand, according to [75] thinning reduces fire risk, in [76,77] the variation of leaf area before and after thinning is 0.008 m^2, and per hectare and year is 0.0001164 $ha^{-1}year^{-1}$. Therefore, the thinning rate acts on the biomass decrease, which could decrease the fire risk and, therefore, there will not be an increase in the burned area, given $\eta = 0.008 \times 0.0001164 = 0.000000093$ m^2 ha^{-1} $year^{-1}$. In addition, biomass also decreases due to the occurrence of forest fires. In Chile, the area of forest plantations affected by forest fires in the last 5 years corresponds to 3,960,000 m^2 [78]. Then, the rate of biomass decrease due to fires is the ratio between the area burned with respect to forest plantation species and the number of total fires recorded in Chile in the last thirty years, giving μ_1 = 218,413/3,960,000 = 0.055 m^{-2}. In [77], they show that the mean annual increment is 29.1 m^3 ha^{-1}. Then, the maximum intrinsic growth rate is the quotient between the mean annual increment and the carrying capacity, and we obtain r_0 = 29.1/400 = 0.07275 $year^{-1}$ (for simulation convenience we leave it at $year^{-1}$). The natural mortality of *P. radita* plantations is almost zero due to the high intensity of silvicultural treatments. However, due to the lack of silvicultural operations in our model, plants may die due to lack of fertilization and plague control which is not included in this model, then ρ = 0.06 $year^{-1}$ is considered. For the contribution of thinning to intrinsic growth, according to [71] 356 trees were thinned out of 1,156 trees that obtained an increase of 3.153 m, also have a volume increase of 5 m^3 ha^{-1} after thinning, leaving ν = (5 m^3 ha^{-1} ha^{-1})/(3.153 m)^3 = 0.159 ha^{-1}. On the other hand, according to [64] the area affected by fires in the last ten years for forest plantations of the genus Pinus on average corresponds to 34,541.98 ha, but in the year 2020 the area affected was 14,152.54 ha $year^{-1}$. Then, the increase in the burned area is the ratio between the burned area of *Pinus* in 2020 and the total area affected by fires in the last ten years is μ_2 = 14,152.54/34,541.98 = 0.4097 $year^{-1}$. In Chile large forestry companies have an estate of approximately 3 million hectares with exotic plantations, with a budget of approximately 20 million USD per year. Therefore, the fire prevention budget is $\theta = 3/20 = 0.15$ ha USD^{-1} year^{-1} [79]. The biomass dry weights were converted to carbon weight and, assuming a carbon content of 50% of the total biomass weight, the carbon capture rate is $\alpha = 0.50$ Tn C m^{-3} year^{-1} [80].

References

1. Pedersen, J.S.T.; Santos, F.D.; van Vuuren, D.; Gupta, J.; Coelho, R.E.; Aparício, B.A.; Swart, R. An assessment of the performance of scenarios against historical global emissions for IPCC reports. *Glob. Environ. Chang.* **2021**, *66*, 102199. [CrossRef]
2. Lam, M.K.; Lee, K.T.; Mohamed, A.R. Current status and challenges on microalgae-based carbon capture. *Int. J. Greenh. Gas Control* **2012**, *10*, 456–469. [CrossRef]
3. Ambardekar, A.A.; Siebenmorgen, T.J.; Counce, P.A.; Lanning, S.B.; Mauromoustakos, A. Impact of field-scale nighttime air temperatures during kernel development on rice milling quality. *Field Crops Res.* **2011**, *122*, 179–185. [CrossRef]
4. UNFCCC. Adoption of the Paris Agreement FCCC/CP/2015/L.9/Rev.1. United Nations Framework Convention on Climate Change. In Proceedings of the Parties Twenty-First Session, Paris, France, 30 November–11 December 2015.
5. Kim, Y.; Tanaka, K.; Matsuoka, S. Environmental and economic effectiveness of the Kyoto Protocol. *PLoS ONE* **2020**, *15*, e0236299. [CrossRef] [PubMed]
6. Karousakis, K. Incentives to reduce GHG emissions from deforestation: Lessons learned from Costa Rica and Mexico. *OECD Pap.* **2007**, *7*, 1–50. [CrossRef]
7. Dixon, R.K.; Solomon, A.; Brown, S.; Houghton, R.; Trexier, M.; Wisniewski, J. Carbon pools and flux of global forest ecosystems. *Science* **1994**, *263*, 185–190. [CrossRef] [PubMed]
8. FAO yPNUMA. El Estado de Los Bosques del Mundo 2020: Los Bosques, la Biodiversidad y las Personas, Roma. Available online: https://www.fao.org/documents/card/en/c/ca8642es (accessed on 8 August 2022).
9. Lewis, S.L.; Wheeler, C.E.; Mitchard, E.T.; Koch, A. Restoring natural forests is the best way to remove atmospheric carbon. *Nature* **2019**, *568*, 25–28. [CrossRef] [PubMed]
10. Nghiem, N. Optimal rotation age for carbon sequestration and biodiversity conservation in Vietnam. *For. Policy Econ.* **2014**, *38*, 56–64. [CrossRef]
11. Kaipainen, T.; Liski, J.; Pussinen, A.; Karjalainen, T. Managing carbon sinks by changing rotation length in European forests. *Environ. Sci. Policy* **2004**, *7*, 205–219. [CrossRef]
12. Fragoso-López, P.I.; Rodríguez-Laguna, R.; Otazo-Sánchez, E.M.; González-Ramírez, C.A.; Valdéz-Lazalde, J.R.; Cortés-Blobaum, H.J.; Razo-Zárate, R. Carbon sequestration in protected areas: A case study of an Abies religiosa (HBK) Schlecht. et Cham Forest. *Forests* **2017**, *8*, 429. [CrossRef]
13. Jörgensen, K.; Granath, G.; Lindahl, B.D.; Strengbom, J. Forest management to increase carbon sequestration in boreal Pinus sylvestris forests. *Plant Soil* **2021**, *466*, 165–178. [CrossRef]
14. Pérez-Cruzado, C.; Mansilla-Salinero, P.; Rodríguez-Soalleiro, R.; Merino, A. Influence of tree species on carbon sequestration in afforested pastures in a humid temperate region. *Plant Soil* **2012**, *353*, 333–353. [CrossRef]
15. Olmedo, G.F.; Guevara, M.; Gilabert, H.; Montes, C.R.; Arellano, E.C.; Barría-Knopf, B.; Gárate, F.; Mena-Quijada, P.; Acuña, E.; Bown, H.E.; et al. Baseline of carbon stocks in *Pinus radiata* and *Eucalyptus* spp. plantations of Chile. *Forests* **2020**, *11*, 1063. [CrossRef]
16. Derwisch, S.; Schwendenmann, L.; Olschewski, R.; Hölscher, D. Estimation and economic evaluation of aboveground carbon storage of Tectona grandis plantations in Western Panama. *New For.* **2009**, *37*, 227–240. [CrossRef]
17. Wakker, E.; Watch, S.; Rozario, J.D. *Greasy Palms: The Social and Ecological Impacts of Large-Scale Oil Palm Plantation Development in Southeast Asia*; AIDEnvironment: Amsterdam, The Netherlands, 2004.
18. Nasir, N.; Abd Aziz, M.I.; Banitalebi, A. Carbon absorption control model of oil palm plantation. *Sains Malays.* **2019**, *48*, 921–925. [CrossRef]
19. Abd Aziz, M.I.; Nasir, N.; Banitalebi, A. The Optimal Felling Rate in the Palm Oil Plantation System. *MATEMATIKA Malays. J. Ind. Appl. Math.* **2019**, *35*, 95–104. [CrossRef]
20. Reed, W.J. The effects of the risk of fire on the optimal rotation of a forest. *J. Environ. Econ. Manag.* **1984**, *11*, 180–190. [CrossRef]
21. Sohngen, B.; Mendelsohn, R. An optimal control model of forest carbon sequestration. *Am. J. Agric. Econ.* **2003**, *85*, 448–457. [CrossRef]
22. Brown, T.J.; Hall, B.L.; Westerling, A.L. The impact of twenty-first century climate change on wildland fire danger in the western United States: An applications perspective. *Clim. Chang.* **2004**, *62*, 365–388. [CrossRef]
23. Ning, Z.; Sun, C. Forest management with wildfire risk, prescribed burning and diverse carbon policies. *For. Policy Econ.* **2017**, *75*, 95–102. [CrossRef]
24. Couture, S.; Reynaud, A. Forest management under fire risk when forest carbon sequestration has value. *Ecol. Econ.* **2011**, *70*, 2002–2011. [CrossRef]
25. Holsten, A.; Dominic, A.R.; Costa, L.; Kropp, J.P. Evaluation of the performance of meteorological forest fire indices for German federal states. *For. Ecol. Manag.* **2013**, *287*, 123–131. [CrossRef]
26. Misra, A.; Verma, M.; Venturino, E. Modeling the control of atmospheric carbon dioxide through reforestation: Effect of time delay. *Model. Earth Syst. Environ.* **2015**, *1*, 24. [CrossRef]

27. Verma, M.; Misra, A. Optimal control of anthropogenic carbon dioxide emissions through technological options: A modeling study. *Comput. Appl. Math.* **2018**, *37*, 605–626. [CrossRef]
28. Verma, M.; Verma, A.K. Effect of plantation of genetically modified trees on the control of atmospheric carbon dioxide: A modeling study. *Nat. Resour. Model.* **2021**, *34*, e12300. [CrossRef]
29. Caetano, M.A.L.; Gherardi, D.F.M.; Yoneyama, T. Optimal resource management control for CO_2 emission and reduction of the greenhouse effect. *Ecol. Model.* **2008**, *213*, 119–126. [CrossRef]
30. Kerdan, I.G.; Giarola, S.; Hawkes, A. A novel energy systems model to explore the role of land use and reforestation in achieving carbon mitigation targets: A Brazil case study. *J. Clean. Prod.* **2019**, *232*, 796–821. [CrossRef]
31. Hudiburg, T.W.; Law, B.E.; Wirth, C.; Luyssaert, S. Regional carbon dioxide implications of forest bioenergy production. *Nat. Clim. Chang.* **2011**, *1*, 419–423. [CrossRef]
32. Jin, X.; Pukkala, T.; Li, F.; Dong, L. Optimal management of Korean pine plantations in multifunctional forestry. *J. For. Res.* **2017**, *28*, 1027–1037. [CrossRef]
33. Altamirano-Fernández, A.; Rojas-Palma, A.; Espinoza-Meza, S. A mathematical model to study the dynamics of carbon capture in forest plantations. *J.Phys. Conf. Ser.* **2022**, *1259*, 012001. [CrossRef]
34. Gaoue, O.G.; Jiang, J.; Ding, W.; Agusto, F.B.; Lenhart, S. Optimal harvesting strategies for timber and non-timber forest products in tropical ecosystems. *Theor. Ecol.* **2016**, *9*, 287–297. [CrossRef]
35. Tahvonen, O. Optimal choice between even-and uneven-aged forestry. *Nat. Resour. Model.* **2009**, *22*, 289–321. [CrossRef]
36. Du, E.; Tang, Y. Distinct Climate Effects on Dahurian Larch Growth at an Asian Temperate-Boreal Forest Ecotone and Nearby Boreal Sites. *Forests* **2021**, *13*, 27. [CrossRef]
37. Favero, A.; Daigneault, A.; Sohngen, B. Forests: Carbon sequestration, biomass energy, or both? *Sci. Adv.* **2020**, *6*, eaay6792. [CrossRef] [PubMed]
38. Lenhart, S.; Workman, J.T. *Optimal Control Applied to Biological Models*; Chapman and Hall/CRC: Boca Raton, FL, USA, 2007.
39. Fleming, W.H.; Rishel, R.W. *Deterministic and Stochastic Optimal Control*; Springer Science & Business Media: Berlin/Heidelberg, Germany, 2012; Volume 1.
40. Huong, V.T. Solution existence theorems for finite horizon optimal economic growth problems. *Optimization* **2021**, *71*, 4243–4263. [CrossRef]
41. Pontryagin, L.S. *Mathematical Theory of Optimal Processes*; CRC Press: Boca Raton, FL, USA, 1987.
42. Kirk, D.E. *Optimal Control theory: An introduction*; Courier Corporation. Dover Publications: New York, NY, USA, 2004.
43. Grüne, L.; Pannek, J. Nonlinear model predictive control. In *Nonlinear Model Predictive Control*; Springer: Berlin/Heidelberg, Germany, 2017; pp. 45–69.
44. Campos, C.; Silva, C.J.; Torres, D.F. Numerical optimal control of HIV transmission in Octave/MATLAB. *Math. Comput. Appl.* **2019**, *25*, 1. [CrossRef]
45. Higham, D.; Higham, N. *MATLAB Guide*; SIAM: Philadelphia, PA, USA, 2016; Volume 150.
46. Navarrete, E.; Bustos, J. Faustmann optimal pine stands stochastic rotation problem. *For. Policy Econ.* **2013**, *30*, 39–45. [CrossRef]
47. Shukla, P.R.; Skeg, J.; Buendia, E.C.; Masson-Delmotte, V.; Pörtner, H.O.; Roberts, D.; Zhai, P.; Slade, R.; Connors, S.; Van Diemen, S.; et al. *Climate Change and Land: An IPCC Special Report on Climate Change, Desertification, Land Degradation, Sustainable Land Management, Food Security, and Greenhouse Gas Fluxes in Terrestrial Ecosystems*; Intergovernmental Panel on Climate Change (IPCC): Geneva, Switzerland, 2019; in press.
48. Diaz-Hormazabal, I.; Gonzalez, M.E. Spatio-temporal analyses of wildfires in the region of Maule, Chile. *Bosque* **2016**, *37*, 147–158.
49. Rodríguez, M.P.R.; Rodríguez, Y.C.; Sierra, C.A.M.; Batista, A.C.; Tetto, A.F. Relación entre variables meteorológicas e incendios forestales en la Provincia Pinar del Río, Cuba. *Floresta* **2017**, 47. [CrossRef]
50. Song, L.; Zhou, Y. The COVID-19 pandemic and its impact on the global economy: What does it take to turn crisis into opportunity? *China World Econ.* **2020**, *28*, 1–25. [CrossRef]
51. Schneider, G.; Troeger, V.E. War and the world economy: Stock market reactions to international conflicts. *J. Confl. Resolut.* **2006**, *50*, 623–645. [CrossRef]
52. Bekessy, S.A.; Wintle, B.A. Using carbon investment to grow the biodiversity bank. *Conserv. Biol.* **2008**, *22*, 510–513. [CrossRef] [PubMed]
53. Burgin, S. BioBanking: An environmental scientist's view of the role of biodiversity banking offsets in conservation. *Biodivers. Conserv.* **2008**, *17*, 807–816. [CrossRef]
54. Meleason, M.A.; Quinn, J.M. Influence of riparian buffer width on air temperature at Whangapoua Forest, Coromandel Peninsula, New Zealand. *For. Ecol. Manag.* **2004**, *191*, 365–371. [CrossRef]
55. Georgi, N.J.; Zafiriadis, K. The impact of park trees on microclimate in urban areas. *Urban Ecosyst.* **2006**, *9*, 195–209. [CrossRef]
56. Peng, G.; Li, J.; Chen, Y.; Norizan, A.P.; Tay, L. High-resolution surface relative humidity computation using MODIS image in Peninsular Malaysia. *Chin. Geogr. Sci.* **2006**, *16*, 260–264. [CrossRef]
57. Kumar, M.; Sheikh, M.A.; Bhat, J.A.; Bussmann, R.W. Effect of fire on soil nutrients and under storey vegetation in Chir pine forest in Garhwal Himalaya, India. *Acta Ecol. Sin.* **2013**, *33*, 59–63. [CrossRef]
58. Du, E.; Terrer, C.; Pellegrini, A.F.; Ahlström, A.; van Lissa, C.J.; Zhao, X.; Xia, N.; Wu, X.; Jackson, R.B. Global patterns of terrestrial nitrogen and phosphorus limitation. *Nat. Geosci.* **2020**, *13*, 221–226. [CrossRef]

59. Du, E. Evidence of soil nutrient availability as the proximate constraint on growth of treeline trees in northwest A laska: Comment. *Ecology* **2016**, *97*, 801–803. [CrossRef]
60. Toro, J.; Gessel, S. Radiata pine plantations in Chile. *New For.* **1999**, *18*, 33–44. [CrossRef]
61. Castillo, E.; Rodriguez, F. Determining response times for the deployment of terrestrial resources for fighting forest fires: A case study: Mediterranean-Chile. *Cienc. Investig. Agrar. Rev. Latinoam. Cienc. Agric.* **2015**, *42*, 97–107. [CrossRef]
62. Apud, E.; Meyer, F. Factors influencing the workload of forest fire-fighters in Chile. *Work* **2011**, *38*, 203–209. [CrossRef]
63. Cartes-Rodríguez, E.; Rubilar-Pons, R.; Acuña-Carmona, E.; Cancino-Cancino, J.; Rodríguez-Toro, J.; Burgos-Tornería, Y. Potential of Pinus radiata plantations for use of harvest residues in characteristic soils of south-central Chile. *Rev. Chapingo Ser. Cienc. For. Ambiente* **2016**, *22*, 221–233. [CrossRef]
64. CONAF. Situación Diaria de Incendios Forestales, Sistema de Información Digital para el Control de Operaciones. 2022. Available online: https://www.conaf.cl/situacion-nacional-de-incendios-forestales/ (accessed on 8 August 2022).
65. Lukes, D.L. *Differential Equations: Classical to Controlled*; Mathematics in Science and Engineering; Academic Press: New York, NY, USA, 1982.
66. Khan, M.; Ali, K.; Bonyah, E.; Okosun, K.; Islam, S.; Khan, A. Mathematical modeling and stability analysis of Pine Wilt Disease with optimal control. *Sci. Rep.* **2017**, *7*, 3115. [CrossRef] [PubMed]
67. Logan, J.D. *An Introduction to Nonlinear Partial Differential Equations*; John Wiley & Sons: Hoboken, NJ, USA, 2008; Volume 89.
68. Soto Aguirre, D. Anuario Forestal. 2021. Available online: https://bibliotecadigital.infor.cl/handle/20.500.12220/31292 (accessed on 1 March 2021).
69. Akinbode, O.; Eludoyin, A.; Fashae, O. Temperature and relative humidity distributions in a medium-size administrative town in southwest Nigeria. *J. Environ. Manag.* **2008**, *87*, 95–105. [CrossRef] [PubMed]
70. Valdivieso, J.P.; Rivera, J.d.D. Effect of wind on smoldering combustion limits of moist pine needle beds. *Fire Technol.* **2014**, *50*, 1589–1605. [CrossRef]
71. Ojeda, H.; Rubilar, R.A.; Montes, C.; Cancino, J.; Espinosa, M. Leaf area and growth of Chilean radiata pine plantations after thinning across a water stress gradient. *N. Z. J. For. Sci.* **2018**, *48*, 10. [CrossRef]
72. Asenjo, S.B. Evolución de las plantaciones forestales en Chile. *For. Reforestación. Cienc. Investig. For.* **2018**, *24*, 89–115. [CrossRef]
73. Gallardo Vera PD, L.; Morales Agoni, R.P.; Sáez, G. Manual de manejo silvícola para coníferas en Aysén. *bibliotecadigital.infor.cl* **2000**, *24*, 1–24. [CrossRef]
74. Alzamora, M.R.; Apiolaza, L.; Ide, S. Physical and economic evaluation of volume losses due to Rhyacionia buoliana (Schiff.) damage in Pinus radiata (D. Don) plantations in Southern Chile. *Bosque* **2002**, *23*, 29–42. [CrossRef]
75. Crecente-Campo, F.; Pommerening, A.; Rodríguez-Soalleiro, R. Impacts of thinning on structure, growth and risk of crown fire in a Pinus sylvestris L. plantation in northern Spain. *For. Ecol. Manag.* **2009**, *257*, 1945–1954. [CrossRef]
76. Fernández, M.P.; Basauri, J.; Madariaga, C.; Menéndez-Miguélez, M.; Olea, R.; Zubizarreta-Gerendiain, A. Effects of thinning and pruning on stem and crown characteristics of radiata pine (Pinus radiata D. Don). *Iforest-Biogeosci. For.* **2017**, *10*, 383. [CrossRef]
77. White, D.A.; Silberstein, R.P.; Balocchi-Contreras, F.; Quiroga, J.J.; Meason, D.F.; Palma, J.H.; de Arellano, P.R. Growth, water use, and water use efficiency of Eucalyptus globulus and Pinus radiata plantations compared with natural stands of Roble-Hualo forest in the coastal mountains of central Chile. *For. Ecol. Manag.* **2021**, *501*, 119676. [CrossRef]
78. CONAF. Incendios. 2022. Available online: https://www.conaf.cl/incendios-forestales/incendios-forestales-en-chile/estadistica-deocurrencia-diaria/ (accessed on 1 June 2022).
79. Hidalgo, M.G. Disciplinamiento de las subjetividades como estrategia de prevención de incendios: El caso de las plantaciones forestales en el sur de Chile. *Perspect. Rural. Nueva Época* **2018**, *16*, 117–141.
80. Raison, R.; Myers, B. The biology of forest growth experiment: Linking water and nitrogen availability to the growth of Pinus radiata. *For. Ecol. Manag.* **1992**, *52*, 279–308. [CrossRef]

forests

MDPI

Article

Modeling Climate Effects on Site Productivity of Plantation Grown Jack Pine, Black Spruce, Red Pine, and White Spruce Using Annual/Seasonal Climate Values

Mahadev Sharma

Ontario Forest Research Institute, Ontario Ministry of Natural Resources and Forestry, 1235 Queen St. East, Sault Ste Marie, ON P6A 2E5, Canada; mahadev.sharma@ontario.ca; Tel.: +1-(705)-992-9775

Abstract: Site index (SI) is a commonly used measure of forest site productivity and is affected by climate change. Therefore, climate effects on site productivity were analyzed and modeled for jack pine (*Pinus banksiana* Lamb.), black spruce (*Picea mariana* (Mill.) B.S.P.), red pine (*Pinus resinosa* Ait.), and white spruce (*Picea glauca* (Moench) Voss) plantations using annual/seasonal values of climate variables. Jack pine and black spruce trees were each sampled from 25 plantations (sites), and red pine and white spruce trees were sampled from 30 and 31 plantations, respectively, from across Ontario, Canada. Stem analysis data collected from 201 jack pine, 211 black spruce, 90 red pine, and 93 white spruce trees were used in this study. To analyze and model climatic effects on site productivity, parameters of the stand height models were expressed in terms of climate variables. A nonlinear mixed-effects modelling approach was applied to fit the stand height models. Climate effects on site productivity was evaluated by predicting stand heights in three areas (the central, eastern/southeastern, and western parts of Ontario) for the period 2021 to 2080 under three emissions trajectories (representative concentration pathways (RCP) 2.6, 4.5, and 8.5 watts m^{-2}). Climate effects on site productivity depended on tree species and location. For jack pine, climate effects were positive and pronounced only in western Ontario under all emissions scenarios. The effects were negative and mild after breast height age (BHA) 50 in central Ontario for black spruce. Similarly, the effects were negative and more pronounced at all areas after BHA 35 for red pine. On the other hand, for white spruce the effects were negative and highly pronounced from a young age under all scenarios, mainly in the southeast. For all species except for jack pine, climate effects were more pronounced under RCP 8.5 than the other two scenarios.

Keywords: height growth functions; dynamic site index models; climatic effects on tree growth; nonlinear height growth models; stand/top height

Citation: Sharma, M. Modeling Climate Effects on Site Productivity of Plantation Grown Jack Pine, Black Spruce, Red Pine, and White Spruce Using Annual/Seasonal Climate Values. *Forests* **2022**, *13*, 1600. https://doi.org/10.3390/f13101600

Academic Editors: Dalmonech Daniela, Alessio Collalti and Gina Marano

Received: 29 August 2022
Accepted: 27 September 2022
Published: 30 September 2022

1. Introduction

Site productivity influences growth, mortality, and recruitment of trees in a stand [1]. It is commonly measured in terms of a site index (SI), with SI defined as stand height (average height of dominant and codominant trees) at a specified stand age [2]. Most stand scale growth and yield models that are used in developing forest management plans are driven by SI. Consequently, accurate SI estimates are fundamental for informed forest management decisions.

The site index depends on species, site, growing environment, and climate [3]. Researchers have used different approaches to examine and model climate effects on site productivity. Some have regressed SI in terms of climate/environmental variables directly [4–7]. However, Ung et al. [5] indicated that linear relationships between the SI and biophysical variables were inadequate for use in growth and yield models.

Similarly, Albert and Schmidt [6] found less than 40% variation in SI explained by biophysical variables for Norway spruce (*Picea abies* (L.) Karst.) and common beech (*Fagus sylvatica* L.) trees in Lower Saxony, Germany. On the other hand, Weiskittel et al. [7]

reported that 68% of the variability in SI was explained by climate-related variables for tree species grown in western U.S. forests.

Bergh et al. [8] used a process-based simulation model, to compute and compare the effect of increased temperature on net primary productivity (NPP) for Norway spruce (Picea abies), Scots pine (Pinus sylvestris), black cottonwood (Populus trichocarpa), and European beech (Fagus sylvatica) growing in the Nordic countries (Finland, Denmark, Norway, Iceland, and Sweden). Their results showed that if the temperature is increased by 4 degrees C, Norway spruce and Scots pine NPP would increase by 24%–37% in spring. In another study, Pedlar and McKenney [9] used published and provenance trial data to assess the estimated growth response of five northern conifers to climate change. They reported that climate warming could have a significant positive effect on cold-origin (northern) populations, but negative effects on warm-origin (southern) populations. Similarly, Guo et al. [10] investigated local adaptation process of bud phenology of five black spruce populations originating from the latitudinal range of boreal forest. They found a relationship between bud phenology and the mean annual temperature at the sites of tree origin.

Since climate is not the sole factor influencing site productivity, SI expressed solely in terms of climate variables would not provide accurate estimates of forest site productivity. To improve model efficiency, other researchers defined the parameters of SI models in terms of climate- and site-related variables [11–13]. However, other than stating that incorporating biophysical variables (including climate) improved fit statistics, these researchers did not quantify the magnitude and nature (positive or negative) of climate effects on site productivity.

Recently, stand height growth/SI models have been developed by incorporating climate variables in stand height growth models for several tree species in Ontario, Canada: plantation grown jack pine (*Pinus banksiana* Lamb.) and black spruce (*Picea mariana* Mill. B.S.P.) [14], red pine (*Pinus resinosa* Ait.) [15], white spruce (*Picea glauca* (Moench) Voss) [16], and white pine (*Pinus strobus* L.) [3] and natural origin mixed stands of jack pine and black spruce [17] and black spruce and trembling aspen (*Populus tremuloides* Michx.) [18].

When climate-sensitive SI models were developed for jack pine, black spruce, red pine, and white spruce plantations, only 30-year average values of climate variables were available for evaluation [14–16]. Therefore, the average values of climate variables over the lifespan of sample trees were used to develop climate-sensitive stand height growth/SI models for these tree species. These models were evaluated using projected average values of climate variables significant in the models for a 30-year period under different climate change scenarios.

However, in the case of white pine plantations and natural origin mixed stands, annual and seasonal values of climate variables were available for past and future growth periods. Past and future annual/seasonal values of index variables derived from climate-related variables (e.g., climatic moisture index) were also available. Therefore, Sharma and Parton [3] and Sharma [17,18] analyzed climate effects on site productivity of these tree species using annual/seasonal values of temperature- and precipitation-related climate and derived variables and developed climate-sensitive stand height growth/SI models for these plantations and mixed stands. In these studies, model evaluation included the use of projected annual/seasonal values of climate variables for a future 80-year growth period under three climate change scenarios.

Since the values of climate variables fluctuate every year/season, it is intuitive to use annual/seasonal values of climate variables to examine and model climate effects on site productivity. Models developed using annual/seasonal values of original and derived climate variables will provide more accurate information about the climate effects on tree growth than those developed using the average values over the period of tree growth. Therefore, the objectives of this study were to derive stand height growth/SI models for jack pine, red pine, black spruce, and white spruce plantations by incorporating yearly/seasonal values of climate variables and to assess the effects of future climate scenarios on stand height growth of these tree species using projected yearly/seasonal values of climate variables.

2. Methods

2.1. Height and Age Data

Data used in this study were collected from jack pine, red pine, black spruce, and white spruce trees grown in plantations. Twenty-five even-aged monospecific plantations were sampled for each of jack pine and black spruce. Similarly, 30 and 31 monospecific plantations were sampled for red pine and white spruce, respectively. These plantations were selected from across the species' range [19] in Ontario (Figure 1). Details of sampling trees and collecting stem analysis data have been provided in studies by Sharma et al. [14] for jack pine and black spruce and Sharma and Parton [15,16] for red pine and white spruce.

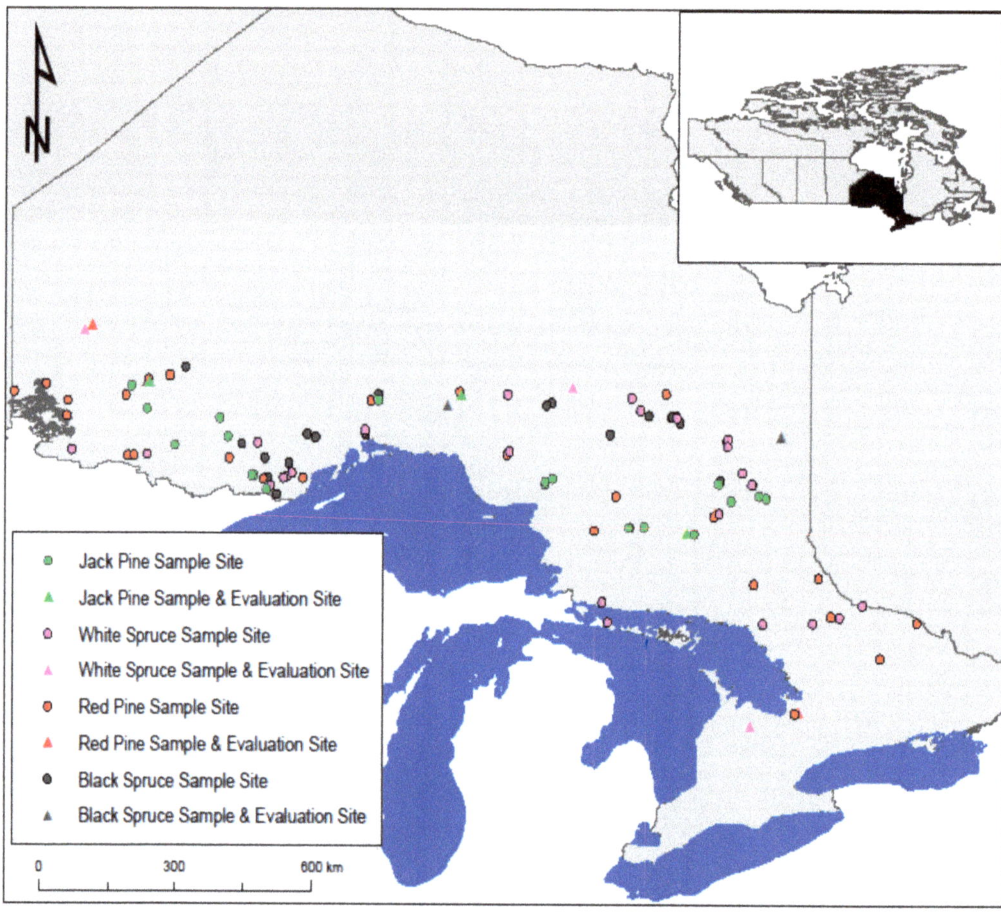

Figure 1. Distribution of jack pine and black spruce plantation sites sampled across Northern Ontario, Canada. Latitude and longitude ranged from 47° N to 50° N and 80° W to 92° W, respectively.

Height growth of all tree species used in this study was erratic before trees reached breast height [14–16]. Therefore, unless otherwise specified, height from breast height and age from breast height (breast height age, BHA) were used for all analyses reported in this study. As a result, tree height refers to height above breast height and age to BHA.

Since height growth below breast height was irregular, BHA of trees sampled from three plots at a site did not necessarily reach a particular BHA during the same calendar year. Combining the growth series from the three plots would result in averaging height growth across years, precluding analysis of climate effects on stand height growth (mean height of three trees sampled from a site) using a climate variable value associated with a specific calendar year. Therefore, in this study growth series from the three plots at each site were not combined to obtain site-scale estimates.

Observed heights were plotted against their ages to form height–age curves for each tree. Trees for which curves indicated possible injuries or early height growth suppression were discarded. This resulted in 201 and 211 jack pine and black spruce trees, respectively, to be used for analysis. None of red pine or white spruce trees sampled had any noticeable defect, so all trees (90 and 93 for red pine and white spruce, respectively) were used in analyzing climate effects and developing stand height/SI models. Summary statistics for total age, total height, diameter at breast height (DBH), and BHA for the trees used in this study are presented in Table 1.

Table 1. Summary statistics (N = number of samples, Std Dev = standard deviation) for total height, diameter at breast height (DBH), total age, and breast height age (BHA) of plantation grown trees in Ontario, Canada and climate variables that best explained the variation in stand height growth of tree species used in this study. (CMI = climatic moisture index, MDTR = mean diurnal temperature range, PWQ = precipitation of warmest quarter, GSMT = growing season mean temperature, AnMxT = annual maximum temperature).

Variable	N	Mean	Std Dev	Minimum	Maximum
		Jack pine			
Height (m)	201	16.71	2.38	11.80	23.02
DBH (cm)	201	19.59	4.02	9.90	34.30
Total age (year)	201	42.22	10.26	26.00	63.00
BHA (year)	201	38.68	10.00	23.00	60.00
CMI (April)	2700	2.92	2.45	−5.62	14.54
CMI (May)	2700	0.58	3.18	−6.85	10.03
MDTR (°C)	2700	12.25	1.31	7.2	16.2
		Black spruce			
Height (m)	211	12.55	2.23	6.82	17.85
DBH (cm)	211	16.39	2.81	10.10	24.80
Total age (year)	211	38.34	8.40	24.00	84.00
BHA (year)	211	32.01	7.10	20.00	46.00
CMI (July)	2700	−1.35	3.39	−10.1	12.28
PWQ (mm)	2700	243.67	47.72	97	406
		Red pine			
Height (m)	90	21.83	4.10	13.25	30.90
DBH (cm)	90	31.39	6.23	21.30	56.20
Total age (year)	90	59.90	18.12	27.00	97.00
BHA (year)	90	55.26	18.07	23.00	93.00
CMI (March)	3456	4.02	2.18	−1.52	14.36
CMI (April)	3456	2.72	2.62	−6.22	14.5
CMI (Oct)	3456	3.68	3.66	−6.22	19.45
CMI (Nov)	3456	5.40	2.73	−0.63	17.89
GSMT (°C)	3456	13.18	0.73	10.87	15.82

Table 1. *Cont.*

Variable	N	Mean	Std Dev	Minimum	Maximum
		White spruce			
Height (m)	93	20.98	2.91	15.25	29.35
DBH (cm)	93	29.97	6.67	16.30	52.90
Total age (year)	93	55.62	10.26	42.00	87.00
BHA (year)	93	49.86	10.18	35.00	82.00
AnMaxT (°C)	2700	7.44	0.52	6.14	8.05

CMI = mean monthly precipitation–monthly potential evapotranspiration.

2.2. Climate Data

All climate variables for each plot location for each tree species were estimated using Canadian climate models [20]. These models were generated from continuous climate grids using ANUSPLINE based on corrected Canadian weather station data [21,22], which includes many stations in Ontario. For each plot location, estimates of average seasonal and yearly values of these variables were calculated for each year, starting when the sampled tree reached breast height, until 2018.

A total of 68 climate-related variables were computed, including minimum, mean, and maximum air temperatures and total precipitation, estimated for each month of the year, for each quarter (consecutive three-month periods), and annually. The 68 variables also included longitude, latitude, and elevation (site-related variables). Details of calculating climate data were documented by Sharma et al. [14] and Sharma and Parton [15,16]. In addition, potential evapotranspiration (PET) was subtracted from mean monthly precipitation (MMP) to estimate climatic moisture index (CMI) for each year (see [23]). These values were also calculated for each plot for each species.

Total or partial CMI values were then computed by summing the 12-month or partial-month (months for which CMI was significant in the regression) values of CMI for each year for each sample site. Estimates for all climate variables were provided by Dan McKenney (Canadian Forest Service, 2018, pers. comm.). Summary statistics of climate variables that explained the variation in stand height growth of jack pine, black spruce, red pine, and white spruce plantations are included in Table 1.

2.3. Stand Height/Site Index Equations

Sharma et al. [14] and Sharma and Parton [15,16] evaluated variants of Chapman-Richards and Hossfeld IV functions for the tree species used in this study and found a variant of the Hossfeld IV function (Equation (1)), also known as McDill–Amateis growth function (see [24] p. 126), provided the best fit statistics (R^2 and MSE). This variant also produced the most consistent and biologically realistic height estimates across productivity classes for all four species studied. The variant (model form) was:

$$H = \frac{\alpha_0}{1 - \left(1 - \frac{\alpha_0}{H_1}\right)\left(\frac{A_1}{A}\right)^{\alpha_1}} + \varepsilon \tag{1}$$

where H and H_1 are stand heights (from breast height) at BHAs A and A_1, respectively, α_0 and α_1 are model parameters and ε is the error term. This model form was used in this study as the base function to examine and model climate effects on stand height growth. In general, α_0 defines the asymptote of the curve, and α_1 determines the shape. α_1 is also called the rate parameter that determines the growth rate. To analyze and model the climate effects on stand height growth, the asymptote and rate parameter (α_0 and α_1, respectively) in Equation (1) were expressed in terms of climate variables.

2.4. Model Fitting and Evaluation

The data used in this study came from stem analysis. These data are hierarchical (i.e., height–age series within sites), resulting in two sources of variation: among sites and within a site. Observations within a site (height–age series) (correlated) are dependent as they originate from the same tree. However, observations among sites are independent. To address the problem of autocorrelation, a mixed-effects modelling approach was used to fit the stand height growth model for all tree species.

To examine and model climate effects on site productivity, climate- and site-related variables were partitioned into three groups (precipitation, temperature, and site) and introduced to Equation (1) successively from each group. The climate- and site-related variables that were significant and resulted in the lowest Akaike information criterion (AIC; [25]) value were selected and incorporated into the stand height models. Quadratic transformations and two-way interactions of the climate variables that were significant in the regression were also introduced one by one in the presence of their original variables. All climate and site variables, their two-way interactions, and transformations that were both significant and improved model fit were selected as climate variables.

Random effects parameters were added successively to the fixed-effects coefficients of climate variables as necessary. Goodness-of-fit criteria such as log-likelihood (twice the negative log-likelihood) ratio, assessment of model residuals, and AIC were used to evaluate the model with random effects. The model with the smallest goodness-of-fit value was considered best. The model form that resulted in the smallest value of AIC was used as the final model for each species.

Estimated values of stand heights of jack pine, black spruce, red pine, and white spruce trees using the models with climate variables were used to evaluate climatic effects on future stand height growth for three areas. These areas were in the center (near Hearst), the eastern (north of Sudbury for jack pine and black spruce and near Barrie for red pine and white spruce), and western (near Dryden for jack pine and black spruce and near Red Lake for red pine and white spruce) parts of Ontario, where the trees were sampled (see Figure 1).

The evaluation of climate effects was performed under three emissions trajectories (2.6, 4.5 and 8.5 Watts/m^2). These trajectories, known as representative concentration pathways (RCPs), produce different levels of warming at the end of the century using the Canadian model [20]. The projected values of climate variables (from [20]) that were significant in expressing the asymptote and rate parameter in Equation (1) were used in evaluating climate effects. Height growth curves were also generated for the 80-year growth period (2021–2100) for all tree species for all emissions scenarios.

3. Results

The base model (Equation (1)) coefficient estimates were provided in studies by Sharma et al. [14], Sharma and Parton [15], and Sharma and Parton [16] for jack pine and black spruce, red pine, and white spruce, respectively. Those estimates remain the same and, hence, are not reported here. Any differences would be associated with the climate variables since annual/seasonal values were used in this study instead of the average values over the trees' past growth periods used in the previous studies.

Climate effects on stand height growth were analyzed by expressing parameters (α_0 and α_1) in Equation (1) in terms of climate variables as described earlier. For jack pine, mean diurnal temperature range (MDTR), its quadratic transformation, and April CMI (CMI_{Apr}) explained the variation in asymptote. Similarly, MDTR and May CMI (CMI_{May}) were significant in explaining the rate parameter for this tree species. For black spruce, precipitation of warmest quarter (PWQ) and July CMI (CMI_{Jul}) explained the variations in the asymptote and rate parameter, respectively.

For red pine, growing season mean temperature (GSMT) and the sum of March, April, October, and November CMIs (CMI_{Sum}) explained the variation in the asymptote and rate parameter, respectively. On the other hand, only one temperature-related variable (annual maximum temperature (AnMaxT)) was significant in explaining the variation in the rate parameter for white spruce.

As stated, random effects parameters were sequentially added to the fixed-effects coefficients, starting with the intercept in the expression for asymptote. Only site-level random effects associated with the intercept in expressions for both asymptote and rate parameter were significant for all species. However, no random effects associated with the fixed-effects coefficients attached to climate variables were significant. Final models with climate variables and random effects for jack pine, black spruce, red pine, and white spruce can be mathematically expressed as:
jack pine

$$H_{ijk} = \frac{\beta_0 + b_{0i} + \beta_1 * CMI_{Apr} + \beta_2 MDTR + \beta_3 MDTR^2}{1 - \left(1 - \frac{\beta_0 + b_{0i} + \beta_1 * CMI_{Apr} + \beta_2 MDTR + \beta_3 MDTR^2}{H_{ijl(l \neq k)}}\right)\left(\frac{A_{ijl(l \neq k)}}{A_{ijk}}\right)^{\beta_4 + b_{4i} + \beta_5 CMI_{May} + \beta_6 MDTR}} + \varepsilon_{ijk} \quad (2)$$

black spruce

$$H_{ijk} = \frac{\beta_0 + b_{0i} + \beta_1 PWQ}{1 - \left(1 - \frac{\beta_0 + b_{0i} + \beta_1 PWQ}{H_{ijl(l \neq j)}}\right)\left(\frac{A_{ijl(l \neq k)}}{A_{ijk}}\right)^{\beta_2 + b_{2i} + \beta_3 CMI_{Jul}}} + \varepsilon_{ijk} \quad (3)$$

red pine

$$H_{ijk} = \frac{\beta_0 + b_{0i} + \beta_1 GSMT}{\left(1 - \frac{\beta_0 + b_{0i} + \beta_1 GSMT}{H_{il(l \neq k)}}\right)\left(\frac{A_{ijl(l \neq k)}}{A_{ijk}}\right)^{\beta_2 + b_{2i} + \beta_2 CMI_{Sum}}} + \varepsilon_{ijk} \quad (4)$$

white spruce

$$H_{ijk} = \frac{\beta_0 + b_{0i}}{\left(1 - \frac{\beta_0 + b_{0i}}{H_{ijl(l \neq k)}}\right)\left(\frac{A_{ijl(l \neq k)}}{A_{ijk}}\right)^{\beta_1 + b_{1i} + \beta_2 AnMaxT}} + \varepsilon_{ijk} \quad (5)$$

where H_{ijk} is the stand height at age A_{ijk} (kth observations of series j and site i), H_{ijl} is the stand height at age A_{ijl} at the same series and site (lth observations of series j and site i and $l \neq k$), b_{0i} is site-scale random effect associated with the intercept expressing the asymptote, and b_{mi} (m = 1, 2, and 4) is also site-scale random effect but associated with the intercept that expressed the rate parameter. Both random effects are independent of ε_{ijk}. Random effects, b_{0i} and b_{mi}, are normally distributed with mean zero and variances ${\sigma_0}^2$ and ${\sigma_1}^2$, respectively, and covariance $\sigma_0\sigma_1$. β_0–β_6 are fixed effects parameters to be estimated. Other variables are as defined earlier.

Estimated parameters and fit statistics are listed in Table 2. For all species, fit statistics (RMSE, log-likelihood, AIC) decreased when climate variables were included in the model. Equations (2)–(5) incorporated climate variables that significantly improved fit statistics. As a result, these equations could be used to explain the effects of climate on stand height growth for jack pine, black spruce, red pine, and white spruce. As mentioned, only site-level random effects associated with the intercept of functions used to express both asymptote and rate parameter were significant for all species.

Estimated parameter values are consistent with biological expectations. For jack pine, the coefficients of April CMI and MDTR are negative. On the other hand, the coefficient for the quadratic transformation is positive. MDTR decreases as the rate of climate change increases. Therefore, decreasing MDTR has positive effect on the asymptote if evapotranspiration exceeds precipitation in April. However, the effect diminishes as the rate of climate change increases because of the opposite sign in the quadratic transformation of MDTR. Similarly, coefficients of both climatic variables (May CMI and MDTR) are negative in the expression for the rate parameter. This finding indicates that growth rate increases with the rate of change in climate if evapotranspiration is higher than total precipitation in May.

Table 2. Parameter estimates, their standard errors (SE), and fit statistics (MSE ($\sigma_e{}^2$), variance of b_0 ($\sigma_0{}^2$), variance of b_1 ($\sigma_1{}^2$), covariance of b_0 and b_1 ($\sigma_0\ \sigma_1$), and the Akaike information criterion (AIC) for the models with climate variables (Equations (2)–(5)) for jack pine, black spruce, red pine, and white spruce trees grown in plantation across Ontario, Canada.

	Jack Pine		**Black Spruce**	
Parameters	Estimates	SE	Estimates	SE
β_0	104.810	30.622	30.002	7.0212
β_1	−0.5135	0.1667	0.0890	0.0319
β_2	−14.8825	5.5186	1.0922	0.0126
β_3	0.8107	0.2499	0.0048	0.0012
β_4	1.3133	0.0740	–	–
β_5	−0.0062	0.0012	–	–
β_6	−0.0191	0.0057	–	–
$\sigma_e{}^2$	0.0184	0.0003	0.0116	0.0002
$\sigma_0{}^2$	83.010	37.304	200.00	83.210
$\sigma_1{}^2$	0.0060	0.0020	0.0026	0.0010
$\sigma_0\sigma_1$	−0.3189	0.2001	−0.3313	0.2165
AIC	−8620	–	−10.492	–
	Red pine		White spruce	
β_0	78.9872	9.9897	68.606	6.9399
β_1	−2.1957	0.6933	1.3575	0.0986
β_2	1.0851	0.0204	−0.0246	0.0109
β_3	0.0025	0.0006	–	–
$\sigma_e{}^2$	0.0066	0.0001	0.0152	0.0003
$\sigma_0{}^2$	260.00	117.27	1232.99	680.89
$\sigma_1{}^2$	0.0078	0.0022	0.0086	0.0028
$\sigma_0\sigma_1$	−1.1933	0.4377	−1.5901	0.9488
AIC	−10452	–	−5829	–

All parameter estimates were statistically significant ($p < 0.05$).

For black spruce, the coefficients of climate variables expressing the variation in asymptote and rate parameters (PWQ and CMI_{July}, respectively) are positive, i.e., the asymptote increases as the precipitation of the warmest quarter increases. Similarly, the growth rate increases if July precipitation exceeds evapotranspiration during that month.

For red pine, the coefficient of GSMT expressing the variation in asymptote is negative. Similarly, the coefficient of the climate variable (CMI_{sum}) that was significant in explaining the variation in the rate parameter is positive. This indicates that the increase in growing season mean temperature negatively affects the asymptote. On the other hand, an increase in the sum of March, April, October, and November CMIs will increase the growth rate of red pine plantations.

For white spruce, no climate variable significantly explained the variation in the asymptote. However, the rate parameter could be expressed in terms of one climate variable (AnMaxT), and the coefficient of this variable was negative. Thus, for this tree species, the asymptote is not affected by the change in climate, but height growth is reduced if the AnMaxT increases.

To evaluate climate effects on site productivity of different tree species, future stand heights were predicted for jack pine, black spruce, red pine, and white spruce trees at three areas in Ontario under three climate change scenarios (Figures 2–5). These estimates were made for an 80-year (2021–2100) growth period using only fixed-effects coefficients in the models. For all species, the average height value at age one BHA (0.35 m for jack pine and black spruce and 0.5 m for red pine and white spruce) was used as the initial height for generating height–age curves. Under all climate change scenarios, projected values of annual/seasonal climate variables were used to estimate future stand heights for all species. Height–age curves were also produced for a no climate change scenario for all species.

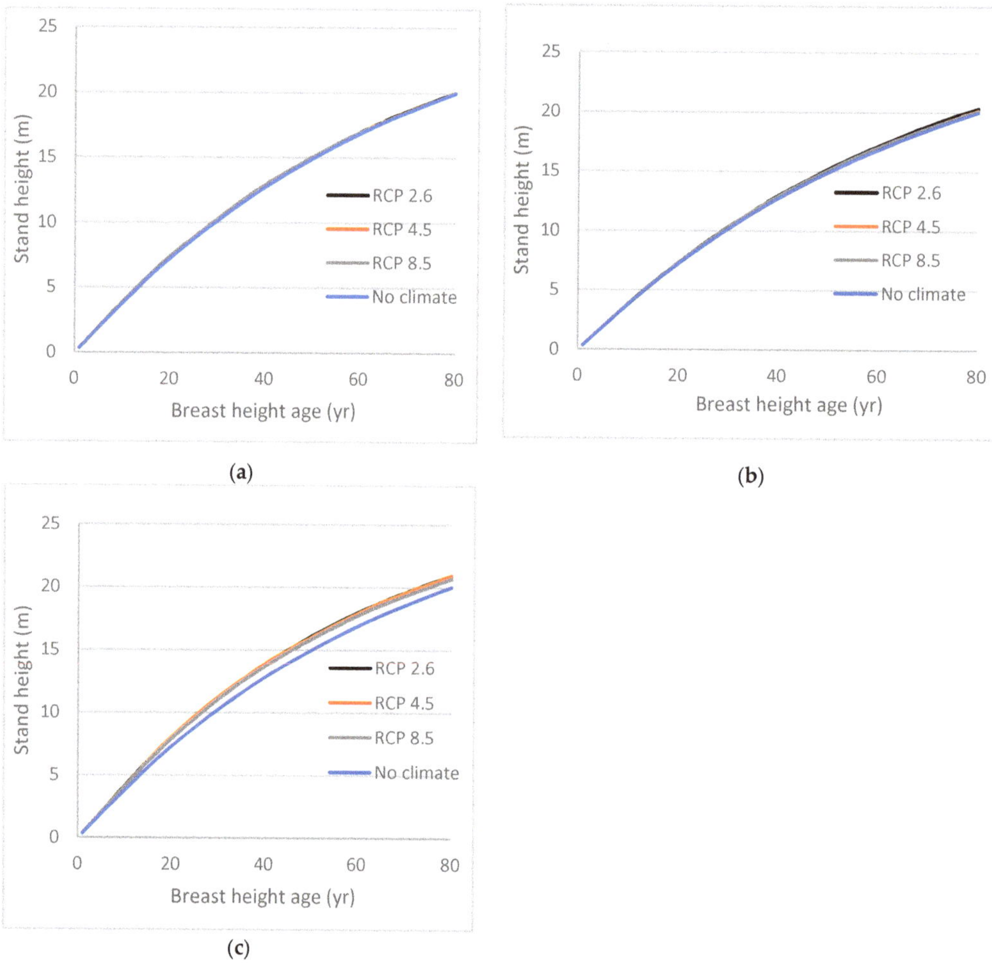

Figure 2. Stand height profiles for jack pine trees generated using projected values of climate variables for the period 2021 to 2080, assuming climate remains stable (no climate) or warms (RCPs 2.6, 4.5, and 8.5), in Equation (2) for (**a**) eastern (near Sudbury), (**b**) central (near Hearst), and (**c**) western (near Dryden) Ontario, Canada. Climate variables were projected for locations close to sample sites using three emissions trajectories known as representative concentration pathways (RCPs).

Jack pine height growth was positively affected by climate change in the west (Figure 2). At BHA 80, stand heights were higher by 4.4, 4.6, and 3.2% under 2.6, 4.5 and 8.5 emission scenarios, respectively, compared to those under the no climate change scenario. However, height growth was not significantly affected by climate change for the east and central areas of Ontario. The positive effect in the west increased from RCP 2.6 to 4.5 but decreased from RCP 4.5 to 8.5. Thus, the positive effect of climate change in the west is not linear but concave down.

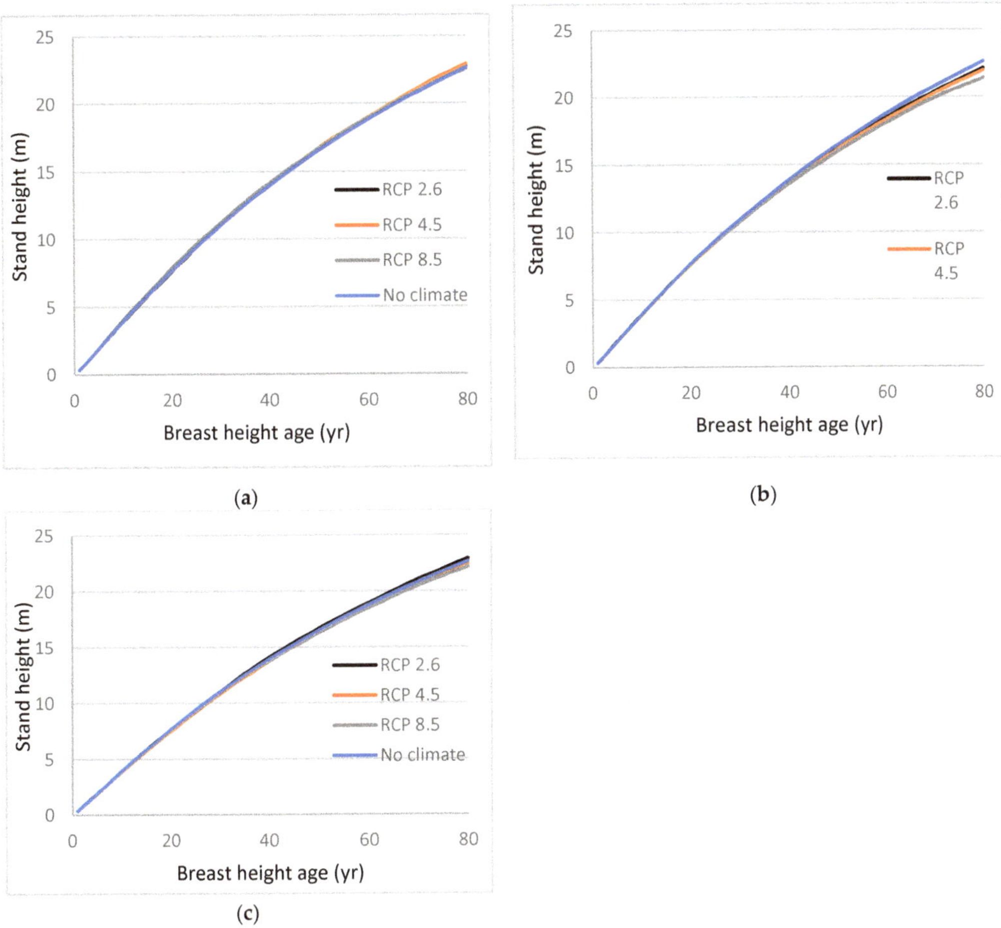

Figure 3. Stand height profiles for black spruce trees generated using projected values of climate variables for the period 2021 to 2080, assuming climate remains stable (no climate) or warms (RCPs 2.6, 4.5, and 8.5), in Equation (3) for (**a**) eastern (near Sudbury), (**b**) central (near Hearst), and (**c**) western (near Dryden) Ontario, Canada. Climate variables were projected for locations close to sample sites using three emissions trajectories known as representative concentration pathways (RCPs).

The climate change effect on black spruce height growth was not pronounced for stands in the east and west (Figure 3). However, the effects on height growth were negative and minimal for stands in central Ontario. The negative effect was more pronounced under RCP 8.5 than the other two (2.6 and 4.5) emission scenarios. At BHA 80, stand heights were lower by 2.2, 2.9, and 5.3% under RCP 2.6, 4.5, and 8.5 emission scenarios, respectively, compared to those under the no climate change scenario.

At all locations, red pine stand height growth was negatively affected by climate (Figure 4). However, the differences in stand heights under RCP 2.6 and 4.5 and the no climate change scenario were not pronounced in southeastern Ontario across stand age. At BHA 80, under RCP 8.5 stand heights were 6.4% shorter than those under the no climate change scenario. For the other two areas, stand heights under all three emission scenarios were affected by climate change. At the end of the same growth period, in central Ontario, stand heights under RCP 2.6, 4.5, and 8.5 were shorter by 3.7, 5.8, and 10.5%, respectively, compared to those under the no climate change scenario. Similarly, in the west, stand heights under RCP 2.6, 4.5, and 8.5 were shorter by 5.4, 7.5, and 9.3%, respectively, relative to the no climate change scenario.

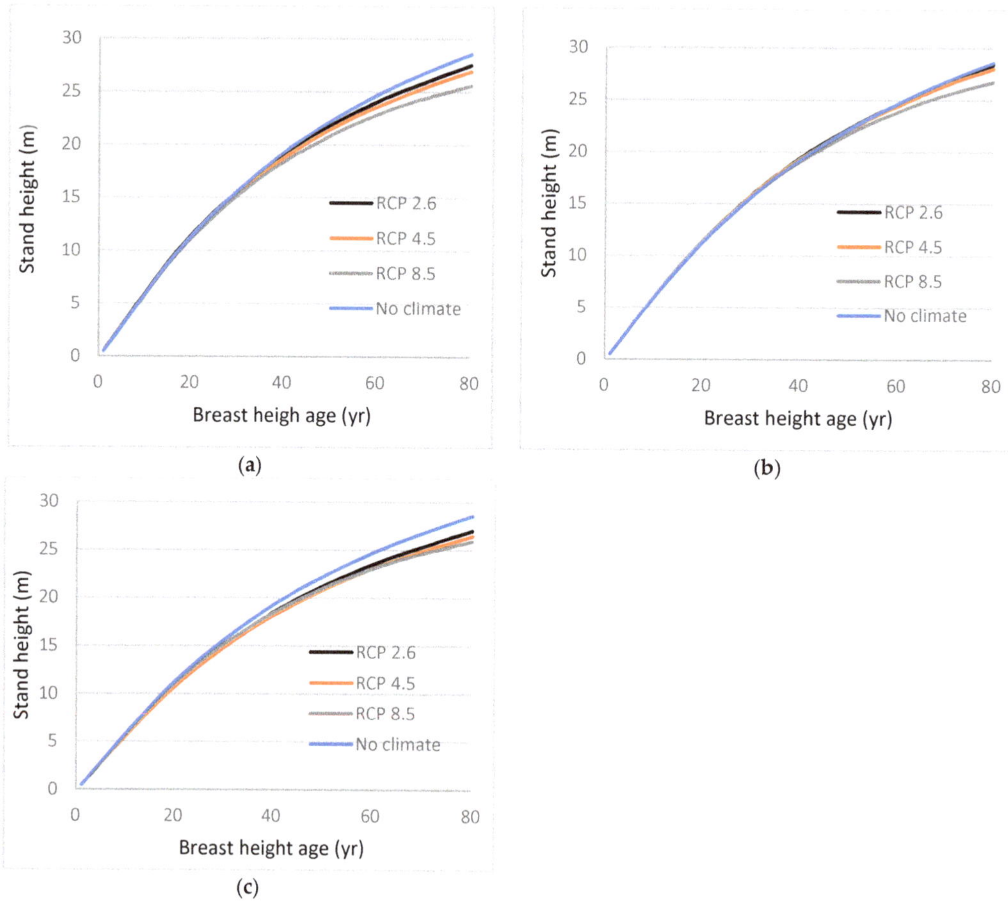

Figure 4. Stand height profiles for red pine trees generated using projected values of climate variables for the period 2021 to 2080, assuming climate remains stable (no climate) or warms (RCPs 2.6, 4.5, and 8.5), in Equation (4) for (**a**) southeastern (near Barrie), (**b**) central (near Hearst), and (**c**) western (near Red Lake) Ontario, Canada. Climate variables were projected for locations close to sample sites using three emissions trajectories known as representative concentration pathways (RCPs).

For white spruce, the difference in stand heights under all emission scenarios relative to the no climate change climate scenario was not pronounced for central and western Ontario across stand age (Figure 5). However, stand heights under all three emission scenarios were lower relative to the current climate scenario in the southeastern part of the province (near Barrie). At BHA 80, stand heights under RCP 2.6, 4.5, and 8.5 in this area were lower by 18.3, 18.3, and 21.0%, respectively.

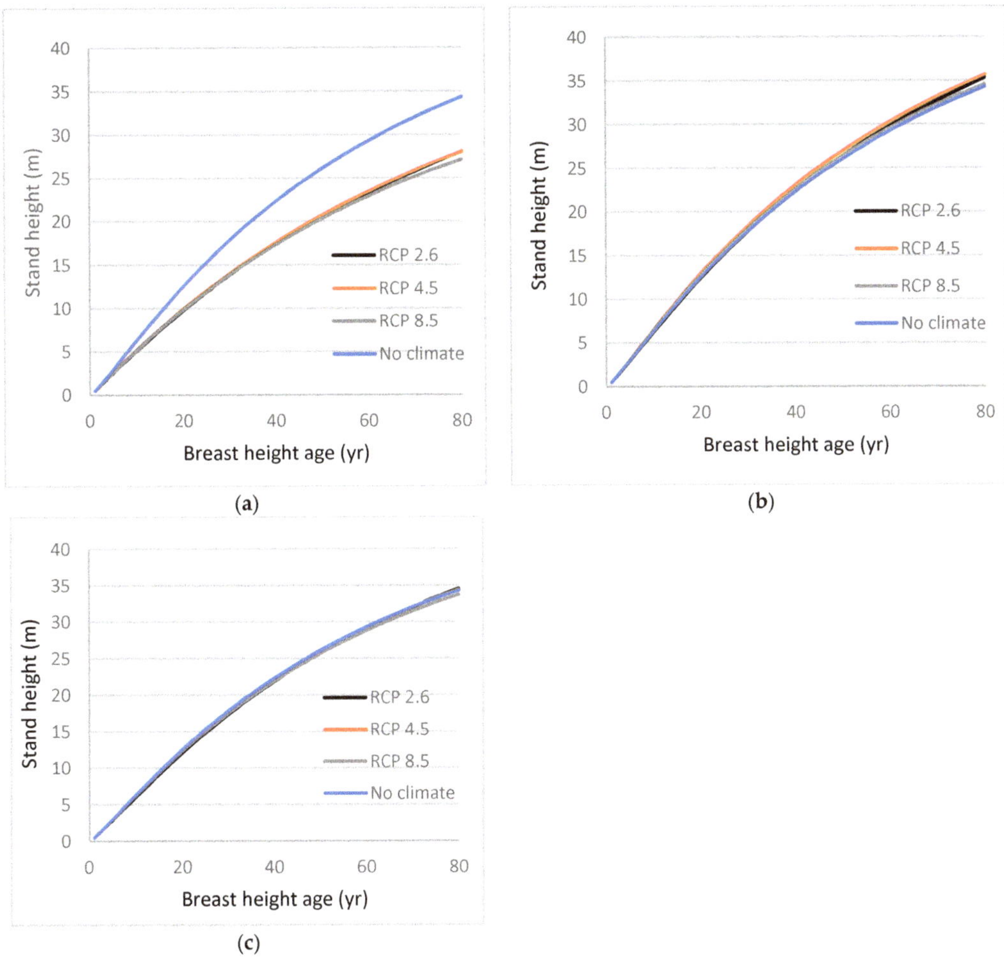

Figure 5. Stand height profiles for white spruce trees generated using projected values of climate variables for the period 2021 to 2080, assuming climate remains stable (no climate) or warms (RCPs 2.6, 4.5, and 8.5), in Equation (5) for (**a**) southeastern (near Barrie), (**b**) central (near Hearst), and (**c**) western (near Red Lake) Ontario, Canada. Climate variables were projected for locations close to sample sites using three emissions trajectories known as representative concentration pathways (RCPs).

4. Discussion

Site productivity is affected by climate and other environmental conditions [26]. Climate effects on site productivity have recently been analyzed and modelled for jack pine, black spruce, red pine, white spruce, and white pine plantations [3,14–16] and for natural origin jack pine and black spruce [17] and black spruce and trembling aspen [18] mixed stands in Ontario. Climatic conditions such as changes in temperature and precipitation regimes were used to analyze climate effects in these studies. The nature and magnitude of effects varied by species and geographic region.

As mentioned, [14–16] used average seasonal and annual values of climate variables over the lifespan of trees to analyze climate effects on site productivity of jack pine, black spruce, red pine, and white spruce plantations. They reported that stand height growth of jack pine and black spruce plantations was affected by both precipitation- and temperature-related variables (growing season total precipitation (GSTP) and growing season mean temperature (GSMT)). Similarly, stand height growth of white spruce plantations was affected by warmest quarter total precipitation (WQTP) and warmest quarter mean temperature (WQMT). However, red pine plantation height growth was affected only by a temperature-related variable (GSMT).

Values of climate variables fluctuate almost every year. Therefore, the effects of climate on tree growth vary from year to year. Now we have the projected annual/seasonal values of climate variables available for a future 80-year growth period to evaluate climate effects. Moreover, derived values of climate variables (e.g., CMI) are also available for use in analyzing and evaluating of climate effects. Therefore, in this study annual/seasonal values of climate variables including CMI were used to reanalyze climate effects on site productivity of jack pine, black spruce, red pine, and white spruce plantations.

In contrast to findings by Sharma et al. [14] that GSTP and GSMT explained the variation in both the asymptote and rate parameters of the jack pine and black spruce height growth models, in this study MDTR, its quadratic transformation, and April CMI explained the variation in the asymptote of the jack pine height growth model and PWQ that of black spruce. Similarly, MDTR and May CMI significantly affected the rate of height growth for jack pine and July CMI affected that of black spruce. The climatic effects found by Sharma et al. [14] were negative for both species, minimal for jack pine and more pronounced for black spruce. In this study, however, the effects were positive for jack pine and negative for black spruce, and for both species, they were minimal where present.

When the average values of climate variables over tree lifespan were used, only GSMT explained the variation in the rate parameter of the red pine height growth model [15]. The effect of climate on site productivity was highly negative in all three areas evaluated. In this study, however, the sum of March, April, October, and November CMI was also significant in the model for red pine. GSMT and the sum of CMIs explained the variations in the asymptote and the rate parameter, respectively. The effects were negative and pronounced only after BHA 30 for all three areas.

In another study, Sharma and Parton [16] reported both temperature- and precipitation-related variables (WQTP and WQMT) explained the variation in both the asymptote and rate parameter of a white spruce height growth model. The effect of climate was negative and more pronounced for white spruce than for jack pine, black spruce, and red pine. In this study, however, only AnMaxT was significant in explaining variations in the rate parameter of the height growth model for white spruce plantations. The effects were negative and highly pronounced in central Ontario. However, the effects were minimal in the other areas.

For white pine plantations, only MDTR affected stand height growth [3]. The effect was mild and positive in central Ontario and negative in the south. It was not pronounced in other areas. For jack pine and black spruce natural origin mixed stands, a temperature-related variable (GSMT) was important in explaining the variation in stand height growth for both jack pine and black spruce trees [17]. The effect was negative and not pronounced for jack pine but positive and pronounced after BHA 35 years for black spruce. Annual/seasonal values of climate variables were used in analyzing and modelling the climate effects in these studies.

Sharma [18] also examined the climate effects on site productivity of black spruce and trembling aspen natural origin mixed stands using annual/seasonal values of climate variables. A temperature-related variable (MDTR) was important in explaining the variation in stand height growth for both black spruce and trembling aspen trees. The effect was positive for both species but not pronounced in three of the four areas evaluated.

Sharma [18] reported that even in natural origin mixed stands, climate variables that explained the height growth of black spruce grown with different tree species were not the same. Stand height growth of black spruce was explained by GSMT and MDTR in the presence of jack pine and trembling aspen, respectively. The climate variable that explained the variation in the stand height growth of trembling aspen grown with black spruce was also MDTR. Although MDTR was the significant climate variable in the stand height growth models for both black spruce and trembling aspen, it explained the variation in the asymptote for black spruce but in the rate parameter for trembling aspen [18]. These findings indicated that climate effects on site productivity depend not only on tree species but also on stand type (plantations vs. natural origin mixed stands) and species mixture (other tree species growing in the stands).

Climate effects on site productivity also depended on the time over which climate variable values were calculated. First, the climate variables significant in the model averaged over trees' lifespan differed from those significant based on annual/seasonal values linked to growth period. Second, the nature and magnitude of the effects differed. Since climate varies annually, climate effects analyzed using the annual/seasonal values would be more accurate than those using average values. Therefore, annual/seasonal values are recommended for analyzing climate effects on tree/forest growth.

A site index expressed in terms of biophysical variables alone does not provide an accurate estimate of site productivity because it is determined by more than climate and other environmental variables. Climate and environmental variables are estimated at landscape scale, but several microsite variables (e.g., soil type, available nutrients) also influence site productivity. Therefore, climate effects on SI should be analyzed by incorporating climate variables in SI/stand height growth models. The effects of microsite variables on SI/stand height growth are reflected by the initial values of stand heights required in the models presented here. Better soil with more available nutrients may produce higher initial height values at a particular stand age.

The results presented here are consistent with other studies conducted in other geographic regions. As Bergh et al. [8] reported, net primary production of Scots pine grown in Nordic countries could increase with the increase in temperature. Similarly, the study by Pedlar and McKenney [9] showed that the growth response of five northern conifers to climate change could be positive on cold-origin (northern) populations, but negative on warm-origin (southern) populations. As mentioned earlier, climate effects on jack pine site productivity were positive in the north and severely negative for white spruce in the south.

The models presented here can be readily applied to statistical growth and yield models to estimate site productivity more accurately under a changing climate. These models characterize not only stand height growth models that can be used under a changing climate but also a means to evaluate the effect of climate on site productivity that depends on tree species and geographic location. For a given tree species, the climatic effect on site productivity can be explained by interpreting the sign and magnitude of the coefficients of the climate variables significant in the models.

The estimates for the coefficients of base model (Equation (1)) have been presented by Sharma et al. [14] for jack pine and black spruce, by Sharma and Parton [15] for red pine, and by Sharma and Parton [16] for white spruce. Those estimates remain the same and, hence, are not reported here.

5. Conclusions

Climate effects on site productivity were reanalyzed and modelled for jack pine (*Pinus banksiana* Lamb.), black spruce (*Picea mariana* (Mill.) B.S.P.), red pine (*Pinus resinosa Aiton*), and white spruce (*Picea glauca* (Moench) Voss) plantations using annual/seasonal values of climate variables. For this analysis, parameters of the stand height growth model were expressed in terms of climate variables for all tree species. A nonlinear mixed-effects approach was applied to fit the models with climate variables. Including climate variables improved the model fit statistics for all four tree species.

Climate effects on site productivity depended on tree species and location. For jack pine, the effects were positive and pronounced only in the west of Ontario under all three emissions scenarios (2.6, 4.5, and 8.5 watts m^{-2}). For black spruce, effects were negative and minimal after BHA 50 in central Ontario. Similarly, the effects were negative and more pronounced in all areas (southeast, central, and west of Ontario) after BHA 35 for red pine. On the other hand, the effects were negative and notable from when white spruce were young under all scenarios in the southeast. The effects under RCP 8.5 were more pronounced than those under other two scenarios for all species except jack pine. The difference between the effects under RCP 2.6 and 4.5, however, was not as pronounced as those between RCP 4.5 and 8.5 for all areas for all species.

The climate effects analyzed using annual/seasonal values of climate variables differed in nature (positive or negative) and magnitude from those estimated using the average values of climate variables over the trees' lifespan. Since climate effects analyzed using the annual/seasonal values of climate variable would be more accurate than those estimated using the average values, the former are best to be used in forest management planning.

Funding: This research received no external funding.

Data Availability Statement: The data used in this study have been presented in Table 1.

Acknowledgments: This study was supported by the Ontario Ministry of Natural Resources and Forestry (MNRF). Support for data collection was provided by the Forestry Futures Trust Enhanced Forest Productivity Science Program. The author is grateful to Daniel McKenney, Canadian Forest Service, for providing estimates of climate variables for study sites and Lisa Buse, Ontario Forest Research Institute, for editing an earlier version of the manuscript.

Conflicts of Interest: The authors declare no conflict of interest.

References

1. Vanclay, J.K. *Modelling Forest Growth and Yield: Applications to Mixed Tropical Forests*; CAB International: Oxon, UK, 1994; 312p.
2. Sharma, M.; Amateis, R.L.; Burkhart, H.E. Top height definition and its effect on site index determination in thinned and unthinned loblolly pine plantations. *For. Ecol. Manag.* **2002**, *168*, 163–175. [CrossRef]
3. Sharma, M.; Parton, J. Modelling effects of climate on site productivity of white pine plantations. *Can. J. For. Res.* **2019**, *49*, 1289–1297. [CrossRef]
4. Hunter, I.R.; Gibson, A.R. Predicting Pinus radiata site index from environmental variables. *N. Z. J. For. Sci.* **1984**, *14*, 53–64.
5. Ung, C.-H.; Bernier, P.Y.; Raulier, F.; Fournier, R.A.; Lambert, M.C.; Regniere, J. Biophysical site indices for shade tolerant and intolerant boreal species. *For. Sci.* **2001**, *47*, 83–95.
6. Albert, M.; Schmidt, M. Climate-sensitive modelling of site-productivity relationships for Norway Spruce (Picea abies (L.) Karst.). *For. Ecol. Manag.* **2010**, *259*, 739–749. [CrossRef]
7. Weiskittel, A.R.; Crookston, N.L.; Radtke, P.J. Linking climate, gross primary productivity, and site index across forests of the western United States. *Can. J. For. Res.* **2011**, *41*, 1710–1721. [CrossRef]
8. Bergha, J.; Freeman, M.; Sigurdsson, B.; Kellomaki, S.; Laitinen, K.; Niinisto, H.P.; Linder, S. Modelling the short-term effects of climate change on the productivity of selected tree species in Nordic countries For. *Ecol. Manag.* **2003**, *183*, 327–340. [CrossRef]
9. Pedlar, J.H.; McKenney, D.W. Assessing the anticipated growth response of northern conifer populations to a warming climate. *Sci. Rep.* **2017**, *7*, 43881. [CrossRef]
10. Guo, X.; Klisz, M.; Puchalka, R.; Silvestro, R.; Faubert, P.; Belien, E.; Huang, J.; Rossi, S. Common-garden experiment reveals clinal trends of bud phenology in black spruce populations from a latitudinal gradient in the boreal forest. *J. Ecology* **2022**, *110*, 1043–1053. [CrossRef]
11. Beaumont, J.F.; Ung, C.H.; Bernier-Cardou, M. Relating site index to ecological factors in black spruce stands: Tests of hypotheses. *For. Sci.* **1999**, *45*, 484–491.
12. Wang, Y.; LeMay, V.M.; Baker, T.G. Modelling and prediction of dominant height and site index of Eucalyptus globulus plantations using a nonlinear mixed-effects model approach. *Can. J. For. Res.* **2007**, *37*, 1390–1403. [CrossRef]
13. Bravo-Oviedo, M.; Bravo, T.F.; Montero, G.; del Rio, M. Dominant height growth equations including site attributes in the generalized algebraic difference approach. *Can. J. For. Res.* **2008**, *38*, 2348–2358. [CrossRef]
14. Sharma, M.; Subedi, N.; TerMikaelian, M.; Parton, J. Modeling climatic effects on stand height/site index of plantation-grown jack pine and black spruce trees. *For. Sci.* **2015**, *61*, 25–34. [CrossRef]
15. Sharma, M.; Parton, J. Climatic effects on site productivity of red pine plantations. *For. Sci.* **2018**, *64*, 544–554. [CrossRef]
16. Sharma, M.; Parton, J. Analyzing and modelling effects of climate on site productivity of white spruce plantations. *For. Chron.* **2018**, *93*, 173–182. [CrossRef]
17. Sharma, M. Climate effects on jack pine and black spruce productivity in natural origin mixed stands and site index conversion equations. *Trees For. People* **2021**, *5*, 100089. [CrossRef]
18. Sharma, M. Climate effects on black spruce and trembling aspen productivity in natural origin mixed stands. *Forests* **2022**, *13*, 430. [CrossRef]
19. Rowe, J.S. *Forest Regions of Canada*; Government of Canada, Department of the Environment, Canadian Forestry Service: Ottawa, ON, Canada, 1972; Publication No. 1300.
20. McKenney, D.W.; Hutchinson, M.F.; Papadopol, P. Customized spatial climate models for North America. *Bull. Am. Meteorol. Soc.* **2011**, *92*, 1161–1622. [CrossRef]

21. Mekis, É.; Vincent, L.A. An overview of the second generation adjusted daily precipitation data set for trend analysis in Canada. *Atmos. Ocean.* **2011**, *49*, 163–177. [CrossRef]
22. Vincent, L.A.; Wang, X.L.; Milewska, E.J.; Wan, H.; Yang, F.; Swail, V. A second generation of homogenized Canadian monthly surface air temperature for climate trend analysis. *J. Geophys. Res. Atmos.* **2012**, *117*, D18110. [CrossRef]
23. Hogg, E.H. Climate and the southern limit of the western Canadian boreal forest. *Can. J. For. Res.* **1994**, *24*, 1835–1845. [CrossRef]
24. Burkhart, H.E.; Tome, M. *Modeling Forest Trees and Stands;* Springer: Dordrecht, The Netherlands, 2012; 457p.
25. Akaike, H. A Bayesian analysis of the minimum AIC procedure. *Ann. Inst. Stat. Math.* **1978**, *30*, 9–14. [CrossRef]
26. Clutter, J.L.; Fortson, J.C.; Pienaar, L.V.; Brister, G.H.; Bailey, R.L. *Timber Management: A Quantitative Approach;* Krieger Publishing Company: Malabar, FL, USA, 1983.

MDPI AG
Grosspeteranlage 5
4052 Basel
Switzerland
Tel.: +41 61 683 77 34

Forests Editorial Office
E-mail: forests@mdpi.com
www.mdpi.com/journal/forests

www.ingramcontent.com/pod-product-compliance
Lightning Source LLC
Chambersburg PA
CBHW061842210926
51850CB00006B/950
9783725817573